Ivosic

Beginning Statistics with Data Analysis

Beginning Statistics with Data Analysis

Frederick Mosteller
Harvard University

Stephen E. Fienberg
Carnegie-Mellon University

Robert E. K. Rourke
Formerly of the St. Stephen's School

ADDISON-WESLEY PUBLISHING COMPANY
Reading, Massachusetts
Menlo Park, California
London
Amsterdam
Don Mills, Ontario
Sydney

This book is in the Addison-Wesley Series in Statistics

Library of Congress Cataloging in Publication Data

Mosteller, Frederick, 1916–
Beginning statistics with data analysis.

Bibliography: p.
Includes index.
1. Statistics. I. Fienberg, Stephen E.
II. Rourke, Robert E. K. III. Title.
QA276.12.M67 1983 519.5 82-16325
ISBN 0-201-05974-6

ISBN 0-201-05974-6
ABCDEFGHIJ-DO-89876543

To Cleo Youtz

Preface

THE WORLD OF STATISTICS

To introduce the student to the world of statistics, this book treats the following areas in turn:

1. exploratory data analysis
2. methods for collecting data
3. formal statistical inference
4. techniques of regression and analysis of variance

Examples are drawn from the broad range of applications in business, government, medicine, social sciences, biology, and everyday life.

Many people associate statistics with mathematics, but very little in the way of formal mathematics is required to understand how to look at and analyze data. The level of mathematics necessary for an understanding of the material in this book is one year of algebra, as found in contemporary texts written for grade 9.

EXPLORATORY VERSUS CONFIRMATORY ANALYSIS

We can divide statistical methods into two categories: exploratory and confirmatory. Exploratory methods offer a systematic set of tools for trying to discover what the data seem to be saying. Confirmatory methods usually employ probabilistic approaches to try to settle specific questions. For example: Will the new burglarproof lock reduce burglary losses?

A common situation arises in which data need analysis, but no clear question was asked before the data were generated. As the world turns, these messy problems come up frequently, and as frequently are handed out with remarks such as "Here, you've had a course in statistics. Tell us what these numbers say." Someone equipped only for formal inference and well-designed studies can become totally lost in a sea of flotsam. Consequently, we believe the student needs some general-purpose tools for data analysis, and we provide them in Chapters 1 through 4. At the same time, because the student will read in the scientific literature the formal inferential analyses that go with designed studies and narrowly posed questions, material on confirmatory analysis must also be available. We provide this in Chapters 5 through 10 and Chapter 16. In more advanced data analysis, the current practice of statistics often blends exploratory and confirmatory ideas. We proceed in this spirit as we discuss regression and analysis-of-variance methods in Chapters 11 through 15.

In carrying out one fundamental job, to control and measure variability and to assign it to its sources, statistics uses design of investigation and analytic methods. Creating hypothetical structures often adds strength and direction to the effort. Such simple ideas as a probability distribution, variation around a line or curve, and the approximate additivity of effects from several sources help us analyze and interpret otherwise sprawling, messy sets of data. Fortunately, we do not have to believe these models exactly. Usually a rough approximation gives us enough to boost our understanding by several levels. When we have no structure in mind in advance, the methods of exploratory data analysis offer us a start with rather simple underlying structures. All these methods, both formal and exploratory, give us ways of summarizing information heavily contaminated with chance variation.

OUR APPROACH TO COMPUTING

In these days, when calculators and high-speed computers do so much of the work, we believe that the student's effort should be oriented toward understanding a method and what it does rather than toward carrying out lengthy calculations. For example, when all arithmetic work was done by hand calculation, students spent days working out the details of a multiple regression analysis. This time spent with the numbers had its value, because each data point became an old friend. Furthermore, as more time was spent on a data set, more thought was given to the work and to the further steps needed. Today, with computers at hand, we can do a similar thing by exploring a variety of analyses, adding and excluding variables and data points, and plotting the data and the residuals in a variety of ways. Thus we have more chance of appreciating the information contained in the data, because we need not be bogged down in the details of calculations.

Access to convenient packages of statistical programs has improved dramatically in recent years, as has the availability of graphics for both high-speed computers and minicomputers. Many desk-top minicomputers offer prepackaged programs ample for working out the details of any problem in this book. For these reasons, we have, especially in the latter part of the book, emphasized the understanding of method and the interpretation of computer output rather than the carrying out of heavy arithmetic calculations.

EXAMPLES AND PROBLEMS

We have provided genuine data for both examples and problems. The tang of reality adds zest to the learning process. New ideas are approached through special, practical examples from which general principles are extracted. These general principles are then used to equip the student with procedures for tackling a wide variety of real-life problems. Students should not have to face their first real-life assignments innocent of the nasty irregularities that commonly occur.

Among the important but not loudly sung themes of science and statistical analysis, one that worries beginners says that we have no one correct way to proceed. Many ways may be appropriate, and some better than others, but unless we deal with a frequently repeated situation, we can rarely be confident that one way is best, though we may know that some ways are likely to be strong under a variety of conditions. Thus the student may have several ways available to attack a specific problem. Appreciating this should be one goal of the student in this course. Thus the reader of this book will find some examples analyzed repeatedly by different methods, sometimes to answer different questions and sometimes much the same question.

Each section of the book ends with a set of problems. These reinforce the ideas and methods developed. Sometimes they fill in the details or extend an example; sometimes they follow up on earlier problems; often they contain new data sets for analysis. Each chapter ends with a set of review problems.

OTHER SPECIAL FEATURES

1. The first four chapters give a brief course in exploratory data analysis. Indeed, the role of data analysis in statistics is stressed throughout.
2. We place special emphasis on data collection, and we associate analysis

with data collection. This helps to motivate formal ideas of estimation and variability.

3. The book ends with a chapter on the design of comparative experiments, linking the basic ideas of design back to the methods of analysis discussed earlier.
4. The chapter on probability can be skipped without losing continuity, because the basic notions required for formal methods of statistical inference appear in other chapters.
5. We offer an elementary treatment of multiple regression, something absent from many books at this level. The examples of computer output in this material allow the teacher to use the ideas of high-speed computers even if they have none for the students to use.
6. A solutions manual is available to the instructor and offers complete, worked solutions to both odd- and even-numbered problems.

When a problem depends upon the student having already solved one or more previous problems, we often indicate this by the word "Continuation" before the problem.

A star (★) next to a problem or a section means that the level of difficulty may be higher than for other material or that the material might be skipped in a first reading or that a problem or project may be especially time-consuming.

OUTLINE FOR COURSES

This book is designed to include the topics an instructor is likely to need for various introductory courses. Chapters 1 through 7 plus Chapter 17 (except Section 17-4) provide the basis for a one-semester course on Data Analysis and Collection. A more traditional one-quarter course on Statistics with Probability can be based on Chapters 1 through 3 and Chapters 5 through 10. Alternatively, the first 12 chapters (with or without Chapter 8, on probability) provide material for a one-semester introductory course. Instructors who wish to teach a one-quarter course in Exploratory Data Analysis can do so by using Chapters 1 through 4 and Chapters 11 through 14, and omitting the sections on the t tests and F tests from the latter chapters. Finally, a second course in statistics focusing on Regression and Analysis of Variance can follow Chapters 3 and 4 and Chapters 11 through 15, plus parts of Chapters 16 and 17.

The diagram showing various routes through the Chapters gives a schematic representation of the connections between chapters and indicates possible chapter sequences for the courses mentioned earlier. The words in the boxes describe the main ideas treated in the chapters.

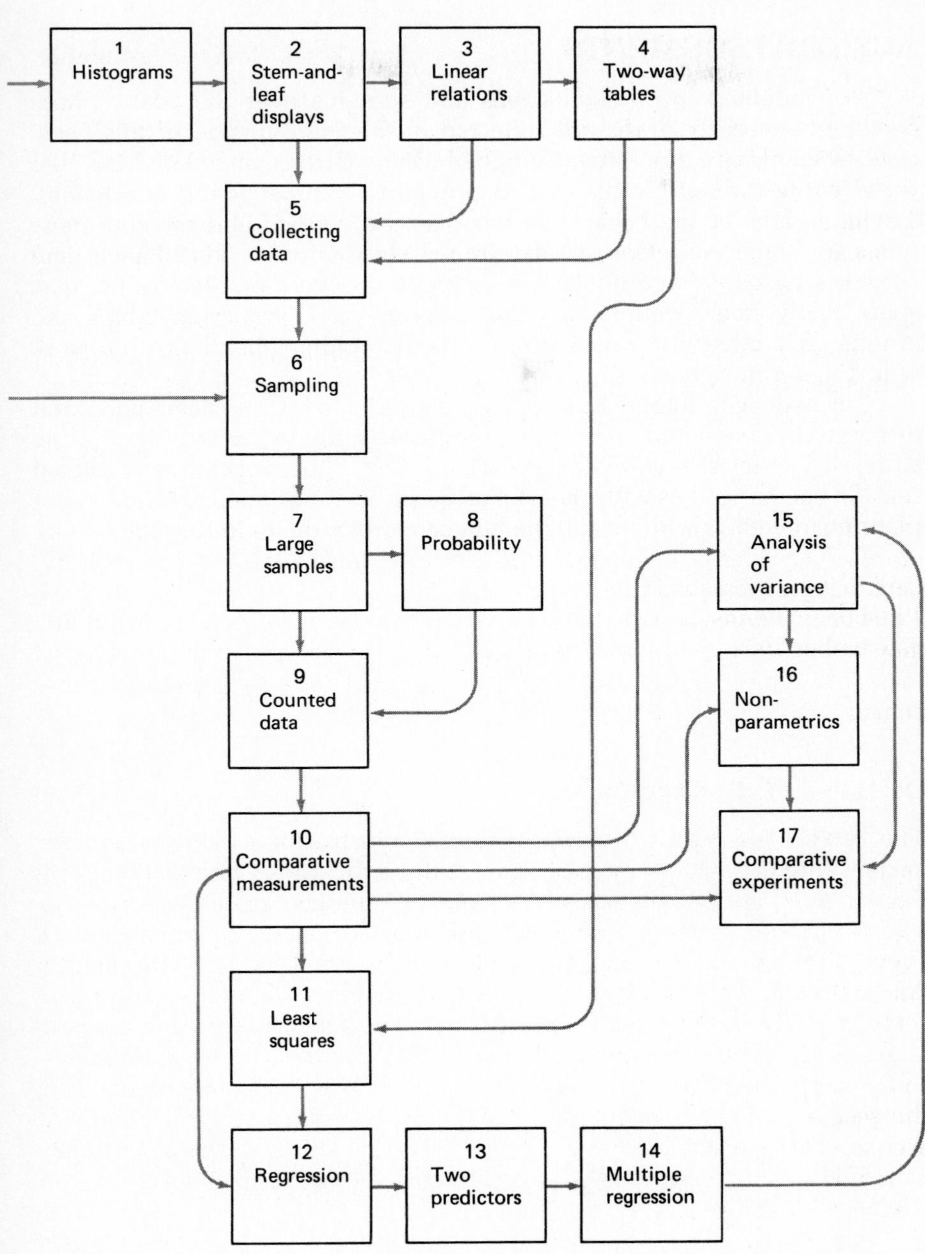

Various routes through the chapters.

ACKNOWLEDGMENTS

We are indebted to our colleagues and students over the years whose comments on early drafts are reflected in the final version of this book. Special thanks are due Kathy Campbell, Diane Griffin, and Gale Mosteller for working through problems and providing comments and corrections. The materials for the book have been carefully typed and retyped many times by Linda Anderson, Holly Grano, Margie Krest, Nina Leech, and Marjorie Olson. We particularly appreciate the work of Cleo Youtz, who spent many hours double-checking computations, preparing tables and graphs, and otherwise organizing the book. Without her help this work would never have been completed.

Our colleague Robert E. K. Rourke participated in the conception and design of the book and contributed enthusiastically to the writing until he suffered a series of severe illnesses. Thereafter, while working at a reduced rate, he encouraged us with cheerful criticism and continued to brighten our planning meetings with his unique humor until his death in July 1981.

Cambridge, Massachusetts F.M.
Pittsburgh, Pennsylvania S.E.F.
November 1982

Contents

1 Frequency Distributions

1-1 Data: An Aid to Action 2
1-2 How to Make a Histogram **8**
1-3 Organizing Data: The Cumulative Frequency Diagram **13**
1-4 Summary of Chapter 1 **22**
Summary Problems for Chapter 1 **23**
Reference **26**

2 Summarizing and Comparing Distributions

2-1 Organizing Data: Tallies and Stem-and-Leaf Diagrams **28**
2-2 More about Stem-and-Leaf Diagrams **34**
2-3 A Quick Picture: Summarizing the Data **38**

2-4 Logarithms for Stem-and-Leaf Diagrams **49**

2-5 Summary of Chapter 2 **55**

Summary Problems for Chapter 2 **55**

References **57**

Finding and Summarizing Relationships 3

3-1 Target for Chapter 3 **60**

3-2 Exact Relations **64**

3-3 Approximate Relations: The Black-Thread Method **73**

3-4 Studying Residuals **77**

3-5 Summary of Chapter 3 **85**

Summary Problems for Chapter 3 **85**

References **87**

Analysis of One-Way and Two-Way Tables 4

4-1 Components, Residuals, and Outliers **90**

4-2 Two-Way Tables of Measurements: Informal Analysis **97**

4-3 Breaking a Table into Its Components **103**

4-4 Summary of Chapter 4 **112**

Summary Problems for Chapter 4 **113**

References **115**

Gathering Data 5

5-1 Gathering Information **118**

5-2 Examples of Questions that Might be Studied by Various Methods **118**

5-3 The Case Study or Anecdote **120**

5-4 Sample Surveys and Censuses 121
5-5 Observational Studies 124
5-6 Controlled Field Studies and Comparative Experiments 127
5-7 Using all Methods 130
5-8 Summary of Chapter 5 133
Summary Problems for Chapter 5 134
References 134

6 Sampling of Attributes

6-1 Samples and Their Distributions 138
6-2 Measures of Variation 142
6-3 Sampling without Replacement from a Finite Population with Two Categories 147
6-4 Variability of $\overline{p}$ in Sampling without Replacement 151
6-5 Infinite Populations and Sampling with Replacement 156
6-6 Frequencies, Proportions, and Probabilities 162
6-7 Summary of Chapter 6 162
Summary Problems for Chapter 6 163
References 164

7 Large Samples, the Normal Approximation, and Drawing Random Samples

7-1 Practical Computation of Probabilities and Methods of Drawing Random Samples 168
7-2 Large Samples from Large Populations with Two Kinds of Elements: Approximate Distribution of $\overline{p}$ when p is not near 0 or 1 168

7-3 Fitting the Standard Normal Distribution to a Binomial Distribution 172
7-4 Random Sampling 178
7-5 Summary of Chapter 7 185
Summary Problems for Chapter 7 186
References 187

8 ★**Probability**

8-1 Introduction 190
8-2 Probability Experiments and Sample Spaces 190
8-3 Independent Events 196
8-4 Dependent Events 203
8-5 Probabilities for Combined Events 206
8-6 Conditional Probability 212
8-7 Assigning Probabilities Using the Chain Rule 219
8-8 Random Variables 223
8-9 Summary of Chapter 8 229
Summary Problems for Chapter 8 230
Reference 232

9 **Comparisons for Proportions and Counts**

9-1 The Need for Comparisons 234
9-2 Matched Pairs— One Winner 237
9-3 Matched Pairs: 0,1 Outcomes 240
9-4 Independent Samples: 0,1 Outcomes 242
9-5 Independence of Attributes 247
9-6 Chi-Square Test for Independence 252
9-7 The $r \times c$ Table 256

9-8 Summary of Chapter 9 261
Summary Problems for Chapter 9 263
References 264

10 Comparisons for Measurements

10-1 Dealing with Measurements 266
10-2 Applying the Normal Approximation to Measurements 272
10-3 Matched Pairs, Measurement Outcomes, and Single Samples 276
10-4 Two Independent Samples: Measurement Outcomes, Variances Known 284
10-5 Two Independent Samples: Measurement Outcomes, Variances Unknown but Equal 286
10-6 Two Independent Samples: Unknown and Unequal Variances 289
10-7 Accepting or Rejecting the Null Hypothesis 292
10-8 Summary of Chapter 10 293
Summary Problems for Chapter 10 294
Reference 295

11 Fitting Straight Lines Using Least Squares

11-1 Plotting Data and Predicting 298
11-2 Aims of Fitting Linear Relations 302
11-3 Criteria for Fitting Straight Lines 309
11-4 Least Squares 314

11-5 Summary of Chapter 11 323
Summary Problems for Chapter 11 324
Reference 325

The Linear Regression Model and Its Use 12

12-1 Fitting Straight Lines to Data 328
12-2 The Linear Regression Model 329
12-3 Estimating Variability 340
12-4 The Proportion of Variability Explained 346
12-5 Tests and Confidence Limits for *m* 352
12-6 Tests and Confidence Limits for *b* 360
★ **12-7** Predicting New *y* Values 364
12-8 Summary of Chapter 12 370
Summary Problems for Chapter 12 372
References 373

Regression with Two Predictors 13

13-1 Predicting with Two Variables 376
13-2 Least-Squares Coefficients for Two Predictors 381
13-3 One versus Two Predictors 384
13-4 Multiplicative Formulas with Two Predictors 391
13-5 Summary of Chapter 13 399
Summary Problems for Chapter 13 399
References 402

Multiple Regression 14

14-1 Many Predictors 404
14-2 Choosing a Multiple Regression Equation 406

14-3 Examples of Estimated Multiple Regression Equations, Their Interpretation, and Their Use **425**
14-4 Summary of Chapter 14 **429**
Summary Problems for Chapter 14 **432**
References **433**

15 Analysis of Variance

15-1 Introductory Examples **436**
15-2 Several Samples **441**
15-3 The F Ratio **451**
15-4 Two-Way Analysis of Variance with One Observation per Cell **462**
15-5 Summary of Chapter 15 **468**
Summary Problems for Chapter 15 **469**
References **471**

16 Nonparametric Methods

16-1 What are Nonparametric Methods? **474**
16-2 The Sign Test **475**
16-3 The Mann-Whitney-Wilcoxon Two-Sample Test **478**
16-4 Analysis of Variance by Ranks: The Kruskal-Wallis Test **486**
16-5 Summary of Chapter 16 **494**
Summary Problems for Chapter 16 **496**
References **496**

17 Ideas of Experimentation

17-1 Illustration of Experiments **500**
17-2 Basic Idea of an Experiment **503**

17-3 Devices for Strengthening Experiments **507**
17-4 Illustrations of Loss of Control **511**
17-5 The Principal Basic Designs **517**
17-6 Analysis of Variance and Experimentation **522**
17-7 Summary of Chapter 17 **526**
Summary Problems for Chapter 17 **527**
References **528**

Appendix I Summations and Subscripts **529**

Appendix II Formulas for Least-Squares Coefficients for Regression with Two Predictors **533**

Appendix III Tables **537**

Short Answers to Selected Problems **557**

Index to Data Sets and Examples **577**

General Index **579**

1

Frequency Distributions

Learning Objectives

1. Reaping the benefits from organizing data and displaying them in histogram form
2. Gaining skill in interpreting frequency histograms
3. Constructing a cumulative frequency diagram
4. Finding quartiles from a cumulative graph

1-1 DATA: AN AID TO ACTION

Much human progress grows from the practice of keeping and analyzing records. Important examples are the records leading to the calendar, an appreciation of the seasons, the credit, banking, and insurance systems, much of modern production processes, and our health and medical systems. We shall, therefore, begin by explaining how to organize, display, and interpret data.

What does this book do? This text primarily equips the reader with the skills

1. to analyze and display a set of data,
2. to interpret data provided by others,
3. to gather data,
4. to relate variables and make estimates and predictions.

ORGANIZING DATA: THE USE OF PICTURES

Data often come to us as a set of measurements or observations along with the number of times each measurement or observation occurs. Such an array is called a **frequency distribution.**

To display a frequency distribution and disclose its information effectively, we often use a type of diagram called a **frequency histogram.** Let us look at some histograms. Examples 1 and 2 suggest some benefits we get from organizing data and displaying them in histogram form.

EXAMPLE 1 *Allocation of police.* The chief of police of New York City has enlisted as many men as his budget permits. He has divided his forces about equally to cover three daily shifts; the first shift runs from midnight until 8 A.M., the second from 8 A.M. until 4 P.M., and the third from 4 P.M. until midnight. This system is bringing many complaints, and it is clear that during certain hours of the day the police calls require more police than are available. The chief selects a certain Sunday in August as a guide to action and makes a histogram (Fig. 1-1) showing the frequency distribution of emergency police calls, hour by hour, in New York City during this Sunday. Using this Sunday in August as a guide for late summer Sundays, what action, short of hiring additional police, does the histogram suggest?

Discussion. The histogram vividly exhibits the changes in the numbers of police calls from hour to hour during the 24-hour period. The day's calls appear to divide into three types of periods:

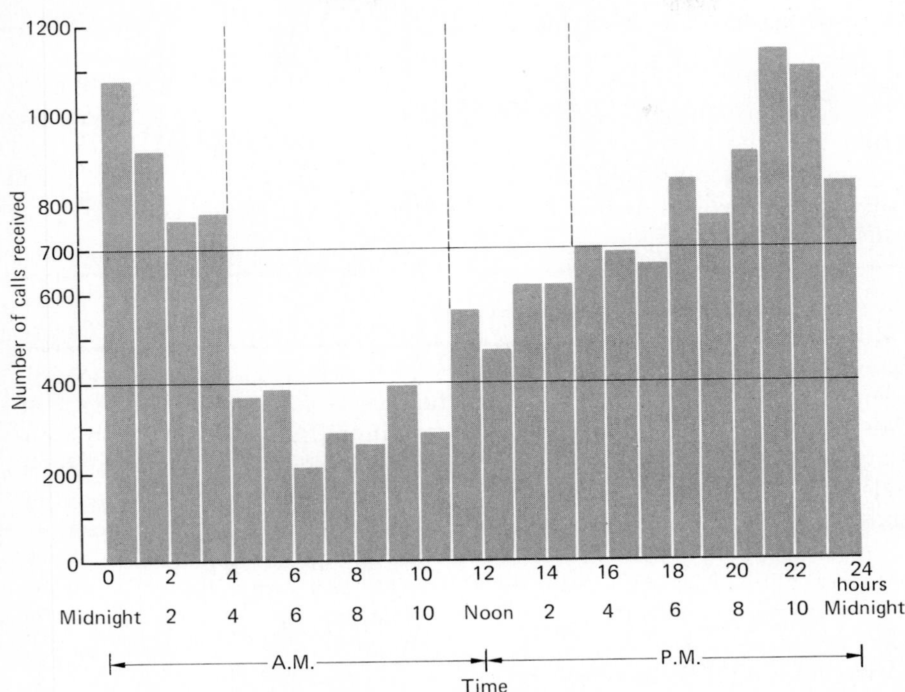

Figure 1-1 Frequency distribution of New York City police calls during a Sunday in August.

Source: Adapted from R. C. Larson (1972). Improving the effectiveness of New York City's 911. In *Analysis of Public Systems*, edited by A. W. Drake, R. L. Keeney, and P. M. Morse, p. 161. Cambridge, Mass.: M.I.T. Press.

1. A connected interval of peak demand, about 700 or more calls per hour from 0 to 4 hours (midnight to 4 A.M.), and another interval from 15 to 24 hours (3 P.M. to midnight); these two intervals join to become one period as we run from one day to the next starting at 15 hours and running through to 4 hours the next day (3 P.M. to 4 A.M.).

2. A period of medium demand, over 400 calls to less than 700 calls per hour from 11 A.M. to 3 P.M.

3. A period of low demand, about 400 or fewer calls per hour from 4 A.M. to 11 A.M.

A study of such histograms over a period of time gives important information about the number of police required for emergency duty during a given shift. Figure 1-1 shows that there are over twice as many calls during the 7-hour period between 6 P.M. and 1 A.M. as during the 7-hour period between 4 A.M. and 11 A.M.

If we assume that the chief does not want to change the shift times, then he might reallocate his force so that the numbers on the shifts are more nearly proportional to the numbers of calls received during the shifts. A practical nonstatistical problem then arises. How can arrangements be made with the union to reschedule the shifts? In the actual event, the rescheduling was worked out. Note that the guide to action came from looking at the whole distribution of calls, not just the average number of calls per hour.

EXAMPLE 2 *A gap in the histogram.* The quality control expert W. E. Deming reported that part of a manufacturing process involves making steel rods. These rods have a lower specification limit (LSL) of 1.000 cm on their diameters; rods smaller than 1.000 cm are too loose in their bearings. Such rods are rejected, or thrown out, which implies losses of labor and material and of the overhead expenses incurred up to this point in the manufacturing process. As an aid to action, the quality-control engineer constructs Fig. 1-2, a histogram displaying the inspectors' measurements for 500 rods. The rods are grouped into intervals by diameter measurements, and each interval is 0.001 cm wide. It is convenient to label the center of each interval on the horizontal axis. Thus the interval centered at 0.998 includes measurements for all rods with reported diameters between 0.9975 and 0.9985. The vertical

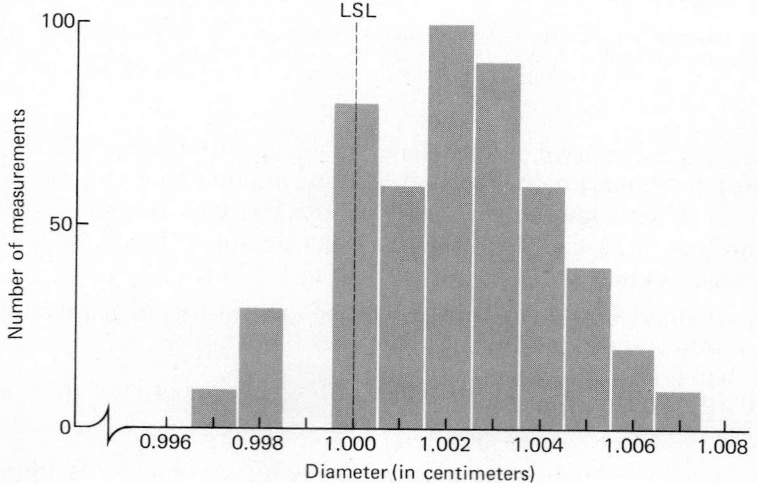

Figure 1-2 Distribution of measurements of the diameters of 500 steel rods.

Source: W. E. Deming (1978). Making things right. In *Statistics: A Guide to the Unknown*, second edition, edited by J. M. Tanur, F. Mosteller, W. H. Kruskal, R. F. Link, R. S. Pieters, and G. R. Rising; special editor, E. L. Lehmann; p. 280. San Francisco: Holden-Day, Inc.

axis shows the number of rods that the inspectors put into each of the 0.001-cm intervals.

Discussion. The histogram shows that the inspectors rejected 30 rods with diameters in the interval centered at 0.998. This histogram definitely has a message for us. The peak at 1.000 cm with the pit at 0.999 raises a question: Are the inspectors passing rods that have diameters only slightly below the LSL? When questioned, the inspectors admitted that they had been passing rods that were only slightly defective. This fault was corrected in the inspection of the next 500 steel rods, and the pit at 0.999 filled up. The corrected inspection process disclosed a basic fault in the production: There were too many rods below the LSL. The machine needed resetting. When the machine setting was corrected and the inspection was correctly carried out, most of the trouble disappeared. The histogram provided the key to appropriate action. Again, note the value in looking at the whole histogram, not just an average.

Why must we organize, display, and interpret data? In light of Examples 1 and 2, we can see reasons that justify our taking these steps. For example, we may want

1. to report something about what is going on, something about the activity or behavior or process that produced the data, or
2. to compare the process under study with other processes, or
3. to take some action with respect to this process, or
4. to detect some quirks, if any, in the behavior of the process.

EXAMPLE 3 *Distribution of length of life for U.S. females.* Table 1-1 gives the age at death for U.S. females based on experience in 1965. Let us construct a histogram to display these data.

Method. The data are given in 5-year age intervals. Because 21 groups are not too unwieldy for graphing, we shall use the ages as they are, that is, grouped into 5-year age intervals. The first group includes all deaths from 0 to less than 5 years of age, the second group includes deaths from 5 to less than 10 years of age, and so on. In the last group we include all deaths at age 100 or over. Why do we not combine pairs of 5-year age intervals, thus getting 11 groups instead of 21? This can be done, but the simplification would have its price. At the extreme, if all the data are assigned to one group, we lose information about the shape of the distribution. We must ask ourselves this question: Does the gain in convenience justify the loss of information? In this example the answer seems to us to be "No." Next we

TABLE 1-1
Distribution of deaths by age at death for U.S. females in 1965

Age at death	*Percent who died in 5-year period*	*Age at death*	*Percent who died in 5-year period*
0–4	2.39	55–59	3.99
5–9	0.18	60–64	5.69
10–14	0.15	65–69	8.41
15–19	0.26	70–74	11.16
20–24	0.34	75–79	14.68
25–29	0.42	80–84	17.19
30–34	0.61	85–89	17.05
35–39	0.89	90–94	8.00
40–44	1.31	95–99	2.14
45–49	1.96	100 & over	0.35
50–54	2.84	Total	100.01

Source: Adapted with the help of N. Keyfitz from N. Keyfitz and W. Flieger (1968). *World Population: An Analysis of Vital Data*, p. 45. Chicago: University of Chicago Press.

choose suitable scales for our axes. On the horizontal scale we choose an interval width that will let us accommodate our 21 age intervals; on the vertical scale, readings up to 20% will handle everything. Finally, the histogram may be drawn and labeled as in Fig. 1-3. Study this figure.

Interpretation. We observe the following:

1. A small spike at the 0–5 age interval. This isolated elevation stems primarily from deaths in infancy.
2. A very slow rise in the percentage of deaths between ages 5 and 40. The lengths of the histogram bars increase very little.
3. After age 40, the bars begin to lengthen rapidly, reaching a peak between 80 and 90 years of age. Thus female deaths increase rapidly after age 40, until between ages 80 and 90 the percentage reaches its maximum.
4. After age 90, the percentage falls rapidly, because few survive past 90 years of age.
5. So few survive beyond age 105 that they have been pooled into the 100–105 interval.

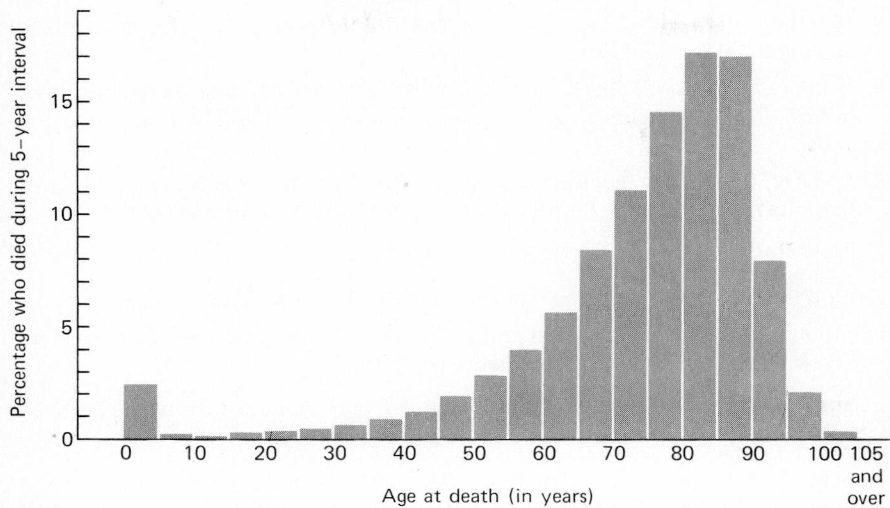

Figure 1-3 Frequency distribution of length of survival for U.S. females based on experience in 1965. Histogram shows percent dying in each 5-year age interval (the interval 100–105 includes all deaths after 100 rather than only those occurring in the interval). Constructed from data in Table 1-1.

Action suggested. The spike at the 0–5 age interval suggests that an investigation into the causes of death of newborn female infants might save lives. Actually, this is constantly being done, and the current spike is small compared with its former value. Because the trend in Fig. 1-3 rises so rapidly after age 65, a public policy inquiry aimed at extending life might ask what diseases are the big killers after age 65, to be followed by efforts to reduce their impact.

Again, note the value of looking at the whole distribution.

PROBLEMS FOR SECTION 1-1

Police Calls. Figure 1-1 plots New York City police calls during each hour of a Sunday in August. Use these data to solve Problems 1 through 6.

1. Estimate the number of police calls during the duty shift from midnight to 8 A.M.

2. Estimate the number of police calls during the duty shift from 8 A.M. to 4 P.M.

3. Estimate the number of police calls during the duty shift from 4 P.M. to midnight.

4. Use the results of Problems 1, 2, and 3 to find the percentage of police calls for each of the three shifts.
5. If the chief of police has 25,000 police available and he decides to assign shift duties using the percentages of Problem 4, how many should be assigned to each shift?
6. Use Fig. 1-1 to show that the number of police calls from 4 A.M. to 6 P.M. is less than the number from 6 P.M. to 4 A.M. Suggest a reason for this result.

Steel Rods. Refer to Fig. 1-2 to solve Problems 7 and 8.

7. Find the number of rods that were rejected because they were too small.
8. Among the 500 steel rods, what group of diameters was produced most frequently? How many rods in this group were produced?

Survival of U.S. Females. Use Table 1-1 and Fig. 1-3 to solve Problems 9 through 11.

9. According to Fig. 1-3, in what 5-year age interval do female deaths occur most frequently? What is the percentage of deaths in this interval?
10. Check your result in Problem 9 by referring to Table 1-1.
11. Use Table 1-1 or Fig. 1-3 to find the percentage of females who survive to age 30 or older.
12. Give four reasons why we should organize, display, and interpret data.

1-2 HOW TO MAKE A HISTOGRAM

If we decide to display the data of a frequency distribution in a histogram, it is well to have a plan in mind. These steps may help:

1. Choose the number of groups into which to divide the data. We ordinarily use from about 6 to 20 groups. (Grouping loses some of the information, but it may make the data more manageable. Sometimes the data may be used directly as given, as in Example 3, without presenting any graphing problems.)
2. Choose convenient whole numbers to bound the groups of measurements, where possible. In Example 3, we used 0, 5, 10, 15, . . . years as group boundaries. Note that the boundaries are equally spaced.
3. Count the number of observations in each group.
4. Choose, for the axes, scales that accommodate all the data.
5. Display the data using vertical bars or, equivalently, plotted points, as in Fig. 1-5. The area of the bar should be proportional to the count in the group. This works out without any additional effort when we use

equally spaced group boundaries because then length can be proportional to the count.

6. Label the figure clearly.

Warning. Do not expect to get a good histogram automatically. It is usually best to rough out the graph first in order to check the scales chosen and assess the effectiveness of the display. Perhaps different scales or a change in grouping will improve the impact of the diagram. For most of us, at least two tries are required. As in art and science, step-by-step improvements advance not just the designing of figures, but all data analysis. The same analysis may have to be done over and over with minor or major variations to try to find out and reveal what the data are saying.

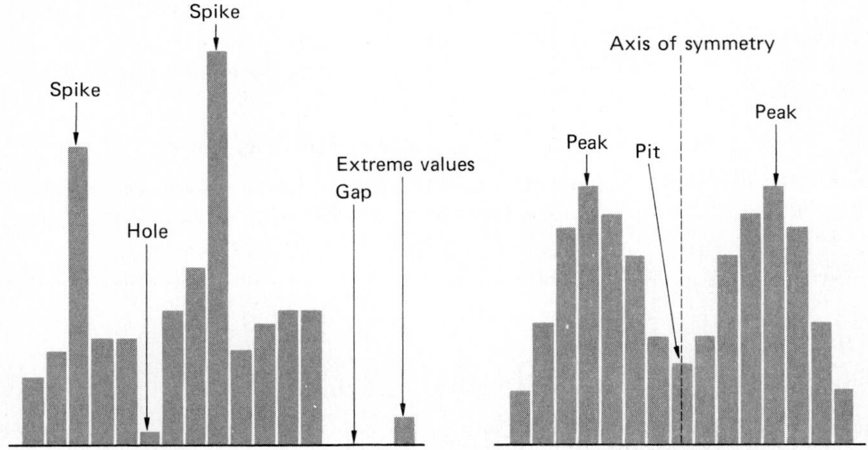

Figure 1-4 Illustrations of spikes, holes, peaks, pits, gaps, extreme values, and symmetry.

What do we look for in a histogram? To interpret the data, we look for any of the following features in the histogram (see Fig. 1-4):

1. *Peaks and pits.* These hilltops and valleys indicate the ranges of measurements or observations that occur most frequently (peaks) or least frequently (pits).
2. *Spikes and holes.* These are isolated high points and low points. Spikes indicate sharp increases, and holes indicate sharp decreases in the frequency. The relative size of the jump distinguishes these from peaks and pits, which occur relatively smoothly rather than jaggedly.
3. *Symmetry.* When the diagram balances about a central vertical line, the data are especially easy to interpret in terms of symmetry.

4. *Extreme values (outliers).* A measurement that strays far from the mean, or center, of the distribution may indicate error or something very special about the object or process being measured.
5. *Gaps.* Often associated with outliers, gaps also may suggest that two or more kinds of measurements are being mixed together.
6. *Other patterns.* We shall discuss other patterns when they arise.

PROBLEMS FOR SECTION 1-2

1. Describe each of the following features of a histogram, and tell what each feature indicates regarding the distribution: (a) a peak; (b) a hole; (c) a pit; (d) a gap.

Football Scores. Figure 1-5 gives the data for Problems 2 through 6. In it we have, for American college football in 1967, the frequencies of score differences from 0 to 40, where

score difference = winner's score minus loser's score.

Recall that in college football the possible scoring plays yield 8 points (a special converted touchdown), 7 points (converted touchdown), 6 points (unconverted

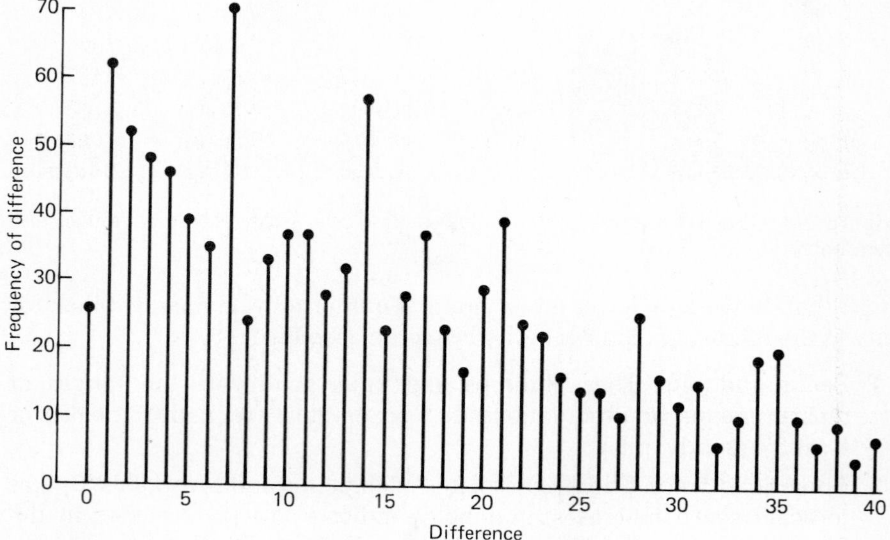

Figure 1-5 Frequency distribution of winner's score minus loser's score for differences not exceeding 40; 1967 college football scores in United States.

Source: Adapted from F. Mosteller (1970). Collegiate football scores, U.S.A. *J. American Statistical Association* p. 45.

touchdown), 3 points (field goal), or 2 points (safety). Total scores are some combination of these scores or zero. The safety and the 8-point touchdown are relatively rare. Use Fig. 1-5 to answer Problems 2 through 6.

2. Which kinds of scores are generally more frequent—close scores or scores with big differences?
3. What differences produce spikes? Try to explain some of these spikes.
4. What explanation might be given for the hole at 8?
5. Is this frequency distribution approximately symmetrical?
★ 6. Differences of 1 seem to occur about twice as often as differences of 0. Why?

Survival of U.S. Males. Table 1-2 gives the age at death for U.S. males based on experience in 1965. Use these data to solve Problems 7 through 10.

TABLE 1-2

Distribution of deaths by age at death for U.S. males in 1965

Age at death	*Percent who died in 5-year period*	*Cumulative percent*	*Age at death*	*Percent who died in 5-year period*	*Cumulative percent*
0–4	3.05	3.05	55–59	7.35	26.21
5–9	0.24	3.29	60–64	9.72	35.93
10–14	0.25	3.54	65–69	12.46	48.39
15–19	0.66	4.20	70–74	13.51	61.90
20–24	0.89	5.09	75–79	13.29	75.19
25–29	0.87	5.96	80–84	11.78	86.97
30–34	1.00	6.96	85–89	8.76	95.73
35–39	1.40	8.36	90–94	3.41	99.14
40–44	2.11	10.47	95–99	0.76	99.90
45–49	3.28	13.75	100 & over	0.10	100.00
50–54	5.11	18.86	Total	100.00	

Source: Adapted with the help of N. Keyfitz from N. Keyfitz and W. Flieger (1968). *World Population: An Analysis of Vital Data*, p. 44. Chicago: University of Chicago Press.

7. Use Table 1-2 to construct a frequency histogram. (Refer to Example 3.)
8. What are the major differences between the histogram from Problem 7 and that for women in Fig. 1-3? What action is suggested by this comparison?
9. Use Table 1-2 or the histogram of Problem 7 to find the 5-year interval in which male deaths occur most frequently. What is the percentage of male deaths in this interval?

10. Use Table 1-2 or the histogram of Problem 7 to find the percentage of males who survive to age 30 or older.

Temperatures. Table 1-3 gives a set of observations of the lowest daily temperatures recorded at the Boston airport weather station in 1969. Use the table to solve Problems 11 through 14.

TABLE 1-3
Distribution of lowest daily temperatures recorded at Boston airport weather station in 1969

Temperature (°F)	*Number of days*				
	Spring	*Summer*	*Autumn*	*Winter*	*Total*
0–4					
5–9				1	1
10–14				6	6
15–19			1	10	11
20–24			6	27	33
25–29	2		7	19	28
30–34	5		14	23	42
35–39	15		11	3	29
40–44	11		11		22
45–49	20	3	17		40
50–54	14	3	13		30
55–59	14	17	9		40
60–64	7	30	1		38
65–69	3	23			26
70–74	2	15			17
75–79		2			2
Total	93	93	90	89	365

Source: Constructed from data in *Climatological Data,* U.S. Department of Commerce, Environmental Science Services Administration, Environmental Data Service, 1969.

11. Display the data for spring in a histogram.

12. Display the data for autumn in a histogram.

13. Interpret the results of the two histograms in Problems 11 and 12.

14. Use Table 1-3 or the histograms of Problems 11 and 12 to find the number of days with temperatures greater than or equal to 50°F (a) in spring, (b) in autumn.

15. Use the *World Almanac* or some other source to locate data for a frequency distribution. Display the data in the form of a histogram, and interpret the results.

1-3 ORGANIZING DATA: THE CUMULATIVE FREQUENCY DIAGRAM

The Real Number Line. It is often convenient to have a geometric representation of the family of real numbers, as shown in the following diagram. We are not primarily concerned here with the method of assigning points on the line to the real numbers. Our main concern is with the following basic assumption:

To every real number there corresponds exactly one point on the line, and, conversely, to every point on the line there corresponds exactly one real number.

When the correspondence has been arranged in the one-to-one manner guaranteed by the assumption, the labeled line is called the **real number line.** It is the horizontal axis that we use in most of our graphing.

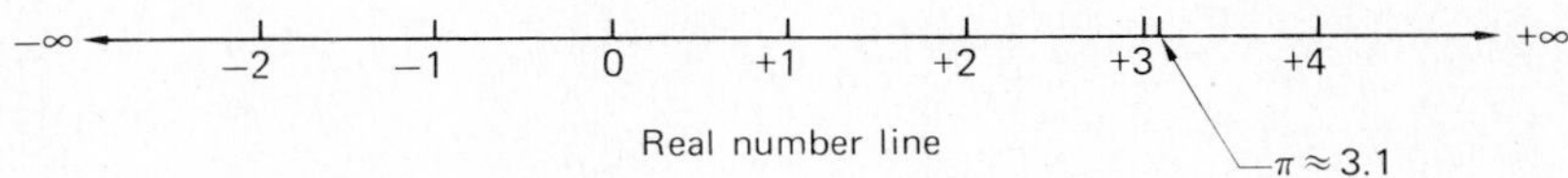

Real number line

Because the numbers on the real number line increase from left to right, we can, with the line in mind, describe many number relations using the words *right* and *left*. Here are some examples:

Algebraic statement	In symbols	Using the real number line
7 is greater than 5	$7 > 5$	7 is to the right of 5
a is less than b	$a < b$	a is to the left of b
all numbers greater than 10	$x > 10$	numbers to the right of 10

Frequency tables and histograms give rise to two kinds of questions:

1. How many measurements are there at a given value or in a given set of values (interval)? We get the answer directly from the table or histogram. This tells us where the measurements are concentrated and where they are sparse.
2. What proportion or percentage of the measurements is less than some given number? We can get the answer by adding group frequencies, but we get it more readily from cumulative frequencies when we have them. These tell us the proportions less than or greater than various numbers: What fraction of students finish at least the 11th grade of secondary school?

To get cumulative frequencies, we add, or accumulate, the group frequencies as we go from the group of smallest measurements to the group of largest measurements. Thus the frequencies of lengths of pipes required for building materials shown in Table 1-4a yield the cumulative frequencies in Table 1-4b.

TABLE 1-4

Frequency distributions for a set of 40 measurements (in feet) of lengths of pipes

TABLE 1-4(a)

Measurement interval	*Frequency*
3.0–3.9	13
4.0–4.9	23
5.0–5.9	3
6.0–6.9	1
Total	40

TABLE 1-4(b)

Measurements less than	*Cumulative frequency*	*Percent*†
3.0	0	0
4.0	13	32.5
5.0	36	90
6.0	39	97.5
7.0	40	100

† Note: We get the percent by dividing the cumulative frequency by the total, 40, and multiplying by 100.

TABLE 1-4(c)

Measurements greater than or equal to	*Cumulative frequency*	*Percent*
7.0	0	0
6.0	1	2.5
5.0	4	10
4.0	27	67.5
3.0	40	100

Cumulative Frequencies and Their Reverses. In Table 1-4b we began the cumulating with the group of smallest measurements and cumulated toward the group of largest measurements. We can get another set of cumulative frequencies by working in the opposite direction, that is, by beginning with the group of largest measurements and cumulating toward the group of smallest measurements. This new cumulative frequency distribution (Table 1-4c) is the complement of the one in Table 1-4b, which is cumulated from the smallest measurements to the largest. For example, the cumulative frequency in Table 1-4b for measurements less than 4 is 13, and in Table 1-4c the cumulative frequency for measurements greater than or equal to 4 is 40 − 13 = 27. The corresponding cumulative percentages are 32.5 and 100 − 32.5 = 67.5. We can call either Table 1-4b or Table 1-4c the **cumulative frequency distribution;** then the other one is the **reverse cumulative distribution.** Hence, if Table 1-4b is called the cumulative frequency distribution, then Table 1-4c gives the reverse cumulative distribution, and vice versa.

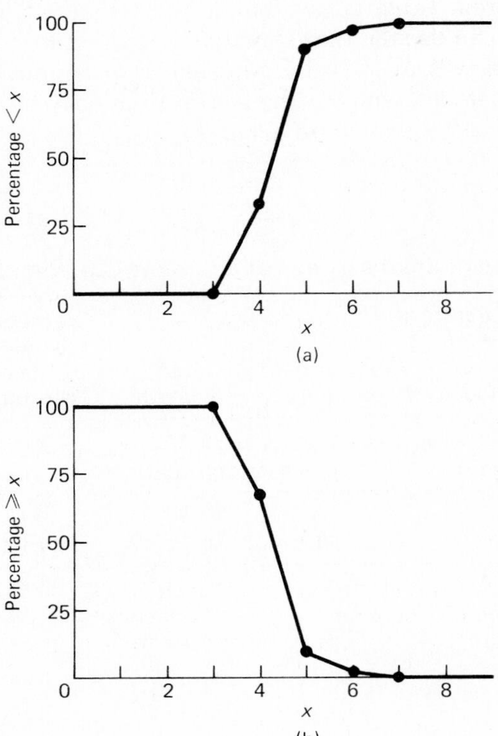

Figure 1-6 (a) A cumulative frequency diagram from Table 1-4(b). (b) Reverse cumulative frequency diagram from Table 1-4(c).

The results of Table 1-4b can be displayed in a **cumulative frequency diagram,** as shown in Fig. 1-6(a) on page 15. Thus 90 percent of the measurements are less than 5.0. The results of Table 1-4c can be displayed in the **reverse cumulative frequency diagram,** as shown in Fig. 1-6(b). Thus 10 percent of the measurements are greater than or equal to 5.0. The decision about which end to accumulate from depends on the type of information that we want to get from the graph. Figure 1-6(b) gives immediate information about percentages greater than or equal to the chosen boundary values. Figure 1-6(a) is handier if we want the number of measurements less than these boundary values.

EXAMPLE 4 Using the data of Table 1-1, construct a cumulative frequency table, and display the results in a cumulative frequency diagram.

SOLUTION. From Table 1-1, we build Table 1-5, the table of cumulative frequencies. We can use the same horizontal scale as in Fig. 1-3. However, the vertical scale will need to be adjusted to accommodate 100 percent instead of 17.19 percent, which is the largest frequency in Fig. 1-3. We then plot the cumulative frequencies to obtain Fig. 1-7(a).

TABLE 1-5†

Distribution of deaths by age at death for U.S. females in 1965

Age at death	*Percent who died in 5-year period*	*Cumulative percent*	*Age at death*	*Percent who died in 5-year period*	*Cumulative percent*
0–4	2.39	2.39	55–59	3.99	15.34
5–9	0.18	2.57	60–64	5.69	21.03
10–14	0.15	2.72	65–69	8.41	29.44
15–19	0.26	2.98	70–74	11.16	40.60
20–24	0.34	3.32	75–79	14.68	55.28
25–29	0.42	3.74	80–84	17.19	72.47
30–34	0.61	4.35	85–89	17.05	89.52
35–39	0.89	5.24	90–94	8.00	97.52
40–44	1.31	6.55	95–99	2.14	99.66
45–49	1.96	8.51	100 & over	0.35	100.01
50–54	2.84	11.35	Total	100.01	

† Constructed from data in Table 1-1.

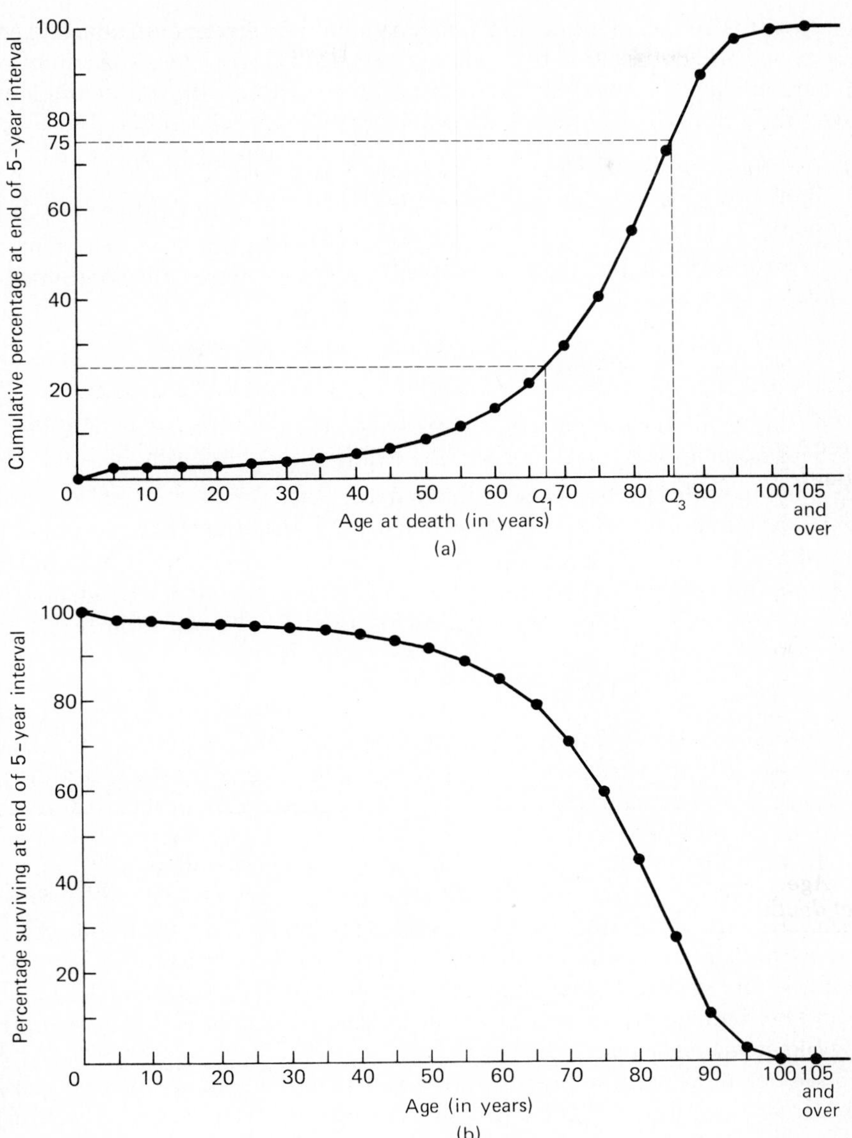

Figure 1-7 (a) Cumulative frequency distribution of age at death for U.S. females based on experience in 1965. Graph shows the percent dead at the end of each 5-year age interval. (b) Reverse cumulative frequency distribution giving survival percentages for U.S. females based on experience in 1965. Graph shows the percent surviving at the end of each 5-year age interval. Constructed from data in Table 1-5.

How to Construct a Cumulative Frequency Diagram. In constructing frequency and cumulative frequency diagrams, we may use either numbers or percentages. We use percentages here. In general, the following steps lead to a cumulative frequency diagram:

1. Get the cumulative frequency to the left of the right-hand boundary in each group. To do this, add the group frequencies, one after another in order, beginning at the leftmost group.
2. If necessary, change the cumulative frequencies to percentages. Use this rule:

$$\text{percentage} = \frac{\text{cumulative frequency} \times 100}{\text{total frequency}}.$$

   Because the frequencies in Table 1-5 were given as percentages, step 2 was not necessary in the preceding example.
3. Choose scales for the axes. Indicate percentages on the vertical axis, and choose a horizontal scale that will accommodate all groups.
4. Plot each cumulative frequency percentage above its corresponding boundary value. Then join the plotted points with line segments. For example, in Fig. 1-7(a), we plot 2.39 percent above the boundary value 5.
5. Label the figure clearly.

Interpretation. Figure 1-7(a) clearly shows that less than 10 percent of deaths occur before age 50, that from age 50 to age 90 there is a marked increase in the percentage of deaths, and that about 90 percent of deaths have occurred by age 90.

By cumulating from the other end, we get a happier view of Table 1-5. Figure 1-7(b) shows the percentage still alive at the end of the interval. Basically, Fig. 1-7(a) and Fig. 1-7(b) give the same information, but which one to present depends a bit on the point we are trying to make. For example, if we were designing old age assistance programs, Fig. 1-7(b) would be likely to be more informative because it would tell the percentage eligible at various ages.

Note that the cumulative diagram does **not** indicate the peaks and pits as vividly as does the histogram in Fig. 1-3. But Figs. 1-7(a) and 1-7(b) much more readily give information about "less than" and "more than," respectively.

Figure 1-8 shows some cumulative distributions and their corresponding frequency distributions or histograms. A good many cumulatives look something like Fig. 1-8(a). We might call the lazy-S shape the standard cumulative shape. In interpreting cumulatives, observe that a flat segment as in Fig. 1-8(b) means "no measurements in the corresponding group or

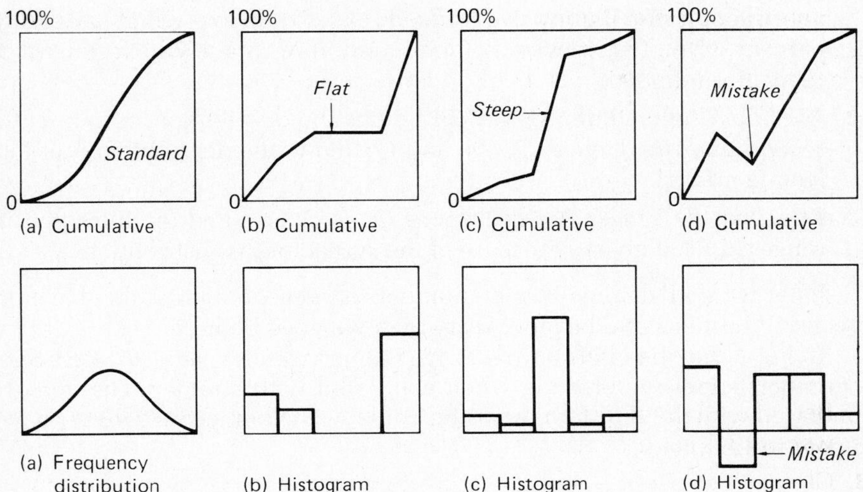

Figure 1-8 Graphs showing cumulative distributions with their corresponding frequency distributions or histograms.

groups." A steep upward slope as in Fig. 1-8(c) means "many measurements in the corresponding group or groups." Finally, a downward slope as in Fig. 1-8(d) means that the person who is constructing the figure has made a mistake.

Cumulative curves that begin cumulating with the group of smallest measurements have no direction to go but up or straight ahead. They cannot descend. Cumulations beginning with the group of largest measurements make cumulative curves that have no direction to go but down or straight ahead. They cannot go up.

Finding quartiles from a cumulative graph. Sometimes it is convenient to divide a distribution of measurements into four equal parts. The numbers that mark these divisions are called quartiles. Look at Fig. 1-7(a). The cumulative graph shows that at about 67 years, 25 percent of the females no longer survive; thus ¼ of the female population dies by age 67. Also, we see from the graph that about 75 percent fail to survive past age 86. We call 67 the first quartile of the distribution and denote it by Q_1; we call 86 the third quartile of the distribution and denote it by Q_3. Thus 50 percent or one-half of females die between ages 67 and 86.

The following are some general notions for large numbers of measurements:

1. *First and third quartiles.* The first quartile of a distribution is a number, Q_1, at or below which at least 25 percent of the distribution is found, and at or above which at least 75 percent is found. Similarly, the third

quartile of a distribution is a number, Q_3, at or below which at least 75 percent of the distribution is found, and at or above which at least 25 percent is found.

2. *Median.* The median, M, which divides the distribution in half, is the second quartile. Hence, Q_1, M, and Q_3 divide the distribution into four equal parts.
3. *Interquartile range.* The difference $Q_3 - Q_1$ is called the interquartile range (IQR). It gives a measure of the spread of the distribution.

Later we shall deal with small numbers of measurements and give more precise definitions and be more fussy than we need be here.

In Fig. 1-7(a) the interquartile range is approximately $86 - 67 = 19$. Such a number is more interesting when compared with another. This may be compared with the IQR for men using Table 1-2. Other uses for the IQR will appear in Chapter 2.

PROBLEMS FOR SECTION 1-3

1. Use Fig. 1-6(a) to find the percentage of measurements less than 5. Use Fig. 1-6(b) to find the percentage of measurements greater than or equal to 5.

Survival of U.S. Females

2. Use both Fig. 1-7(a) and Fig. 1-7(b) to answer the following questions:
   a) What percentage of females died before age 70?
   b) What percentage of females died at age 70 or later?
   c) During what 20-year age interval did the largest percentage of females die?
   d) Why did such a small percentage die after age 95?
   e) Check Q_1 and Q_3 and the IQR.
3. Use the cumulative frequency graph in Fig. 1-6(a) or 1-6(b)
   a) to estimate the median number of observations, M,
   b) to find Q_1, Q_3, and IQR.

Temperatures. Use the data in Table 1-3 for spring and autumn temperatures to answer Problems 4 through 8.

4. Make a cumulative frequency table for spring temperatures.
5. Make a cumulative frequency table for autumn temperatures.
6. Make a cumulative diagram for spring temperatures.
7. Make a cumulative diagram for autumn temperatures.
8. In spring, what temperatures occur less than (a) 25 percent of the time, (b) 75 percent of the time?

Baseball Scores. In the American League in 1968, differences in scores (winner's score minus loser's score) for the 810 games played were those listed with their frequencies in Table 1-6. Use these data to solve Problems 9 through 11.

TABLE 1-6
Frequency distribution of differences in baseball scores (American League, 1968)

Difference in scores	*Frequency*	*Percent*	*Cumulative frequency*	*Cumulative percent*
1	282			
2	165			
3	124			
4	89			
5	58			
6	28			
7	24			
8	15			
9	11			
10	5			
11	3			
12	4			
13	1			
14	1			
Total	810			

Source: Constructed from data in *Official Baseball Guide for 1969*, C. Roewe and P. MacFarlane, editors; C. C. Johnson Spink, publisher. St. Louis: *The Sporting News*.

9. Complete the last three columns in this table.

10. Construct a cumulative frequency diagram.

11. Use the table or the diagram to draw some conclusions about score differences in baseball in 1968.

Small World. The following information is for Problems 12 through 16. Travers and Milgram (1969) conducted an experiment to find out how many intermediaries were needed to reach one person in the United States from another person through a chain of personal acquaintances. Among 296 people, 196 from Nebraska and 100 from Boston, 64 chains were completed. Each person was asked to mail a packet to the "target" person if the sender knew the target person on a first-name basis, or to the person the sender knew on a first-name basis and believed most likely to so know the target person. Table 1-7 gives the distribution of the lengths of the completed chains.

12. Make a cumulative frequency table from Table 1-7.

13. Make a cumulative frequency graph.

14. Use the cumulative frequency graph to estimate the median length of the chains of acquaintances required to get from one person to another in the nation. Find Q_1, Q_3, and the IQR.

TABLE 1-7
Lengths of completed chains of acquaintances

Length	*Frequency*
1	0
2	0
3	2
4	3
5	8
6	14
7	8
8	16
9	6
10	2
11	2
12	3
≥13	0
Total	64

Source: J. Travers and S. Milgram (1969). An experimental study of "the small world problem." *Sociometry* 32:435.

15. Had it been possible to persuade the people who interrupted the chains to continue, and if all chains had gone to completion, do you think the median would have been larger or smaller? Explain why. (Fact: The median stopping point for interrupted chains was 2.)

16. Considering the evidence in Table 1-7, do you think the median length of the chains of acquaintances to reach the president of the United States would be longer or shorter than the median length of chains to reach a random person? Explain why.

1-4 SUMMARY OF CHAPTER 1

In this chapter we have met the following ideas:

1. The collection, organization, and interpretation of data provide an important basis for action in people's everyday affairs.

2. Frequency histograms offer one method of organizing data and displaying them for interpretation. Looking at the whole distribution often provides clues that can guide action.

3. For large numbers of measurements or observations, the method of grouping the data into classes is useful. We lose some of the information in making such groupings, but we make the data more manageable and more revealing.

4. Questions about percentages of measurements are conveniently answered from a cumulative frequency diagram, obtained by adding the frequencies in successive classes.

5. After looking at a distribution, one may want summary statistics. The first quartile, Q_1, the median, M, and the third quartile, Q_3, divide the distribution into four approximately equal parts. Values for Q_1, M, and Q_3 can be readily obtained from a cumulative frequency diagram. The interquartile range ($IQR = Q_3 - Q_1$) is a measure of the spread of a distribution.

SUMMARY PROBLEMS FOR CHAPTER 1

Small World Revisited. In a study similar to the one described for Problems 12 through 16 of Section 1-3, Korte and Milgram (1970) selected 600 starting persons in Los Angeles, each of whom was asked to mail a packet to a preselected "target." Table 1-8 gives the distributions of the lengths of both completed and uncompleted chains. Each uncompleted chain is listed according to the last report on its progress. Use the data from Table 1-8 to solve Problems 1 through 7.

1. Make a cumulative frequency table for the completed chains.
2. Make a cumulative frequency table for the uncompleted chains.
3. Make a cumulative frequency graph for the completed chains.
4. Make a cumulative frequency graph for the uncompleted chains.
5. Use the cumulative frequency graph in Problem 3 to estimate the median length of chains of acquaintances required to get from one person to another in Los Angeles. Find Q_1, Q_3, and the IQR.
6. Compare the answers for Problems 1, 3, and 5 with those for Problems 12, 13, and 14 of Section 1-3 for the original small-world study. Keep in mind that the Los Angeles study involved more than twice as many people as the original one.

TABLE 1-8
Lengths of completed and uncompleted chains for Los Angeles study

Length	Frequency Completed	Frequency Uncompleted
1	0	0
2	0	90
3	2	88
4	4	78
5	14	51
6	22	37
7	32	38
8	18	31
9	11	15
10	8	14
11	4	7
12	3	6
13	4	5
14	1	5
15	2	2
$\geq$16	1	7
Totals	126	474

Source: Adapted from: C. Korte and S. Milgram (1970). Acquaintance networks between racial groups: application of the small world method. *Journal of Personality and Social Psychology*, 15:101–108; and from H. C. White (1970), Search parameters for the small world problem. *Social Forces*, 49:259–264.

7. Does the information on the uncompleted chains in your answers to Problems 2 and 4 affect what you think the median would be if all chains had gone on to completion? Explain your answer.

Sentence Lengths. For Problems 8–12, choose a work of fiction.

8. Turn to page 89 of a work of fiction and count the number of words in the first full sentence. Record this number. Then continue the process until you have 50 successive sentences.

9. Use your data to make a frequency table.

10. Construct a frequency histogram.

11. Construct a cumulative frequency histogram.

12. If, in Problem 11, x is the number of words in a sentence, use the results of Problem 11 to find the median M of x, and also Q_1, Q_3, and the IQR.

13. Figure 1-9 shows a "living" histogram formed by arranging a group of college students, males and females, by their heights. Describe the special patterns that appear.

Figure 1-9 Histogram of a sample of college students arranged by height.
Source: Brian L. Joiner (1975). Living histograms. *International Statistical Review*, Vol. 43, No. 3, p. 339.

Figure 1-10 Histogram of a sample of female college students arranged by height.
Source: Brian L. Joiner (1975). Living histograms. *International Statistical Review*, Vol. 43, No. 3, p. 340.

14. Figure 1-10 on page 25 shows a "living" histogram for only the females from Fig. 1-9. Note the features in this histogram, and use them to explain the special patterns in your answer to Problem 13.

REFERENCE

F. Mosteller, W. H. Kruskal, R. F. Link, R. S. Pieters, and G. R. Rising (editors) (1973). *Statistics by Example: Exploring Data.* Reading, Mass.: Addison-Wesley.

2 Summarizing and Comparing Distributions

Learning Objectives

1. Organizing data by using tallies and stem-and-leaf diagrams
2. Acquiring skill in using stem-and-leaf diagrams to summarize data
3. Calculating and interpreting two locators of the central value of measurements in a distribution: the mean and the median
4. Using logarithms to examine very wide-ranging positive values

2-1 ORGANIZING DATA: TALLIES AND STEM-AND-LEAF DIAGRAMS

Frequency histograms offer vivid displays of data, especially pointing to the presence of peaks and pits, spikes and holes. But it takes a lot of time to make a histogram. Let us therefore look at two faster, informative methods of arranging data.

EXAMPLE 1 *Counties in Vermont.* The areas of the 14 counties in the state of Vermont are as follows (unit: 10 square miles):

78, 67, 61, 53, 66, 66, 8, 47, 69, 72, 93, 71, 78, 96

How can we arrange these data in order to bring out their information and give us some feeling for the data as a whole?

Method 1: Tallies. To make a tally:

1. We divide the data into convenient nonoverlapping groupings such as 0–9, 10–19, 20–29, and so on.
2. We indicate group frequencies with tally marks:

|	indicates 1 observation
||	indicates 2 observations
|||	indicates 3 observations
||||	indicates 4 observations
卌	indicates 5 observations
卌 |	indicates 6 observations
	. . .
卌 卌	indicates 10 observations

3. We separate the grouped measurements from their frequencies with a vertical bar, as shown in Fig. 2-1. This tally gives us some useful information:

   a) *Separation.* The data separate into three groups: the main group of counties with areas in the 40–79 range, a single county with area in the 0–9 range, and two larger counties with areas in the 90–99 range. Thus, a legislature wishing to pass laws dealing with counties by area might find it convenient to make groupings consistent with these three. An attempt to make a division in the 60–69 interval would likely lead to rancorous discussion and would require strong justification.

   b) *Centering.* The distribution seems to be centered in the 60–69 group.

   c) *Spread.* The total spread, or range, of the areas is less than 100 units, and about 80 percent fall in an interval of length 40 units.

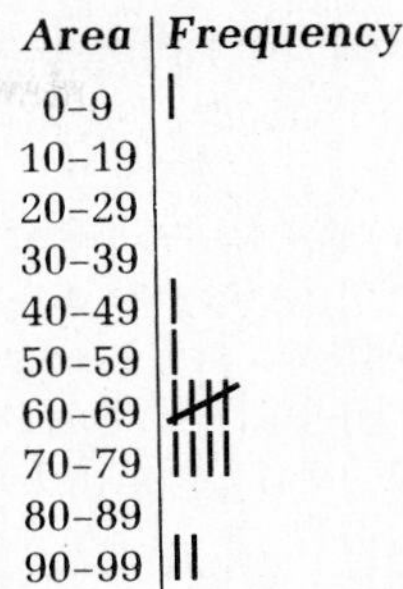

Area	Frequency
0–9	|
10–19	
20–29	
30–39	
40–49	|
50–59	|
60–69	~~||||~~
70–79	||||
80–89	
90–99	||

Figure 2-1 Tallying data into groups. Areas of counties in Vermont (unit: 10 square miles).

Source: Constructed from data in *The World Almanac & Book of Facts 1980*, edited by G. E. Delury, p. 244. New York: Newspaper Enterprise Association.

Method 2: Stem-and-leaf diagrams. Modifications of the tally give us a display that yields still more information. We proceed by the following steps:

1. Instead of listing the group intervals on the left of the vertical bar, we use the tens' digits, as in Fig. 2-2(a).
2. For each measurement, we place on the right side of the vertical bar its units' digit opposite its listed tens' digit. For example, 78 is entered thus: 7|8.
3. We may rearrange the units' digits on the right in ascending order of magnitude to get Fig. 2-2(b).

Figures 2-2(a) and 2-2(b) are called **stem-and-leaf diagrams.** Each row is called a **stem;** the numbers on the left of the vertical bar are **stem ends,** and

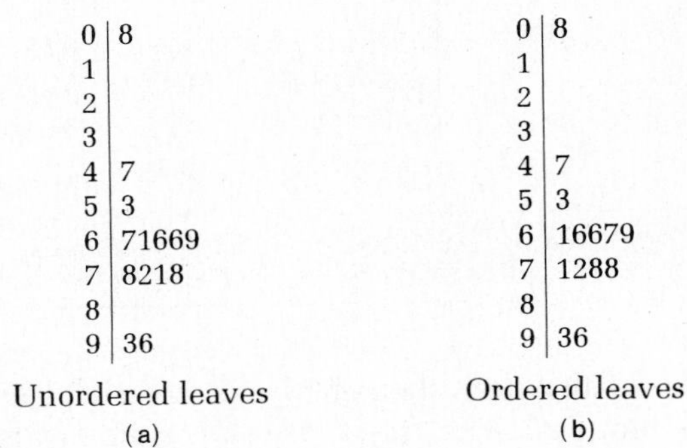

(a) Unordered leaves

0	8
1	
2	
3	
4	7
5	3
6	71669
7	8218
8	
9	36

(b) Ordered leaves

0	8
1	
2	
3	
4	7
5	3
6	16679
7	1288
8	
9	36

Figure 2-2 Stem-and-leaf diagram for areas of counties in Vermont (unit: 10 square miles).

Source: Constructed from data in the *World Almanac & Book of Facts 1980*, edited by G. E. Delury, p. 244. New York: Newspaper Enterprise Association.

the numbers on the right of the bar are called **leaves.** It is usually a good idea to use one-digit leaves. This may require the use of stem ends that have two or more digits.

The stem-and-leaf diagram is more informative than the tally. For example, we can recover to two digits the actual measurements from the stem-and-leaf diagram by inspection. The measurements in the 70–79 range are seen at once to be 71, 72, 78, 78. Indeed, the stem-and-leaf diagram offers most of the features of a histogram plus detailed information about the measurements.

To see the full effectiveness of stem-and-leaf diagrams in organizing data, let us look at a larger set of measurements, a historical set that helped create the field of industrial quality control.

EXAMPLE 2 *Cost of electrical insulation.* To reduce the cost of electrical insulation, a substitute insulating material was extensively tested for its electrical resistance. The first 50 measurements are given in Table 2-1. Reading from left to right beginning at the top row gives the order in which measurements were made. Use these data to make (a) a tally and (b) a stem-and-leaf diagram.

TABLE 2-1

Electrical resistances of insulation (unit: 100 megohms)

50.5	43.5	43.5	39.8	42.9	44.3	44.9	42.9	39.8	39.3
36.5	37.6	33.0	36.9	34.6	52.0	51.0	46.4	51.0	54.5
46.4	47.2	48.1	45.7	44.1	40.7	45.7	51.9	47.3	46.4
46.4	49.0	47.9	48.5	47.0	46.0	41.1	44.1	41.8	47.9
47.9	43.4	49.0	57.5	47.4	50.0	49.0	42.6	41.7	38.5

Source: W. A. Shewhart (1931). *Economic Control of Quality of Manufactured Product*, p. 20. Princeton, N.J.: D. Van Nostrand. By permission of copyright holder, Brooks/Cole Publishing Company.

SOLUTION. We use two-digit stem ends and one-digit leaves and follow the steps used in Example 1. The results are shown in Fig. 2-3 and Fig. 2-4. We can see from either display that the distribution is centered around 45 to 47. The figure is rather asymmetrical, with a long tail toward the small resistances, unusual because long tails usually go toward high measurements. The distribution spreads very widely, and the manufacturer proba-

```
33.0–33.9 | |
34.0–34.9 | |
35.0–35.9 |
36.0–36.9 | ||
37.0–37.9 | |
38.0–38.9 | |
39.0–39.9 | |||
40.0–40.9 | |
41.0–41.9 | |||
42.0–42.9 | |||
43.0–43.9 | |||
44.0–44.9 | ||||
45.0–45.9 | ||
46.0–46.9 | ||||/
47.0–47.9 | ||||/ ||
48.0–48.9 | ||
49.0–49.9 | |||
50.0–50.9 | ||
51.0–51.9 | |||
52.0–52.9 | |
53.0–53.9 |
54.0–54.9 | |
55.0–55.9 |
56.0–56.9 |
57.0–57.9 | |
```

Figure 2-3

```
33 | 0
34 | 6
35 |
36 | 59
37 | 6
38 | 5
39 | 388
40 | 7
41 | 178
42 | 699
43 | 455
44 | 1139
45 | 77
46 | 04444
47 | 0234999
48 | 15
49 | 000
50 | 05
51 | 009
52 | 0
53 |
54 | 5
55 |
56 |
57 | 5
```

Figure 2-4

Figure 2-3 Tally of electrical resistances given in Table 2-1.
Figure 2-4 Stem-and-leaf diagram of electrical resistances given in Table 2-1.

bly is displeased with this and wants action taken. Finally, there are outliers at each end. Again, these suggest the need for remedial action in the manufacturing process. Wild observations suggest that the process is not "in control," not behaving properly.

PROBLEMS FOR SECTION 2-1

1. Use Fig. 2-4 to answer the following:

a) What is the spread of the measurements?
b) What measurements seem most popular?
c) Estimate the center of the distribution.
d) Does the distribution separate into isolated groups?

2. Make a stem-and-leaf diagram for the 50 measurements shown in Table 2-2.

3. Continuation. For the stem-and-leaf diagram in Problem 2, answer the four questions in Problem 1.

TABLE 2-2
Electrical resistances of insulation (unit: 100 megohms)

44.5	46.5	41.7	42.6	41.7	43.8	41.8	45.5	44.5	28.6
29.2	43.8	43.8	43.6	40.9	50.0	43.4	50.0	46.4	43.4
50.0	46.2	42.2	42.8	42.8	50.0	46.2	47.4	42.2	47.0
47.0	47.0	47.0	41.0	41.0	39.4	37.0	36.5	44.5	40.0
48.5	50.0	45.6	47.0	43.1	43.1	50.0	45.8	47.0	44.3

Source: W. A. Shewhart (1931). *Economic Control of Quality of Manufactured Product*, p. 20. Princeton, N.J.: D. Van Nostrand. By permission of copyright holder, Brooks/Cole Publishing Company.

TABLE 2-3
How property tax burdens compare in 88 cities: typical property tax paid in 1972 on a single-family home with a market value of $40,000

	Dollars			*Dollars*	
W	640	Albuquerque, N.M.	C	1480	Madison, Wis.
S	680	Atlanta, Ga.	N	1240	Manchester, N.H.
N	1320	Baltimore, Md.	S	720	Memphis, Tenn.
S	160	Baton Rouge, La.	S	600	Miami, Fla.
W	800	Billings, Mont.	C	1640	Milwaukee, Wis.
S	320	Birmingham, Ala.	C	840	Minneapolis, Minn.
W	760	Boise, Idaho	S	280	Mobile, Ala.
N	1680	Boston, Mass.	S	560	Nashville, Tenn.
N	720	Bridgeport, Conn.	N	2320	Newark, N.J.
N	1600	Buffalo, N.Y.	N	1120	New Haven, Conn.
C	1280	Cedar Rapids, Ia.	S	240	New Orelans, La.
S	520	Charleston, S.C.	N	680	New York City, N.Y.
S	280	Charleston, W.Va.	S	520	Norfolk, Va.
S	680	Charlotte, N.C.	W	1040	Oakland, Calif.
C	1000	Chicago, Ill.	W	600	Ogden, Utah
C	760	Cleveland, Ohio	W	600	Oklahoma City, Okla.
W	880	Colorado Springs, Col.	C	1000	Omaha, Nebr.
S	480	Columbia, S.C.	C	1000	Peoria, Ill.
C	560	Columbus, Ohio	N	800	Philadelphia, Pa.
W	640	Dallas, Texas	W	600	Phoenix, Ariz.
W	760	Denver, Col.	N	920	Pittsburgh, Pa.
C	1240	Des Moines, Ia.	N	1120	Portland, Me.
C	960	Detroit, Mich.	W	960	Portland, Ore.

Property Taxes. To compare property tax burdens in the United States, the property tax paid on a single-family home with market value $40,000 was listed by *U.S. News and World Report* for 88 cities, as shown in Table 2-3. These cities divide into four geographic groups:

Northeast (N)

Southeast (S)

Central (C)

West (W)

The letter N, S, C, or W to the left of the city name indicates the group into which the city falls. Use Table 2-3 to solve Problems 4 through 8.

TABLE 2-3
(Cont.)

	Dollars			*Dollars*	
C	840	Fargo, N.D.	N	1080	Providence, R.I.
S	440	Fort Smith, Ark.	W	480	Reno, Nev.
C	1040	Fort Wayne, Ind.	S	640	Richmond, Va.
C	840	Grand Rapids, Mich.	C	720	St. Louis, Mo.
W	920	Great Falls, Mont.	C	720	St. Paul, Minn.
S	560	Greensboro, N.C.	W	1120	Salem, Ore.
N	1560	Hartford, Conn.	W	600	Salt Lake City, Utah
W	360	Honolulu, Hawaii	W	560	San Antonio, Texas
S	240	Huntington, W.Va.	W	760	San Diego, Calif.
C	1000	Indianapolis, Ind.	W	880	San Francisco, Calif.
S	40	Jackson, Miss.	W	520	Seattle, Wash.
S	440	Jacksonville, Fla.	C	1000	Sioux Falls, S.D.
N	1800	Jersey City, N.J.	W	600	Spokane, Wash.
C	1040	Kansas City, Kans.	C	560	Springfield, Mo.
W	560	Las Vegas, Nev.	S	480	Tampa, Fla.
S	400	Lexington, Ky.	W	560	Tulsa, Okla.
C	1040	Lincoln, Nebr.	W	640	Tucson, Ariz.
S	440	Little Rock, Ark.	N	600	Washington, D.C.
W	1000	Los Angeles, Calif.	N	880	Wilmington, Del.
S	360	Louisville, Ky.	C	1120	Wichita, Kans.
S	600	Macon, Ga.	N	1640	Worcester, Mass.

Source: Basic data from U.S. Department of Commerce, *U.S. News and World Report*, November 26, 1973, p. 110.

4. Make a stem-and-leaf diagram of the amounts listed for the Northeast.
5. Make a stem-and-leaf diagram of the amounts listed for the Southeast.
6. Make a stem-and-leaf diagram of the amounts listed for the Central region.
7. Make a stem-and-leaf diagram of the amounts listed for the West.
8. Continuation. Comment on the results obtained in Problems 4 through 7.

Sentence Lengths. A book was chosen, and in each of 50 sentences taken from the book the number of words was counted and recorded. The following counts were obtained for the 50 sentences:

41, 36, 72, 21, 36, 10, 34, 18, 29, 29,
24, 26, 21, 08, 19, 22, 11, 15, 24, 37,
15, 24, 15, 15, 22, 17, 19, 24, 28, 25,
20, 08, 14, 16, 14, 21, 28, 30, 24, 19,
35, 10, 28, 09, 24, 17, 35, 28, 14, 32

Use these data to solve Problems 9 through 11.

9. Construct a stem-and-leaf diagram for these data. (Use one-digit stem ends and one-digit leaves.)
10. If you did Problem 8 in the Summary Problems for Chapter 1, construct a stem-and-leaf diagram for the counts of the 50 sentences you obtained. Compare your diagram with the one from Problem 9 in this set.
11. Use the stem-and-leaf diagram of Problem 10 to answer the four questions in Problem 1.

Antelope Bands. E. T. Seton, in *The Arctic Prairies,* listed the numbers of antelopes in 26 bands seen along the Canadian Pacific Railroad in Alberta, within a stretch of 70 miles, as follows:

8, 4, 7, 18, 3, 9, 14, 1, 6, 12, 2, 8, 10,
1, 3, 6, 4, 18, 4, 25, 4, 34, 6, 5, 16, 4

Use these data to solve Problems 12 and 13.

12. Construct a stem-and-leaf diagram for these data.
13. Use the diagram of Problem 12 to answer the four questions in Problem 1.

2-2 MORE ABOUT STEM-AND-LEAF DIAGRAMS

Sometimes data cover a wide range of values, say from 10 to 10,000. For economy of space in our stem-and-leaf diagram, it is convenient to indicate the extra digits by extra vertical lines. For example:

we indicate 24 by 2 |4,
we indicate 240 by 2 ||4,
we indicate 2400 by 2|||4.

All numbers having the same number of bars belong to a "cycle." The preceding display illustrates three cycles—that is, numbers of two digits, three digits, and four digits. If we had the numbers

4	8	12
25	85	96
342	445	679

in our distribution, we could tabulate them as shown in Fig. 2-5. We have made the numbers 4 and 8 into 4.0 and 8.0 to produce three cycles here. The number 0 is ordinarily not used as a stem when we have several orders of magnitude, although we did use it in Fig. 2-2, which had just one measurement not in the same cycle.

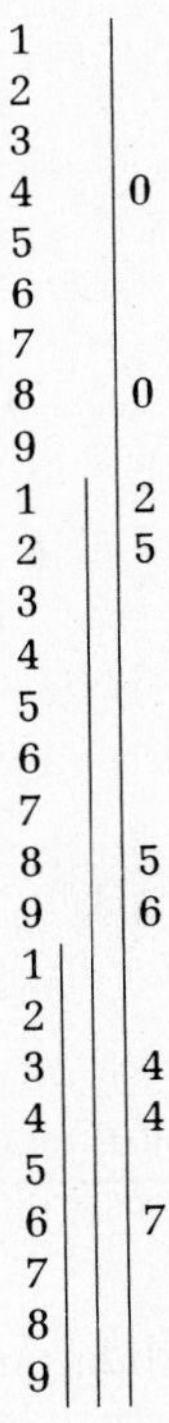

Figure 2-5 Stem-and-leaf diagram for wide-ranging values.

Note that we "cut" 679 to 6|||7 instead of rounding up to 6|||8. Sometimes rounding is preferable, but cutting usually is speedier in making stem-and-leaf diagrams and ordinarily is about as useful.

EXAMPLE 3 *Awards.* In 1973, under the Emergency School Aid Act, various U.S. school districts were given awards for special educational projects. In Tennessee the sizes of the awards in dollars were as shown in Table 2-4. Figure 2-6 shows a stem-and-leaf diagram for these amounts.

TABLE 2-4
Awards for projects in Tennessee

Dollar amount	*Logarithm*†
428,975	5.63
149,508	5.17
141,648	5.15
142,028	5.15
135,490	5.13
191,598	5.28
70,602	4.85
2,212,460	6.34
1,271,019	6.10
202,286	5.31
226,261	5.35
207,449	5.32
167,170	5.22
64,856	4.81
12,165	4.09
83,001	4.92
19,308	4.29
37,684	4.58

† These logarithms will be needed later.

Discussion. The stem-and-leaf diagram shows that the distribution was asymmetrical; most awards were less than $200,000. The spike at $100,000 suggests that there may have been a special reason for awards with this stem. The display also emphasizes that two of the awards were exceptionally large, one over $1 million and another over $2 million.

The distributions of the sizes of things occurring in nature, such as city sizes, heights of mountains, and lengths of rivers, tend to have such shapes, often with long tails like this one. Note that the stem-and-leaf diagram

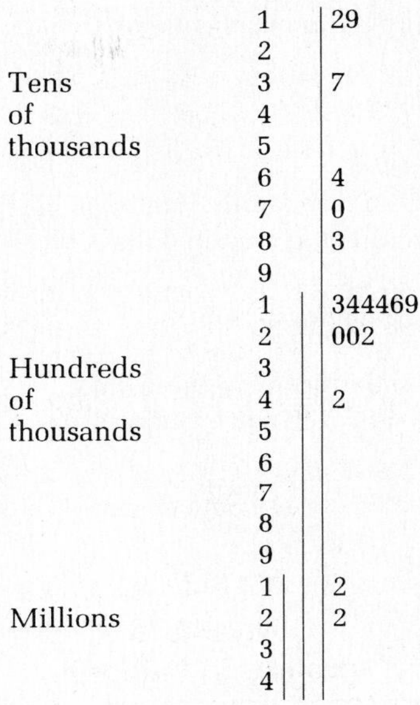

Figure 2-6 Stem-and-leaf diagram for the distribution of the sizes of monetary awards to compensatory educational projects in 1973 for the state of Tennessee. Constructed from data in Table 2-4.

collapses the numbers, with each additional vertical | corresponding to one more power of 10.

It is a feature of a stem-and-leaf diagram running over several cycles that the counts tend to pile up at the low end of each cycle. This is a bit troublesome in thinking about the shape of the distribution. In Fig. 2-6 note that each cycle has counts in the 1-stem. Out of 18 awards, 9 had 1-stems, and 4 had 2-stems. Later we shall describe a way to make a smoother stem-and-leaf diagram.

PROBLEMS FOR SECTION 2-2

1. Indicate on a stem-and-leaf diagram the following numbers:

34,670, 721,950, 5,250,000

2. Exhibit the following numbers on a stem-and-leaf diagram:

3	6	9
34	67	92
351	665	937

3. Make a stem-and-leaf diagram using the following numbers:

47	56	65
472	567	658
4725	5673	6584

4. *Stereo headphones.* In a comparative study of stereo headphones, 23 different models were purchased. The prices, in dollars, were as follows:

37, 50, 60, 69, 50, 75, 40, 45, 70, 60, 30, 50,
25, 28, 30, 30, 35, 39, 40, 75, 29, 35, 60

a) Make a stem-and-leaf diagram of these data.
b) Comment on the shape of the distribution.

School Aid. Table 2-5 displays California awards made in 1973 under the Emergency School Aid Act. Use these data to solve Problems 5 and 6.

TABLE 2-5
Awards for projects in California

111,734	278,329	153,741
544,197	58,590	486,019
431,985	256,156	827,917
875,318	665,939	2,224,778
229,195	358,442	161,237
470,920	59,656	48,523
90,662	498,000	12,720
50,369	44,684	33,066
61,155	218,870	71,806
16,956	122,578	329,306

5. Make a stem-and-leaf diagram of the California awards.
6. Comment on the similarities and differences in the shapes of the distributions for California (given in Problem 5) and for Tennessee (given in Example 3).

2-3 A QUICK PICTURE: SUMMARIZING THE DATA

In life, we are faced with a vast array of information. Many sources teem with sets of measurements and observations that, on occasion, require analysis. For convenience, the working statistician has developed methods of putting the information into capsule form. A short, informative summary saves both space and time.

What information do we want in a summary? Two vague ideas are especially important:

1. the location of the **central value** of the measurements in the distribution
2. the **spread,** or extent to which the measurements are spaced out

We shall describe two commonly used locators of the central value of a distribution, the mean and the median, which are often called **measures of location.**

To describe the spread of the measurements, we sometimes use the **range,** which is the difference between the two extreme measurements in the distribution. Alternatively, and frequently, we arrange the measurements in ascending order, from left to right, on a line. We then find three numbers, Q_1, M, and Q_3, that divide the distribution into four nearly equal numbers of measurements as shown.

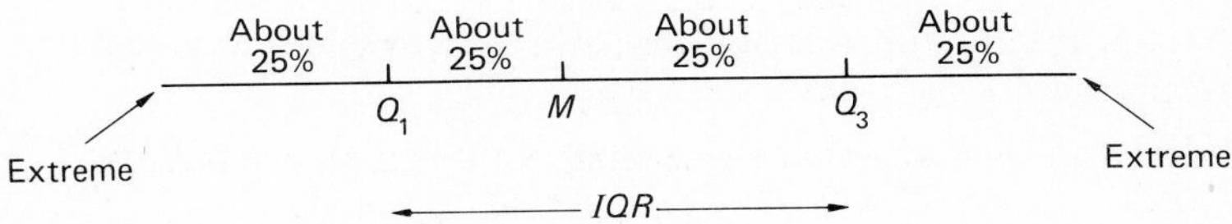

The IQR is the distance $Q_3 - Q_1$, which contains about 50 percent of the measurements. It is a very useful measure of spread, as we stated in Section 1-3. Note that $M - Q_1$ need not be close in size to $Q_3 - M$.

With the aid of some simple examples, we shall now show how to calculate the values of Q_1, M, and Q_3. We use n to denote the number of measurements, and we assume that the n measurements have been arranged in ascending order on a line.

Locators of the Central Value of Measurements in a Distribution. The most common locator of the center is the well-known *mean*, or average, used in arithmetic. If we have a set of measurements,

$$7 \quad 9 \quad 10 \quad 15 \quad 16, \tag{1}$$

then the mean of the measurements is

$$\frac{7 + 9 + 10 + 15 + 16}{5} = 11.4.$$

Add the measurements and divide by the total number of measurements to get the mean.

A second locator for the center is the middle-sized number among the measurements. This number is called the *median*. When the numbers are ordered from least to greatest in set (1), the middle measurement, or median, is 10. If the number of measurements, n, is *odd*, as in set (1), the median is the $(n/2 + 1/2)$th measurement from either end. In (1), $n/2 + 1/2 = 5/2 + 1/2 = 3$, so we take the 3rd measurement, 10. It is convenient to think of the measurements as being ranked from least to greatest, with the least having rank 1 and the largest rank n.

If n is *even*, we get two measurements in the middle, as in set (2):

$$7 \quad 9 \quad 10 | 15 \quad 16 \quad 20. \tag{2}$$

Then we use the average of the two middle measurements,

$$\frac{10 + 15}{2} = 12.5,$$

as our median. For even n, we take the average of the $n/2$th and $(n/2 + 1)$st measurements for the median.

We shall frequently use medians as our center locators because:

a) medians divide the distribution into halves, and thus have a ready interpretation,

b) medians are less affected by extreme values than are means.

EXAMPLE 4 Find the mean and the median for the set of measurements in Example 1, Section 2-1.

SOLUTION.

$$\text{mean} = \frac{\text{sum of measurements}}{n}$$

$$= \frac{78 + 67 + 61 + 53 + 66 + 66 + 8 + 47 + 69 + 72 + 93 + 71 + 78 + 96}{14}$$

$$= \frac{925}{14} = 66.1.$$

Because $n = 14$, which is even, we have two middle measurements—the 7th (14/2) and the 8th (14/2 + 1). In Fig. 2-2(b), where the 14 measurements are arranged in ascending order, we find the 7th and 8th measurements to be 67 and 69, respectively. Hence,

$$\text{median} = \frac{67 + 69}{2} = 68.$$

Note that in both the odd case and the even case we are finding the measurement with rank $\frac{1}{2}(n + 1)$. In the odd case we get one of the measurements in the distribution, and in the even case we interpolate to a number in the interval between the two middle measurements. What is going on is that we are really working on the intervals between the measurements rather than the measurements themselves. One measurement divides the line into two pieces, two measurements divide it into three pieces, and, more generally, n measurements divide it into $n + 1$ intervals. For these data, which come as individual measurements, rather than in groups, we shall be thinking of the intervals when we compute quartiles.

A Measure of Spread. We want a number Q_1 such that 25 percent of the measurements are less than Q_1. Just as $\frac{1}{2}(n + 1)$ helped us with the median, $\frac{1}{4}(n + 1)$ helps us with Q_1. In set (3),

	7	9	10	15	16	20	22	(3)
				↑ M				
Rank	1	2	3	4	5	6	7	

the number 15 is the median, M. There are three numbers to its left. It is natural that the first quartile will be the median of these three numbers, 9. If we compute

$$\frac{1}{4}(n + 1) = \frac{1}{4}(7 + 1) = 2$$

we shall pick the measurement with rank 2, which agrees with our choice of 9.

Similarly, Q_3 has rank $\frac{3}{4}(n + 1)$. For set (3), this gives us $\frac{3}{4}(8) = 6$, or $Q_3 = 20$.

When $n + 1$ is not divisible by 4, we can interpolate if we are being careful, or round off if we are being rough and ready. In set (4),

$$2 \quad 10 \quad 13 \quad 15 \quad 16 \quad 19 \quad 23 \quad 30 \qquad (4)$$

$$\tfrac{1}{2}(n + 1) = \tfrac{1}{2}(9) = 4\tfrac{1}{2}, \quad \tfrac{1}{4}(n + 1) = \tfrac{1}{4}(9) = 2\tfrac{1}{4},$$

$$\tfrac{3}{4}(n + 1) = \tfrac{27}{4} = 6\tfrac{3}{4},$$

$$M = \frac{15 + 16}{2} = 15.5.$$

To get Q_1, we interpolate $\frac{1}{4}$ of the way from the 2nd to the 3rd measurement; the interval has length $13 - 10 = 3$:

$$Q_1 = 10 + \tfrac{1}{4}(13 - 10) = 10\tfrac{3}{4} = 10.75.$$

Similarly, to get Q_3, we interpolate ¾ of the way from the 6th to the 7th; the interval has length $23 - 19 = 4$:

$$Q_3 = 19 + \tfrac{3}{4}(23 - 19) = 19 + 3 = 22.$$

The interquartile range is

$$IQR = Q_3 - Q_1 = 22 - 10.75 = 11.25.$$

If we had been trying to avoid interpolation, we would have rounded 2¼ to 2 and 6¾ to 7 and used the 2nd measurement for Q_1, the 7th for Q_3.

Our summary of the measurements in set (4) is

Q_1	M	Q_3
10.75	15.5	22

$IQR = 11.25$

Note: the foregoing **summary statistics,** mean, median, and quartiles, are numbers marked on a scale. They may or may not coincide with measurements in the distribution. We expect the IQR to include about one-half the measurements and, for ungrouped data, half the $n + 1$ intervals.

EXAMPLE 5 *How is the weather?* The Jones family is moving from Arizona to Boston, Mass. Ms. Jones wants to know about the weather in Boston so that she can plan the family wardrobes. How cold is winter? Is spring colder than autumn? How warm does it get in summer? As an aid to action, Ms. Jones writes to the weather station at the Boston airport and gets the information in Table 2-6, which lists the distribution of daily minimum temperatures, season by season, for the year 1969. Find the summary statistics for spring and autumn, and use the results to answer one of Ms. Jones's questions.

SOLUTION. In this example our data are grouped, and so we cannot locate Q_1, M, and Q_3 in the same way we did for ungrouped data. There are two ways to get these summary statistics: We can interpolate within an interval or we can make a cumulative graph and read off the information as in Section 1-3. To illustrate both methods, we interpolate to get the summary statistics for spring and use the cumulative diagram to get the statistics for autumn.

When we have grouped data, we use n instead of $n + 1$ to locate Q_1, M, and Q_3. We have $n = 93$, $n/4 = 23\tfrac{1}{4}$, $n/2 = 46\tfrac{1}{2}$, and $3n/4 = 69\tfrac{3}{4}$. Cumulating

TABLE 2-6
Distribution of lowest daily temperatures recorded at Boston airport weather station in 1969

Temperature (°F)	*Number of days*				
	Spring	*Summer*	*Autumn*	*Winter*	*Total*
0–4					
5–9				1	1
10–14				6	6
15–19			1	10	11
20–24			6	27	33
25–29	2		7	19	28
30–34	5		14	23	42
35–39	15		11	3	29
40–44	11		11		22
45–49	20	3	17		40
50–54	14	3	13		30
55–59	14	17	9		40
60–64	7	30	1		38
65–69	3	23			26
70–74	2	15			17
75–79		2			2
Total	93	93	90	89	365

Source: Constructed from data in *Climatological Data*, U.S. Department of Commerce, Environmental Science Services Administration, Environmental Data Service, 1969.

the number of days for spring from Table 2-6, we get Table 2-7. We see that Q_1 falls in the interval 40–44, M in the interval 45–49, and Q_3 in the interval 55–59. We think of the observations within an interval as being spread continuously over the interval.

To get Q_1 we interpolate in the interval 40–44, which has 11 observations. There are 22 observations up to 39.5°. We want to go to 23¼, or 1¼ observations beyond 39.5°, so we have

$$Q_1 = 39.5 + \frac{23\frac{1}{4} - 22}{11} \times 5 = 39.5 + \frac{1.25}{11} \times 5 = 39.5 + \frac{6.25}{11} \approx 40.1.$$

(The symbol $\approx$ means "approximately equal to.")

TABLE 2-7†

Distribution of lowest daily temperatures recorded at Boston airport weather station, spring 1969

	Temperature interval (°F)	*Right boundary*	*Frequency*	*Cumulative frequency*
	25–29	29.5	2	2
	30–34	34.5	5	7
	35–39	39.5	15	22
$Q_1 \rightarrow$	40–44	44.5	11	33
$M \rightarrow$	45–49	49.5	20	53
	50–54	54.5	14	67
$Q_3 \rightarrow$	55–59	59.5	14	81
	60–64	64.5	7	88
	65–69	69.5	3	91
	70–74	74.5	2	93
	Total		93	

† Constructed from the data in Table 2-6.

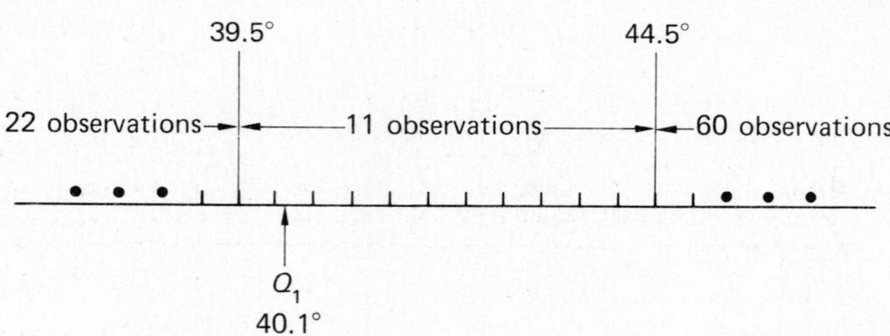

To get M we interpolate in the interval 45–49, with 20 observations:

$$M = 44.5 + \frac{46\frac{1}{2} - 33}{20} \times 5 = 44.5 + \frac{13.5}{20} \times 5 \approx 44.5 + 3.4 = 47.9.$$

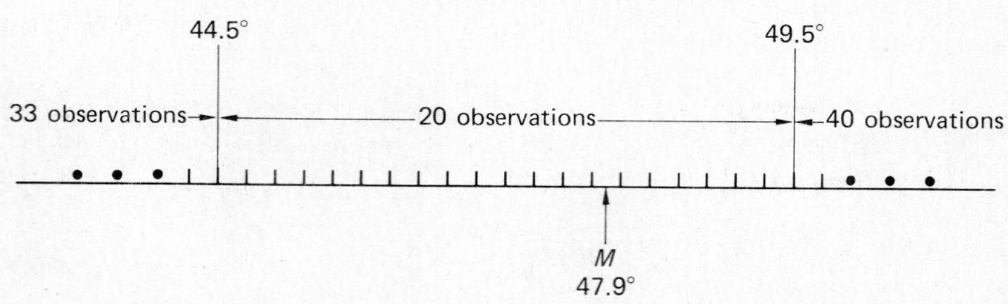

To get Q_3 we interpolate in the interval 55–59, with 14 observations:

$$Q_3 = 54.5 + \frac{69\frac{3}{4} - 67}{14} \times 5 = 54.5 + \frac{2.75}{14} \times 5 = 54.5 + \frac{13.75}{14} \approx 55.5.$$

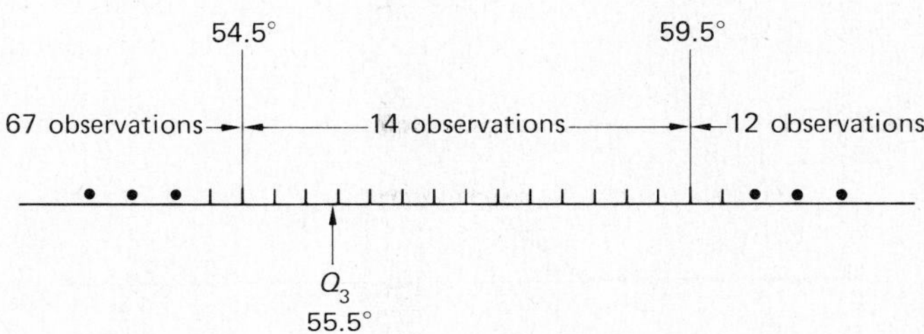

To get the summary statistics for autumn, we shall make and use a cumulative distribution diagram. First, use the data in Table 2-6 to construct the cumulative data shown in Table 2-8. The cumulative graph is shown in Fig. 2-7. From this figure we see that 25 percent or fewer of the observations record temperatures less than or equal to 32.5°, 50 percent of the observations are less than 42.2°, approximately, and 75 percent are less than 49.7°.

TABLE 2-8†

Cumulative percents of minimum temperatures for autumn 1969

Temperature interval (°F)	*Frequency*	*Cumulative frequency*	*Cumulative percent*
15–19	1	1	1.1
20–24	6	7	7.8
25–29	7	14	15.6
30–34	14	28	31.1
35–39	11	39	43.3
40–44	11	50	55.6
45–49	17	67	74.4
50–54	13	80	88.9
55–59	9	89	98.9
60–64	1	90	100.0
Total	90		

† Constructed from the data in Table 2-6.

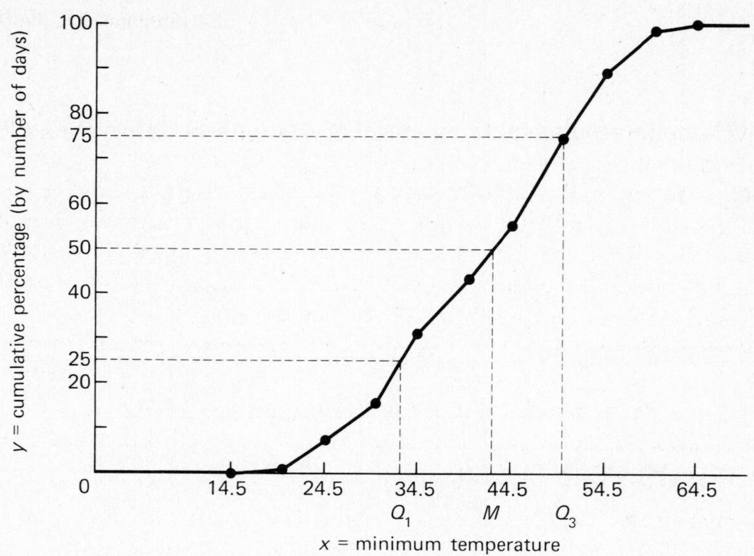

Figure 2-7 Cumulative distribution of minimum daily temperatures (°F), autumn 1969, Boston airport weather station. Constructed from data in Table 2-8.

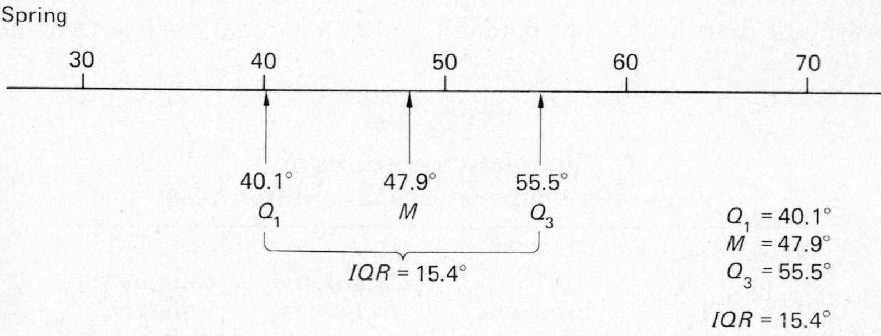

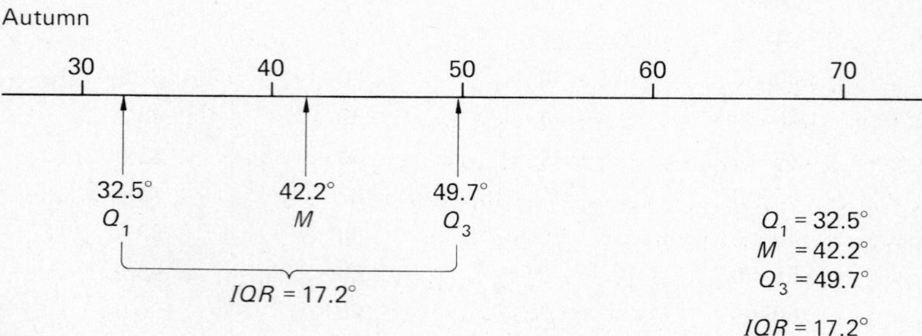

Figure 2-8 Summary of minimum daily temperatures at Boston airport weather station, spring and autumn 1969.

The summary statistics for spring and autumn are shown graphically in Fig. 2-8 on page 46. We can thus report to Ms. Jones tht autumn temperatures in Boston are about 6° colder than spring temperatures, if we take 1969 as a typical year, and autumn temperatures are more variable than spring temperatures. Figure 2-8 also shows each set of summary statistics in a vertical list, which is a usual and convenient form of report. Investigations of winter and summer temperatures are left as problems.

PROBLEMS FOR SECTION 2-3

Temperatures. Table 2-6 displays summer and winter minimum temperatures in Boston in 1969. Use these data to solve Problems 1 through 3.

1. Find the summary statistics for summer minimum temperatures.

2. Find the summary statistics for winter minimum temperatures.

3. Assuming that 1969 was a typical year, tell Ms. Jones what kind of weather she may expect in Boston in the extreme seasons.

4. *Stereo headphones.* In a comparative study of stereo headphones, 23 different models were purchased. The prices, in dollars, were as follows:

37, 50, 60, 69, 50, 75, 40, 45, 70, 60, 30, 50,
25, 28, 30, 30, 35, 39, 40, 75, 29, 35, 60

a) Make a stem-and-leaf diagram of these data, or use the one made for Problem 4, Section 2-2.

b) Find the median and quartiles, and make a schematic summary.

Heroin Purchases. Table 2-9 gives the data for Problems 5 through 18. They were gathered for the Drug Abuse Council by undercover investigators purchasing heroin in Manhattan during June 1971. The data have been rounded to the nearest gram and dollar.

5. Make a stem-and-leaf diagram for the amounts purchased.

6. Use the diagram from Problem 5 to obtain Q_1, M, Q_3, and the IQR for the amount of purchase.

7. Make a stem-and-leaf diagram for potency of heroin purchased.

8. Find the upper quartile Q_3 of potency, that is, the potency percentage above which about one-quarter of the cases lie.

9. Make a stem-and-leaf diagram for the price per raw gram of heroin.

10. Find Q_1, M, Q_3, and the IQR for price per raw gram.

11. Make a stem-and-leaf diagram for the price per pure gram of heroin.

12. Find Q_1, M, Q_3, and the IQR for price per pure gram.

TABLE 2-9
Data from 29 heroin purchases during June 1971

Purchase number	*Amount purchased (grams)*	*Potency (percent)*	*Price per raw gram (dollars)*	*Price per pure gram (dollars)*
1	23	67	55	83
2	66	43	60	140
3	375	12	1	11
4	20	8	30	365
5	20	15	25	163
6	16	11	22	203
7	12	50	53	106
8	24	28	46	162
9	19	44	64	144
10	130	34	31	90
11	14	3	9	327
12	8	26	79	310
13	27	64	48	74
14	21	57	56	99
15	20	73	64	87
16	35	45	34	77
17	112	51	33	65
18	20	11	30	276
19	16	20	40	194
20	26	61	45	75
21	22	58	56	96
22	13	61	43	70
23	27	6	44	799
24	20	12	33	279
25	6	16	26	167
26	26	46	50	108
27	30	35	33	95
28	28	14	46	324
29	30	38	41	106

Source: G. F. Brown, Jr., and L. Silverman (1973). *The Retail Price of Heroin: Estimation and Applications.* Public Research Institute, Center for Naval Analyses, Arlington, Va.

13. Find out whether the price of raw heroin per gram is less variable than the price of pure heroin per gram, using the results of Problems 10 and 12.

14. Perhaps price changes are multiplicative. Compute Q_3/Q_1 for both raw and pure heroin prices, and see which is more variable in a multiplicative sense.

15. Break purchases by the amount purchased in grams into two parts, the 7 largest purchases and the 22 smallest. List the pure gram prices for the 22 smallest purchases and for the 7 largest.

16. Find the median pure gram price for the 22 smallest purchases in Problem 15.

17. Find the median pure gram price for the 7 largest purchases in Problem 15.

18. From the results of Problems 16 and 17, are large or small purchases cheaper per gram?

Property Taxes. Use the data in Table 2-3 to solve Problems 19 through 23. Stem-and-leaf diagrams of these data were drawn for Problems 4 through 7 of Section 2-1.

19. Compute Q_1, M, Q_3, and the IQR for the Northeast (N).

20. Compute Q_1, M, Q_3, and the IQR for the Southeast (S).

21. Compute Q_1, M, Q_3, and the IQR for the Central region (C).

22. Compute Q_1, M, Q_3, and the IQR for the West (W).

23. Do the values calculated in Problems 19 through 22 adequately summarize the information in the corresponding stem-and-leaf diagrams?

24. *Project.* Obtain some data from an almanac or other source. Then make a stem-and-leaf diagram. Comment on the shape of the distribution.

2-4 LOGARITHMS FOR STEM-AND-LEAF DIAGRAMS

As we have seen, some sets of data have values ranging over numbers of very different sizes. For example, a list of governmental dollar expenditures could include such magnitudes as

$$10,\ 100,\ 10{,}000,\ 1{,}000{,}000,\ 1{,}000{,}000{,}000.$$

Such numbers can be expressed more compactly as powers of 10:

$$10^1,\ 10^2,\ 10^4,\ 10^6,\ 10^9.$$

Indeed, if we make it clear that we are talking about powers of 10, we can merely report the exponents:

$$1,\ 2,\ 4,\ 6,\ 9.$$

These exponents are called the **logarithms** of the numbers, or **logs** for short. The number 10 is called the **base** of this system of logarithms.

Another frequently used base is e ($e = 2.71828\ldots$). (Other bases are sometimes used, such as the base 2.) Many hand calculators offer the choice of 10 or e as the base. When 10 is used, the abbreviation for logarithm is log; when e is used, the abbreviation is ln. On some calculators, using e saves a step; on others, using 10 saves a step.

The range of our original numbers is about 10^9, or a billion, whereas the range of their logarithms to the base 10 is $9 - 1 = 8$. If we try to plot the original numbers on a line, the first four cluster together at one end, with the 10^9 at the other; their logarithms plot nicely. Or if we try to put the data into a few class intervals of equal length, the original numbers all clump together, but their logs distribute nicely.

More generally, if a number N is expressed as a power of 10, so that

$$N = 10^n,$$

we say

$$\log_{10} N = n.$$

(Read: The logarithm of N to the base 10 is n.)

DEFINITION When N is expressed as a power of 10, the exponent of 10 is the common logarithm of N.

If it is understood that we use base 10, we merely write

$$\log N = n.$$

Logarithms of numbers to base 10 are called **common logarithms.** Logarithms of numbers to base e are called **natural logarithms.**

EXAMPLE 6 *Awards.* Figure 2-6 shows a stem-and-leaf diagram for the distribution of the sizes of monetary awards to compensatory educational projects in 1973 for the state of Tennessee. In this figure the smallest award is about \$10,000 and the largest about \$2,200,000. The stem-and-leaf diagram is badly crowded at the low end, and the numbers straggle at the high end. Recall that the vertical lines tell how large the numbers are. One vertical line for numbers in the 10,000s, two vertical lines for numbers in the 100,000s, three for numbers in the millions. Thus the stem-and-leaf diagram takes a step toward taking a logarithm of the number by indicating the number of digits. It does not do this smoothly. Numbers from 10,000 to 99,000 are spread over the same distance as numbers from 100,000 to 990,000. This

40	9
41	
42	9
43	
44	
45	8
46	
47	
48	15
49	2
50	
51	3557
52	28
53	125
54	
55	
56	3
57	
58	
59	
60	
61	0
62	
63	4

Figure 2-9 Stem-and-leaf diagram using logarithms for the distribution of the sizes of monetary awards to compensatory educational projects in 1973 for the state of Tennessee. Constructed from data in Table 2-4.

has a tendency to make the stem-and-leaf diagram clump up at the small end of each cycle, as we have discussed.

The two peaks artificially arise because of the jerky arrangement of the stem-and-leaf diagram. The logarithms of the numbers would make a smooth transformation and eliminate the artificial piling-up that ruins our view of the shape of the data.

In Fig. 2-9 we have made a stem-and-leaf diagram of the logarithms of the awards, and we get a picture much more like the ones we are used to—piled up a bit in the middle, straggled at the ends, and nearly symmetrical. For many purposes, Fig. 2-9 is more instructive than Fig. 2-6. To get Fig. 2-9, we used the logarithms of the original numbers shown in Table 2-4.

Before we discuss the meaning of logarithms in this context, let us answer two objections that some people raise: (a) Some believe that the original measurements are inviolable and must not be changed. (b) Some believe that introducing logarithms or square roots or other transformations leads to complications they do not want to cope with. They want to keep things "simple."

Let us treat argument (b) first. If we want to force a body of data to reveal its hidden information, we have to use whatever tools we can find or make. Generally speaking, the processes of science are not simple, although sometimes they lead to answers that seem simple when compared with the complexity we perceived because of our former ignorance. That is the simplicity we seek. We know, therefore, that argument (b) often is incompatible with getting good information, and so we can set that argument aside. Life is not simple, but we try to find ways to simplify it.

Argument (a) is a little different. It overlaps argument (b), but it may stem from a misunderstanding. Some of these people may think such maneuvers are a form of cheating. To analyze our data, we may study any numbers we like. We are simply looking for better ways of describing the data and of understanding them, usually to find out more about why they are as they are. Numbers that may be complicated in the form of their original measurements may be simple on another scale. For example, if we are given the areas of squares, the area, A, makes a curvilinear relation with the length of the side, s; indeed, $A = s^2$. But if we analyze square roots of areas, we are analyzing lengths of sides, an analysis that is for some purposes preferable and even simpler. The moral is that quantities should be measured on scales that will inform us about them, not necessarily on their recorded scales.

"But," asks the skeptic, "if we allow these transformations, can't you make the data look any way you like?" If "anything goes" were the rule, that would be true. But the transformations and other devices we use are actually part of a systematic set of operations used for exploratory data analysis. Different fields have different devices in common use. Their value has been proved over and over in many problems, so that not using them becomes the exception rather than the rule. For example, economists analyzing monetary data tend to take logarithms unless they are primarily interested in totals. They think that money changes by percentages, and the idea of interest rate provides one basis for this belief.

Similarly, people who study the effectiveness of medications frequently study the logarithm of the size of the dose rather than the size of the dose itself, again with the idea of a multiplicative effect. People in the field of weather modification often study the cube root of the amount of rainfall.

What is right about the skeptic's objection is that we cannot, on the basis of one analysis with one set of data treated on a no-holds-barred basis, reach a general conclusion about the good way to analyze data of the given kind. The skeptic's mistake lies in thinking that we should tie our hands behind our backs just because we will not be able to reach rock-hard conclusions.

The use of logarithms compresses the scale of the larger numbers and spreads out the scale of the smaller ones.

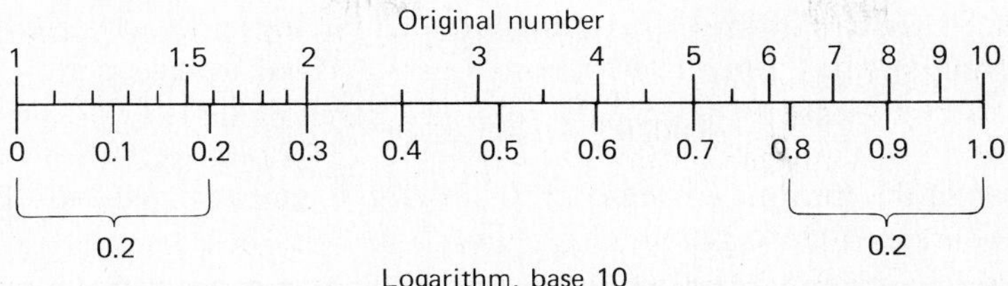

Figure 2-10 Relation between original numbers and the logarithmic scale. Note that the original numbers from 1 to 1.6 and 6.3 through 10 have about the same range in the logarithm, 0.2

EXAMPLE 7 A statistical use of logarithms. In comparing incomes of people from different occupations, we may note that *A*'s income is 1.5 times that of *B* and 10 times that of *C*. The hint is already there that logarithms may be useful. On the other hand, if it is the differences that are important, as in "*A*'s income is \$100 greater than *B*'s and \$300 greater than *C*'s," then logarithms of the original incomes may not be a good choice for analysis.

One person may say that *A*'s income is 10 times *B*'s, and another may say that *B*'s income is 0.1 of *A*'s. This leads to an awkward asymmetry. When we treat this matter with logarithms, we get the following:

$$\text{first person:} \quad \log \frac{A\text{'s income}}{B\text{'s income}} = \log 10 = 1$$

$$\text{second person:} \quad \log \frac{B\text{'s income}}{A\text{'s income}} = \log 0.1 = \log 10^{-1} = -1,$$

because

$$\log \frac{x}{y} = \log x - \log y$$

and

$$\log \frac{y}{x} = \log y - \log x.$$

Thus, on the log scale, we get the same magnitude, and it is just a matter of sign.

This reminds us that numbers less than 1 have negative logarithms. If $x > y > 0$, then $\log x > \log y$. The value y/x is less than 1, and $\log y - \log x$ will be negative. For stem-and-leaf purposes, we shall ordinarily be interested in the shape of the distribution, and we can always recover its original position. So we can evade the problem of negative logarithms by multiplying all our numbers by a power of 10. For example, if our numbers are

0.00200, 0.0362, 0.1567, . . .

and none of the numbers is smaller than 0.00200, we might just multiply all our numbers by 1000 to get rid of the less-than-1 problem.

TABLE 2-10

Populations of states in 1970 (in thousands)

	Log			*Log*	
3,444	3.54	Alabama	694	2.84	Montana
302	2.48	Alaska	1,484	3.17	Nebraska
1,772	3.25	Arizona	489	2.69	Nevada
1,923	3.28	Arkansas	738	2.87	New Hampshire
19,953	4.30	California	7,168	3.86	New Jersey
2,207	3.34	Colorado	1,016	3.01	New Mexico
3,032	3.48	Connecticut	18,241	4.26	New York
548	2.74	Delaware	5,082	3.71	North Carolina
757	2.88	District of Columbia	618	2.79	North Dakota
6,789	3.83	Florida	10,652	4.03	Ohio
4,590	3.66	Georgia	2,559	3.41	Oklahoma
770	2.89	Hawaii	2,091	3.32	Oregon
713	2.85	Idaho	11,794	4.07	Pennsylvania
11,114	4.05	Illinois	950	2.98	Rhode Island
5,194	3.72	Indiana	2,591	3.41	South Carolina
2,825	3.45	Iowa	666	2.82	South Dakota
2,249	3.35	Kansas	3,924	3.59	Tennessee
3,219	3.51	Kentucky	11,197	4.05	Texas
3,643	3.56	Louisiana	1,059	3.03	Utah
994	3.00	Maine	445	2.65	Vermont
3,922	3.59	Maryland	4,648	3.67	Virginia
5,689	3.76	Massachusetts	3,409	3.53	Washington
8,875	3.95	Michigan	1,744	3.24	West Virginia
3,805	3.58	Minnesota	4,418	3.65	Wisconsin
2,217	3.35	Mississippi	332	2.52	Wyoming
4,677	3.67	Missouri			

Source: Constructed from data in *The World Almanac & Book of Facts 1980*, edited by G. E. Delury, p. 193. New York: Newspaper Enterprise Association.

PROBLEMS FOR SECTION 2-4

Use a pocket calculator to solve the following problems:

1. Compute log 8.58, log 85.8, and log 0.858.
2. Compute log 6.1, log 61.0, log 610, and log 6100.
3. Make a stem-and-leaf diagram of the logarithms of the areas of Vermont counties (Section 2-1, Example 1).
4. Make a stem-and-leaf diagram of the logarithms of the nine numbers given at the beginning of Section 2-2. Compare this stem-and-leaf diagram with the one shown in Fig. 2-5.
5. Make a stem-and-leaf diagram of the logarithms of the California awards in Table 2-5 (Section 2-2, Problem 5). Compare with Fig. 2-9 for Tennessee.
6. Use the data in Table 2-10 to make a stem-and-leaf diagram of the logarithms of the 1970 populations of states in the United States.

2-5 SUMMARY OF CHAPTER 2

1. Tallies and stem-and-leaf diagrams provide fast, informative methods of arranging data.
2. Summary statistics save space and time; they give a short, informative picture of the data. Some useful summary statistics are the median, M, of the distribution, and the first and third quartiles, Q_1 and Q_3. The difference $Q_3 - Q_1 = IQR$, the interquartile range, which is a useful measure of the spread of the distribution.
3. The common logarithm of a number is the exponent of that number when it is expressed as a power of 10. For example:

$$\text{if } N = 10^n, \text{ then } \log N = n,$$
$$\text{if } N = e^n, \text{ then } \ln N = n.$$

4. The logarithm of a number, N, can be found using a pocket calculator (or a logarithm table).
5. We sometimes get more helpful information by examining logarithms of the data, rather than by studying the raw data.

SUMMARY PROBLEMS FOR CHAPTER 2

Real Estate. Prices for real estate vary considerably from area to area, even within or around a single city. The values of all real estate transfers for two wards in Pittsburgh and for two nearby suburbs, Fox Chapel and

TABLE 2-11

Real estate transfers for two Pittsburgh wards and two adjacent suburbs, 1980 (in thousands)

Pittsburgh			
14th Ward	*19th Ward*	*Fox Chapel*	*Wilkinsburg*
125	52	437	25
41	34	286	26
79	27	340	62
135	45	195	50
68	287	200	30
72	20	280	42
66	23		65
200	33		58
100	63		16
75	32		383
110	28		166
164	20		2
77	41		25
88	42		13
50	41		55
56	45		
57			
60			
95			

Source: Constructed from data in the *Pittsburgh Post-Gazette*, Saturday, December 13, and Saturday, December 20, 1980.

Wilkinsburg, officially recorded in a 2-week period in 1980, are listed in Table 2-11. Use these values to solve Problems 1 through 9.

1. For each of the four areas, construct a stem-and-leaf diagram of the amounts listed. Use the same scale for all four diagrams.
2. Using the diagrams from Problem 1, compute M, Q_1, Q_3, and the IQR for each area. Summarize this information and the extremes using the "quick picture" of Section 2-3.
3. Compare the prices for real estate in these four areas of Pittsburgh.
4. Compare the mean and median real estate values for the four areas. For which areas is the mean strongly influenced by extreme values?

5. Use a pocket calculator to take logarithms of the prices in Table 2-11.

6. Use the logarithms of the prices from Problem 5 to construct a stem-and-leaf diagram for each of the four areas.

7. Find Q_1, M, Q_3, and the IQR for the logarithms of the prices for each of the areas.

8. Compare your values of Q_1, M, and Q_3 with the logarithms of the values of Q_1, M, and Q_3 in Problem 2.

9. Can you suggest one or two reasons for the extreme values in the 19th ward and in Wilkinsburg?

REFERENCES

D. C. Hoaglin, F. Mosteller, and J. W. Tukey (editors) (1982). *Understanding Robust and Exploratory Data Analysis.* New York: Wiley.

D. R. McNeil (1977). *Interactive Data Analysis: A Practical Primer.* New York: Wiley.

P. F. Velleman and D. C. Hoaglin (1981). *Applications, Basics, and Computing of Exploratory Data Analysis.* Boston, Mass.: Duxbury.

3

Finding and Summarizing Relationships

Learning Objectives

1. Finding out what an equation of a straight line looks like
2. Transforming some variables so that an equation yields a straight-line graph instead of a curve graph
3. Introducing the "black-thread" method to get an equation of a straight line
4. Defining residuals and outliers from straight-line equations fitted to a set of plotted points

3-1 TARGET FOR CHAPTER 3

Related Variables and Graphs. In elementary mathematics we have met the idea of two variables, say x and y, being related so that when x takes a value, then y has a corresponding value. For example, x might be the length in inches of the side of a square, with y its area in square inches. Each pair of values (x, y) gives rise to a point on a graph, and the set of these points makes up the graph of the relation between x and y. We get the pairs of (x, y) values usually from a table or from a formula like that relating the length of side x to the area y of the square:

$$y = x^2. \tag{1}$$

Here, if $x = 3$, then $y = 3^2 = 9$, and we have the value-pair $(3, 9)$, which yields a point on the graph of $y = x^2$. We put a smooth curve through the plotted points to get the graph. The simplest graph is a straight line; it is the graph of choice. One reason is that interpolation and extrapolation are especially easy for the line; another reason is that the line is the easiest curve to describe or understand.

We now list the aims of this chapter:

Aim 1: *We find out what an equation of a straight line looks like.* For the present, we state, without proof, some facts about equations that have straight lines for graphs. The following example illustrates these facts.

EXAMPLE 1 *The case of the elusive equation.* Soon the metric system of measurements may be in force in the United States. For example, temperatures will be reported in degrees Celsius instead of degrees Fahrenheit as in the past. To help people adjust to the new temperature readings, a bank printed and distributed a short table (Table 3-1) showing temperatures in both Fahrenheit (y) and Celsius (x) degrees.

TABLE 3-1

Temperature readings in degrees Fahrenheit, F, and degrees Celsius, C, from 0°C to 100°C at intervals of 10°C

y: °F	32	50	68	86	104	122	140	158	176	194	212
x: °C	0	10	20	30	40	50	60	70	80	90	100

A student, who knew all about graphing an equation, decided to plot the given data as a guide to interpolating between values in the table. He got the result shown in Fig. 3-1 and immediately noted that all of the plotted points

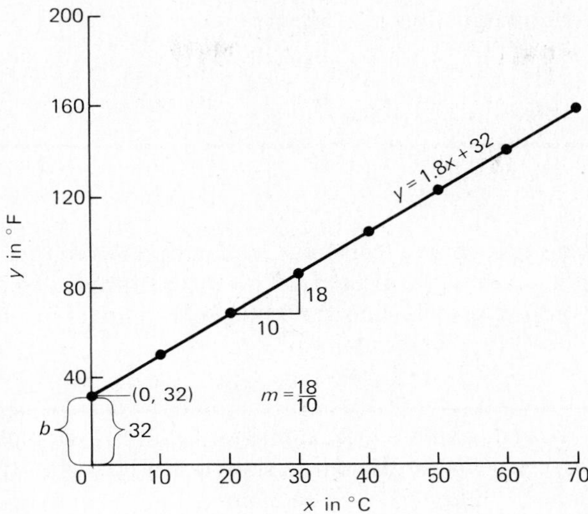

Figure 3-1 Graph of temperature data in Table 3-1 showing a straight line with slope 1.8 and y intercept 32.

were in a straight line. He reasoned that this straight line must have an equation connecting x and y, and he wondered how to find this equation. A friend inspected the table and graph and said: "Of course this line has an equation. One form is

$$y = 1.8x + 32.\text{''} \qquad (2)$$

He checked and found that all of the plotted points indeed satisfied Eq. (2). When he asked her how she got the equation so quickly, she told him three facts:

1. The **slope of a line,** denoted by m, gives

$$m = \frac{\text{change in } y}{\text{change in } x} = \frac{\text{rise}}{\text{run}}.$$

   For this example, the line, using the points (10, 50) and (20, 68), gives slope

$$m = \frac{68 - 50}{20 - 10} = 1.8.$$

2. The y **intercept of a line,** denoted by b, is the y coordinate of the point at which the line cuts the y axis. In this example, $b = 32$.

3. A formula for an equation of a line is

$$y = mx + b. \tag{3}$$

In this example, $y = 1.8x + 32$.

Aim 2: *We show how to transform some variables so that an equation yields a straight-line graph instead of a curved graph.* This procedure enables us to apply straight-line methods to relationships that are not originally expressed in a linear manner.

EXAMPLE 2 *Straightening* $y = x^2$. Consider Eq. (1). If we plot y against x in the usual way, we can get the curve shown in Fig. 3-2(a). If we change variables and let $x^2 = X$, then Eq. (1) becomes

$$y = X \quad \text{or} \quad y = 1 \cdot X + 0, \tag{4}$$

which has the form of Eq. (3), with $m = 1$ and $b = 0$. Now plot y and X, and we get the straight line shown in Fig. 3-2(b).

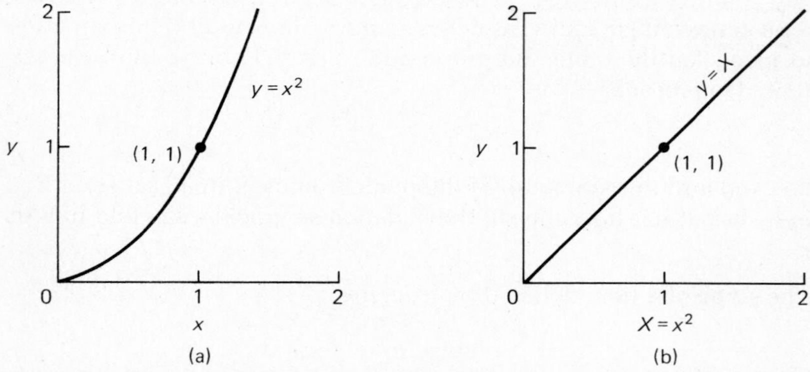

Figure 3-2 The original curve with y plotted against x as shown in 3-2(a) can be straightened by changing the scale on the x axis to $X = x^2$ as shown in 3-2(b).

An alternative method for straightening would take the square root of both sides and then plot $\sqrt{y}$ against x.

To make transformations such as the foregoing, we must know how the equation of a straight line looks. This requires us to have achieved Aim 1.

Note: Our aim in making transformations at this point is different from the aim in Chapter 2:

1. Our present aim is to straighten out the curve, and so find a linear relation.
2. In Chapter 2, our aim was to make a distribution more symmetric and reduce straggling.

Aim 3: *We introduce the "black-thread" method to get an equation of a line that approximately fits the set of plotted points.* Sometimes the corresponding values of x and y are displayed in a table. Then we can use pairs (x, y) obtained from the table to plot a set of points on a graph. The black-thread method may then be used to get an approximate formula for the relation displayed in the table. Section 3-3 offers several examples of this procedure.

Aim 4: *We define residuals and discuss outliers and illustrate their roles in the interpretation of data.*

PROBLEMS FOR SECTION 3-1

1. What are the aims of this chapter?
2. State two techniques for straightening out the graph of the curve $y = x^2$ when $x > 0$.
3. How might you straighten out the graph of the curve for $y = \sqrt{x}$?
4. Copy the following table, and for the line through *A* and *B*, fill in the blanks:

	A	B	Rise: $y_2 - y_1$	Run: $x_2 - x_1$	$m = \frac{\text{rise}}{\text{run}}$
a)	(2,3)	(4,6)	6 − 3 = 3	4 − 2 = 2	3/2
b)	(1,4)	(2,8)	. . .	. . .	. . .
c)	(0,2)	(1,4)	. . .	. . .	. . .

5. Represent the data of Problem 4b on a graph like that given at the top of page 64 for Problem 4a.
6. Find an equation of the line passing through the points (0, 10) and (3, 12).
7. Find an equation of the line passing through the points (0, 0) and (4, 3).
8. Find an equation of the line that passes through (5, 4), (6, 8), and (0, −16).

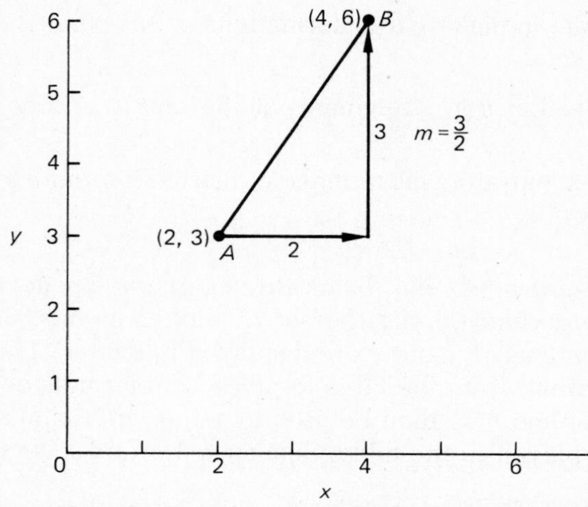

3-2 EXACT RELATIONS

In elementary mathematics we meet exact formulas representing relations between variables. For example:

$$A = \pi r^2, \quad V = \pi r^2 h, \quad d = rt, \quad c^2 = a^2 + b^2,$$

and so on. In the examples that follow, we need to have clearly in mind the idea of a linear function. Figure 3-3 shows the graph representing the linear

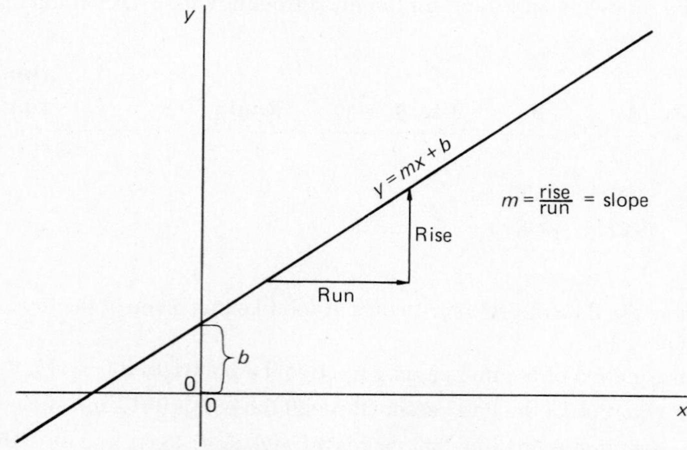

Figure 3-3 Graph of the equation $y = mx + b$.

function given by the equation

$$y = mx + b. \tag{5}$$

To be linear, a functional relation must be such that it can be put into the form of Eq. (5). If, in Eq. (5), the y intercept $b = 0$, then we have a special case, because Eq. (5) is satisfied by $x = 0$ and $y = 0$, which will mean that the line in Fig. 3-3 will pass through the origin (0, 0).

If we are given an equation of a line, then m is the increase in y that results when x is changed by one unit. More generally, m is the change in y divided by the corresponding change in x:

$$\frac{\text{change in } y}{\text{change in } x} = \frac{\text{rise}}{\text{run}} = \text{slope} = m.$$

EXAMPLE 3 *Distance as a function of time.* The distance d traveled by a body moving at a rate of r feet per second for time t seconds is d feet, where

$$d = rt.$$

This distance d can be regarded as a linear function of the time t and the slope r, a constant, because we have

$$d = rt + 0,$$

which has the form of Eq. (5), with $m = r$ and $b = 0$. In physics, this exact linear relation is called a law. The graph of $d = 5t$ is shown in Fig. 3-4. In

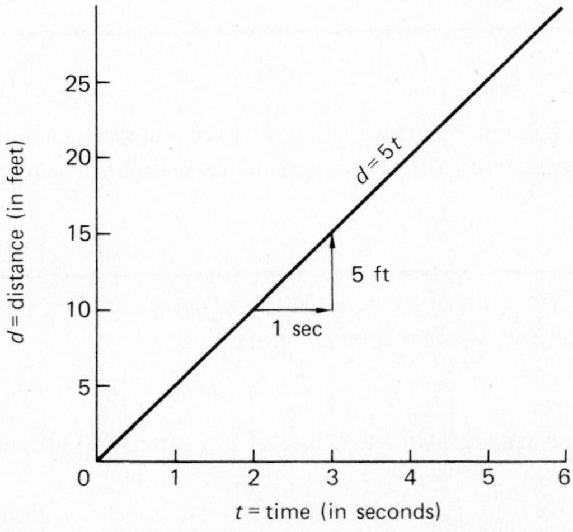

Figure 3-4 Distance as a linear function of time ($r = 5$ ft/sec).

observations, errors of measurement may produce deviations from this law. An experimenter may have measured the following distances:

t	Formula d	Observed Distance	Deviations = Observed − Formula
1	5	5.2	5.2 − 5 = 0.2
2	10	10.1	10.1 − 10 = 0.1
3	15	14.8	14.8 − 15 = −0.2

We see that the observed values of d do not fall precisely on the straight line $d = 5t$. We call these deviations residuals.

DEFINITION A **residual** is the difference between the observed value and the value estimated from an equation of the fitted line.

EXAMPLE 4 *Line not through origin.* If the body in Example 3 starts 8 feet from a point and goes directly away from it at the 5-ft/sec rate, the distance away at time t is $d = 5t + 8$, again a linear function, but one that does not pass through the origin.

At times we have a relation between two variables that is not linear, but often we can transform one of the variables to get a linear relation.

EXAMPLE 5 *The area of a circle.* If the radius of a circle is r units, then the area of the circle is A square units, where

$$A = \pi r^2 \quad (\pi = 3.14159\ldots). \tag{6}$$

We say that A is a **quadratic function** of r. Figure 3-5 shows the curvilinear relation.

It may be more convenient to deal with a linear relation than with a quadratic one. As we know from our work in Section 3-1, to get a linear

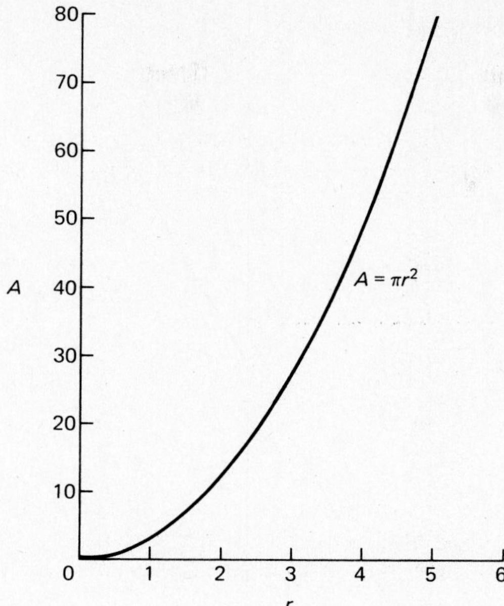

Figure 3-5 Area, A, of circle as a quadratic function of r, the length of the radius.

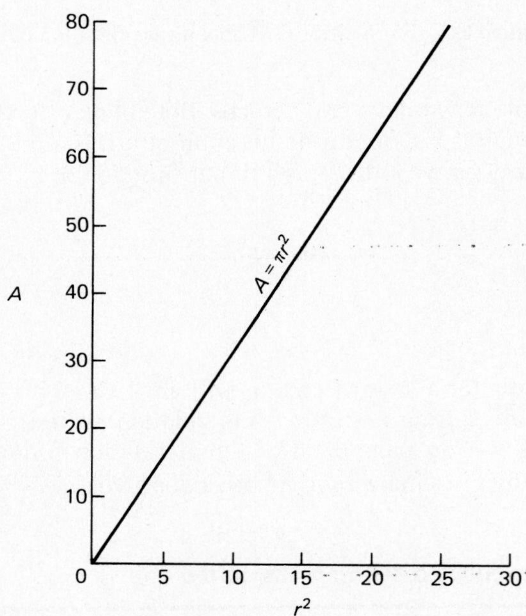

Figure 3-6 Area, A, of a circle as a linear function of r^2.

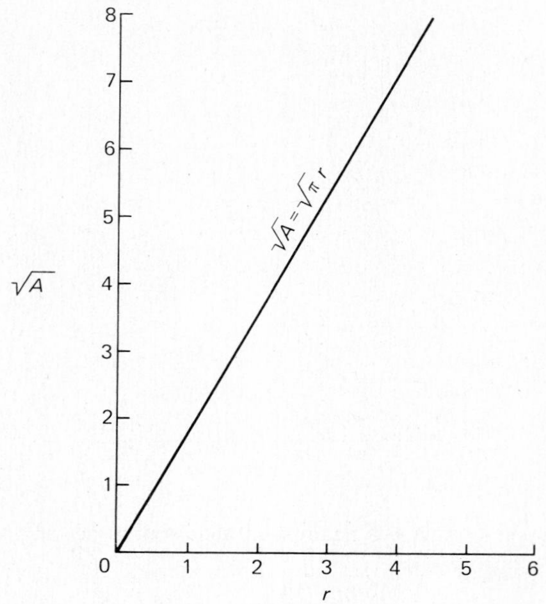

Figure 3-7 For a circle, $\sqrt{A}$ as a linear function of r.

relation in Example 5, we have two choices. Both choices involve a change of variable. Although A is a quadratic function of r, it is a linear function of r^2. See Fig. 3-6. How do we know A is a linear function of r^2? Let $x = r^2$ and $y = A$, and we have

$$A = \pi r^2$$

transformed into

$$y = \pi x,$$

which is the formula for a linear function with $m = \pi$ and $b = 0$.

Another way of getting a straight-line relation would be to plot $\sqrt{A}$ against r. See Fig. 3-7 on page 67. The linear relation follows because if $A = \pi r^2$, then taking the square root of both sides gives $\sqrt{A} = \sqrt{\pi} r$, or

$$\sqrt{A} = \sqrt{\pi} r + 0,$$

which has the form $mr + b$, with $m = \sqrt{\pi}$ and $b = 0$.

Sometimes we suspect that there is a functional relation between two variables, x and y, and we wish to find a formula for it. If the relation is linear or approximately linear, a careful graph of y plotted against x usually helps to choose or "fit" a straight line for the points that we have.

Fitting a Line. When several points fall on the same straight line, the line clearly fits them. When points do not fall on the same straight line, we may wish, nevertheless, to summarize the positions and relations of the points by a straight line that passes among them. Choosing the line is the problem. Often mathematical criteria are available to measure how well the line and points fit together. For example, we might want equal numbers of points above and below the line. Or we might want the sum of the distances of the points from the line to be small. A frequently used criterion is the sum of the squares of the vertical distances of the points from the line—minimizing the sum gives the "least-squares line." Drawing a straight line "by eye" to pass among several points can often be as helpful as these formal methods. One such approach, the black-thread method, lets each person choose his or her own line *subjectively*, as we shall discuss in Section 3-3.

If the points lie along a curve, we still may fit them using a straight-line relation. One possibility is to plot the variable y against some simple function of x, such as x^2, $\sqrt{x}$, $1/x$, log x, or e^x, where $e = 2.71828. \ldots$ If necessary, we can also plot simple functions of y. If any of these plottings turns out to yield a straight line, we have found a linear relation.

Getting Linearity for Various Functions. Let us emphasize the linearity of some relations.

1. If

$$y = 3x^2 + 2 \quad (x \geq 0),$$

then we can set $t = x^2$ and get the transformed equation

$$y = 3t + 2,$$

which gives y as a linear function of t, because we have the form $y = mt + b$, with $m = 3$ and $b = 2$.

2. Again, if

$$y = \frac{2}{\sqrt{x}}$$

and we set

$$t = \frac{1}{\sqrt{x}},$$

then

$$y = 2t,$$

which represents a straight line through the origin of the t–y plane with slope 2 ($m = 2$, $b = 0$).

3. Finally, if $y = 3(10^x)$, we might take logarithms to the base 10 and get

$$\begin{aligned} \log y &= \log [3(10^x)] \\ &= \log 3 + \log (10^x) \\ &= \log 3 + x \log 10 \\ &= \log 3 + x. \end{aligned}$$

Then, if we set $z = \log y$, we get the linear relation

$$z = x + \log 3.$$

Because $z = 1 \cdot x + \log 3$, z is a linear function of x ($m = 1$, $b = \log 3 \approx 0.477$).

EXAMPLE 6 *The planets and their distances from the sun.* Astronomers have long been interested in law-like relations for interplanetary distances. In Table 3-2 we list the known planets and their true distances, d_n, from the sun. (Here the distances are measured in units of $\frac{1}{10}$ times the Earth's distance from the sun.) If we number planets by their order, moving outward

TABLE 3-2

Ordered planets (n) counting outward from the sun; their distances, d_n, from the sun; logarithms of the distances (distances measured in units of $\frac{1}{10}$ times the earth's distance from the sun)

Planet	n	d_n	$\log_{10} d_n$
Mercury	1	3.87	0.588
Venus	2	7.23	0.859
Earth	3	10.00	1.000
Mars	4	15.24	1.183
Asteroids	5	29.00	1.462
Jupiter	6	52.03	1.716
Saturn	7	95.46	1.980
Uranus	8	192.0	2.283
Neptune	9	300.9	2.478
Pluto	10	395.0	2.597

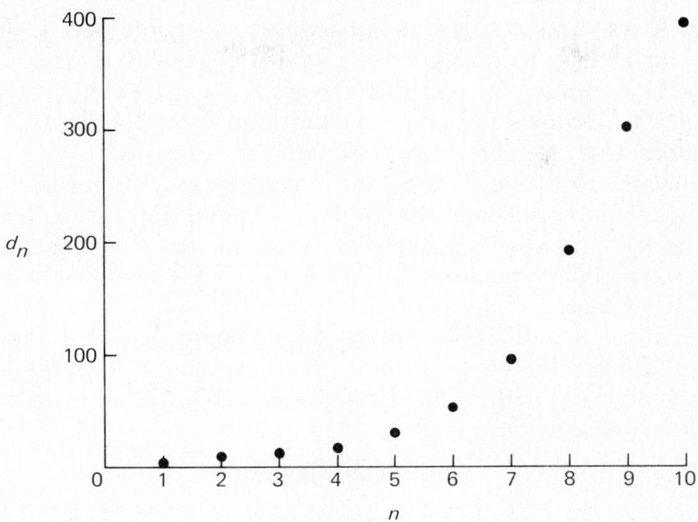

Figure 3-8 Distance d_n of planet from the sun plotted against n, the order number of the planet from the sun. Constructed from data in Table 3-2.

from the sun, then Mercury is 1, Venus 2, and so on. In Fig. 3-8 we plot d_n, the distance from the sun to the planets, against n, the order number of the planet from the sun.

A quick look at Fig. 3-8 is all we need in order to see that the relation between distance and order number is not well described by a straight line. The distances increase at a much more rapid rate than do the order

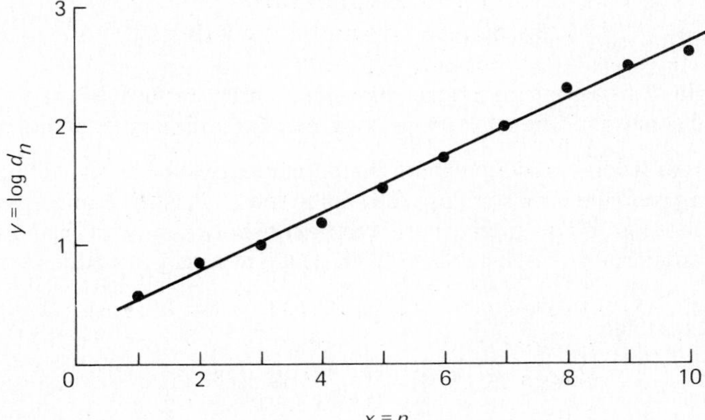

Figure 3-9 Plot of $\log d_n$ against order number of the planet. The straight line has equation $y = 0.330 + 0.233x$. Constructed from data in Table 3-2.

numbers. If we draw straight lines between successive points, the run is always 1, but the rise increases rapidly with n. We could nevertheless draw a smooth curve through the points on the graph. In order to convert this type of relation into a straight line, we can either transform the order numbers or the distances. Here we choose to work with the distances.

Whenever successive changes of the same amount in one variable lead to increasingly larger changes in another variable, statisticians like to take logarithms of the rapidly increasing one. Thus if we take $y = \log_{10} d_n$ (as given in Table 3-2) and plot y against n, the order number is as shown in Fig. 3-9 on the preceding page.

From Fig. 3-9 and Table 3-2 we immediately see that the relation between the logarithm of the distance from the sun and the planet order number is very close to that of a straight line. In Fig. 3-9 we have drawn the straight line with equation

$$y = 0.330 + 0.233x. \qquad (7)$$

All of the points lie very close to this straight line.

PROBLEMS FOR SECTION 3-2

1. Which of the following formulas define y as a linear function of x (m and b are constants)?

(a) $y = 3^x + 2$ (b) $y = \frac{1}{2}x - 5$ (c) $y = \dfrac{m}{x} + b$

(d) $y = m + x + b$ (e) $y = 3x^3 + 2$ (f) $y = \pi 10^x$

(g) $y = (x^2 - 2x)/(\sqrt{x})^2$

2. A straight line passes through the points (3, 5) and (4, 7). Use these points to find a rise and corresponding run, the slope of the line, and the y intercept b.

For convenience in distinguishing the dependent variable (usually called y) in the following problems, we shall use the capital letters B, C, D, E and F instead of y. In each of these problems a precise relation exists between two variables. Find the relation. Transform the x variable in each problem to get the points on a straight line.

3. Given these data:

B	3	6	11	18	27	38	51
x	1	2	3	4	5	6	7

Find a way to plot the relation between B and x to get the points on a straight line.

4. Given these data:

C	5	11	21	35	53	75	101
x	1	2	3	4	5	6	7

Find a way to plot the relation between C and x to get the points on a straight line.

5. Given these data:

D	1024	512	256	128	64	32
x	1	2	3	4	5	6

Find a way to plot the relation between D and x so that the points fall on a straight line.

6. Given these data:

E	120	60	40	30	24
x	1	2	3	4	5

Find a way to plot the relation between E and x so that the points fall on a straight line.

7. Find the slope when F is plotted against x^2 for the following data, and then give the formula relating F and x^2.

F	0.283	0.785	1.54	2.54
x	0.3	0.5	0.7	0.9

For each of the following equations, give a transformation that will put the equation into the form $y = mx + b$ (not necessarily using these letters).

8. $y = 4\pi x^2$

9. $y = 5x^3 - 10$

10. $y = 3/\sqrt{x}$

11. $y = 5(10^x)$

3-3 APPROXIMATE RELATIONS: THE BLACK-THREAD METHOD

Sometimes the relations between the variables are approximate because of measuring errors, or approximate linearity, as we shall next illustrate; at other times a relation is a statistical one, such as that between height and weight for human beings.

EXAMPLE 7 *Temperatures below the earth's surface.* Temperatures have been taken at various depths in an artesian well at Grenoble, France.

The following table, derived from these data, uses a point about 28 meters below the earth's surface as the origin for both temperature and depth. The number of meters below the origin is x, and y is the number of degrees Celsius above the temperature at the origin.

Number of meters below origin, x	40	150	220	270
Number of degrees above the temperature at the origin, y	1.2	4.7	9.3	10.5

These data are plotted in Fig. 3-10. Fit a line through the origin to them.

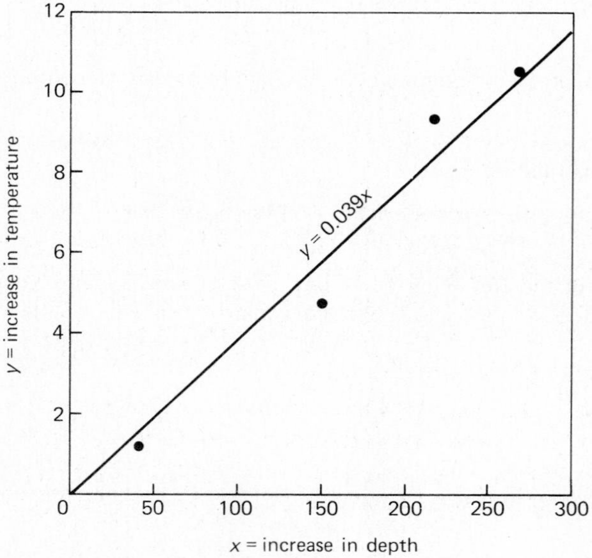

Figure 3-10 Temperature changes with increased depth in an artesian well at Grenoble, France.

Source: Adapted from F. Mosteller, R. E. K. Rourke, and G. B. Thomas, Jr. (1970). *Probability with Statistical Applications*, second edition, p. 396.

SOLUTION. We use the black-thread method, as follows: Take a black thread and stretch it taut so that it passes over the origin and close to the points. Then find the slope of the line it makes. When we did this, we got a slope of 0.039, and so an equation for the fitted line is

$$y = 0.039x.$$

Another observer might get a slightly different slope. Suppose we round it to $y = 0.04x$. We note that our observations do not necessarily fall on our fitted line. For example, for $x = 150$, the corresponding point on the line has $y = 0.04(150) = 6.0$. We call the difference that results, $4.7 - 6.0 = -1.3$, the **residual.**

Black-Thread Method. Lines not going through the origin can also be fitted by the black-thread method. We put the thread down so that it looks like it fits the points, and we read off two points on the thread. These can be used to compute the slope, m, of the line. If one of the points is the y intercept, we also have b; if not, then more effort is required, and

$$b = y_1 - mx_1,$$

where (x_1, y_1) is one of the points we have read off.

EXAMPLE 8 *Voting registration and turnout.* Figure 3-11 shows, for large U.S. cities, the relationship between voting registration and voting turnout, each expressed as a percentage of the adult population. Apply the black-thread method to Fig. 3-11 to get a line relating turnout percentage (T) to registration percentage (R).

SOLUTION. Method A: Two points. To illustrate the method just described, let us choose two values of R and read off values of T from our black thread. For $R_1 = 40$ we read $T_1 = 20$, and for $R_2 = 80$ we read $T_2 = 64$ (you may read something else). Then, for

$$T = b + mR,$$

we compute

$$m = \frac{T_2 - T_1}{R_2 - R_1} = \frac{64 - 20}{80 - 40} = \frac{44}{40} = 1.1$$

$$b = T_1 - mR_1 = 20 - 1.1(40) = -24.$$

Thus we approximate the line by the equation

$$T = 1.1R - 24.$$

Method B: Slope-intercept. On an independent fit, we extend the black thread so that it crosses the y axis. We read the intercept as -16 and the

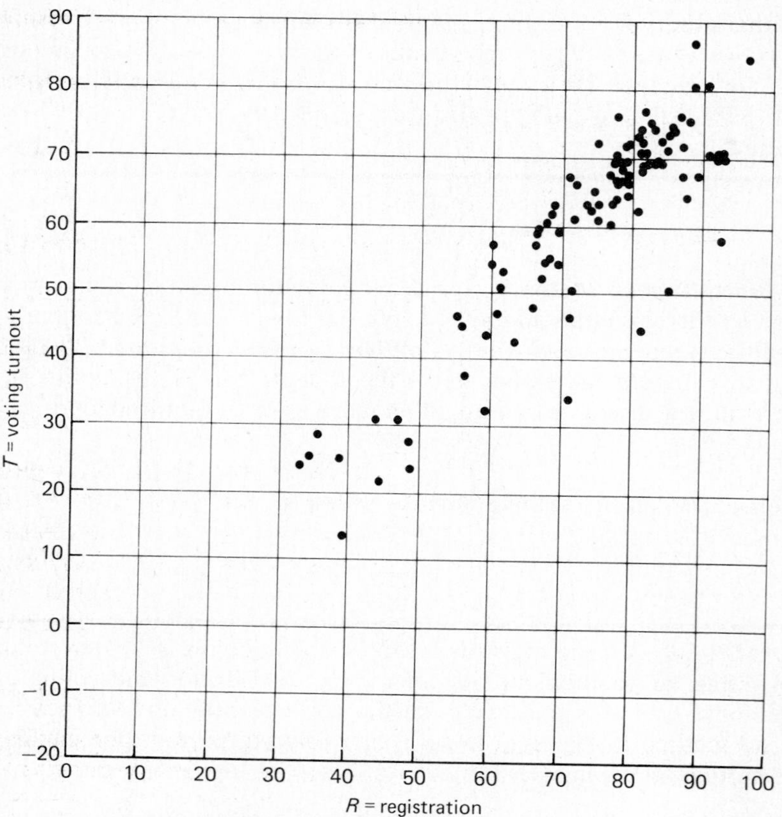

Figure 3-11 City voting registration and election-turnout percentages.
Source: Constructed from data in E. R. Tufte (1978). Registration and voting. In *Statistics: A Guide to the Unknown*, second edition, edited by J. M. Tanur, F. Mosteller, W. H. Kruskal, R. F. Link, R. S. Pieters, and G. R. Rising; special editor, E. L. Lehmann; p. 200. San Francisco: Holden-Day, Inc.

slope as 1 and get the equation

$$T = R - 16.$$

That we get two different straight lines using the two methods should not surprise us, because different people using the same method will almost always get somewhat different answers. This is why we say that the black-thread method is subjective. The important point in this example is that both of our lines provide a fairly good fit to the points, and they do not differ very much. We shall take up more formal methods of fitting lines later in this book, but approximate methods are adequate to get the main idea here.

PROBLEMS FOR SECTION 3-3

1. In Fig. 3-11, choose $R_1 = 30$ and $R_2 = 70$, and read off the corresponding values of T from your black-thread line. Compute m and b and get an approximate equation for the black-thread line.
2. Use the formula obtained in Example 8 (Method A) to find T when $R = 60$.
3. Use the line obtained in Problem 1 to find a value of T when $R = 60$, and compare it with the value obtained in Problem 2.
4. Figure 3-11 shows the relation between voting registration and voting turnout for large U.S. cities. Why are there no points above the 45-degree line running from (0,0) to (100,100)?

3-4 STUDYING RESIDUALS

When we fit a curve to points, we often wish to study the departures from the curve. As we have seen, these departures are often called **residuals,** or errors. *Residuals* may be the better term, because *error* tends to suggest that someone made a mistake and that we have found it. There may be no mistake, and even if there is, we may not have found it. So the more neutral term, *residual,* is preferable.

EXAMPLE 9 *Measuring the coefficient of gravitational attraction.* A body falling freely from rest in a vacuum obeys the law $s = \frac{1}{2}gt^2$, where s is the distance in feet, t is the time in seconds, and g is the gravitational constant whose value we want to estimate. An experimenter measured the time t as the body fell 1 foot, 2 feet, . . . , 5 feet. He then computed t^2 to get the following data:

s	1	2	3	4	5
t^2	0.06	0.12	0.16	0.25	0.31

Plot the points and fit a line through the origin to estimate g. Compute the residuals for each of the points, and comment on them.

SOLUTION. Figure 3-12 shows the points. Four of the five points lie almost on a straight line, but the middle point sits well off the line. We have a choice with our black thread. We can suppose that the errors of measurement combine in a straight line accidentally, or we can suppose that one of the measurements is in substantial error. Our own leaning in this particular problem where theory so clearly offers s as a linear function of t^2 is toward four good measurements and one poor one. Therefore we have put the line

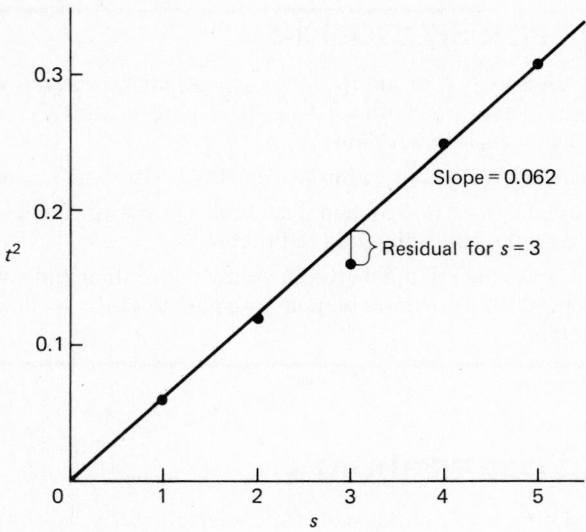

Figure 3-12 Relation between t^2 and s to estimate g. Note the departure of the middle point from the line.

close to four of the points. Because we are fixing s and observing t, we write the approximate straight line as $t^2 = 0.062s$. Writing the equation the other way, we have $s = 16.1t^2$, and so we estimate g to be 32.2 ft/sec². (Note that ft/sec is a unit of velocity, whereas ft/sec², an abbreviation for feet per second per second, is a unit of acceleration.)

The residuals from our line for the five points in this example are

t^2	0.06	0.12	0.16	0.25	0.31
0.062s	0.062	0.124	0.186	0.248	0.310
$t^2 - 0.062s$ = residual	−0.002	−0.004	−0.026	0.002	0.000

Note how the one residual for s = 3 stands out. The observation here seems to be an outlier, resulting from a bad mesaurement.

EXAMPLE 10 *Submarine sinkings.* Table 3-3 shows the results of a historical study of the numbers of German submarines actually sunk each month by the U.S. Navy in World War II, together with the Navy's reports of

TABLE 3-3
Sinkings of German submarines by U.S. Navy in World War II, by months†

Month	*Actual number,* y	*Guesses by U.S. (reported sinkings),* x	*1.14x*	y − *1.14x*
1	3	3	3.42	−0.42
2	2	2	2.28	−0.28
3	6	4	4.56	1.44
4	3	2	2.28	0.72
5	4	5	5.70	−1.70
6	3	5	5.70	−2.70
7	11	9	10.26	0.74
8	9	12	13.68	−4.68
9	10	8	9.12	0.88
10	16	13	14.82	1.18
11	13	14	15.96	−2.96
12	5	3	3.42	1.58
13	6	4	4.56	1.44
14	19	13	14.82	4.18
15	15	10	11.40	3.60
16	15	16	18.24	−3.24
Totals	140	123	140.22	−0.22

† $n = 16$ (months); $\bar{x} = 7.69$; $\bar{y} = 8.75$.

the numbers sunk. Figure 3-13 shows a graph of these data. Use the black-thread method or some other method to fit a line through the origin. Compute for each point

$$\text{residual} = \text{actual} - \text{estimate from line}$$

and plot the residuals against the estimates.

SOLUTION. One way of estimating the slope of a line through the origin is from $\bar{y}/\bar{x}$. The bars over y and x denote means. This gives $8.75/7.69 \approx 1.14$. Table 3-3 shows y, x, 1.14x, and the residual $(y - 1.14x)$ in the final column. Figure 3-14 shows the graph of the residual against the estimated value. Generally speaking, the pattern is that residuals at the left are smaller in absolute magnitude than those at the right. If we take absolute values of residuals and average them for the 8 leftmost points, we get 1.28; the average

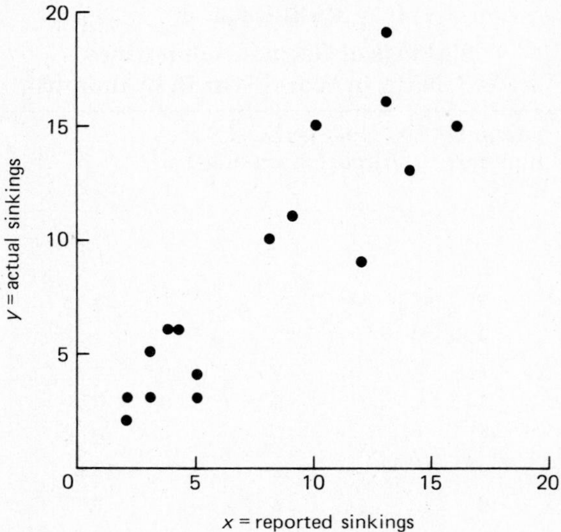

Figure 3-13 Actual and reported sinkings of German submarines in World War II. Constructed from data in Table 3-4.

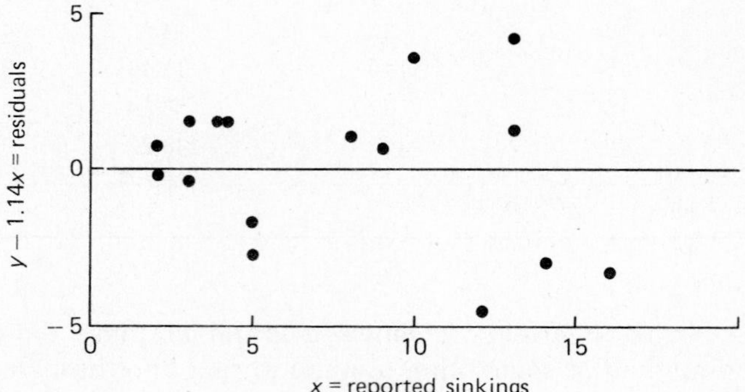

Figure 3-14 Graph of residuals plotted against reported sinkings to illustrate increasing variability with increasing reports. Constructed from data in Table 3-4.

for the 8 rightmost points is 2.68. Thus the residuals on the right average about twice the size of those on the left.

It is a general tendency of measurements of positive quantities for *the larger measurements to be more variable than the smaller ones.* An exception is test scores where there is a maximum score.

There can be various quirks in the data displayed by residuals. One is that they may bring out curvilinearity more clearly after the linear part has been removed. Another is that they may show up patterns of dependence among the observations fitted by a straight line. We shall discuss the various uses of residuals in more detail later, after we take up formal techniques for fitting linear relations.

PROBLEMS FOR SECTION 3-4

1. Maple saplings of various ages are planted in a nursery. The ages of the plants are 2 months, 1 year and 2 months, 2 years and 2 months, 5 years and 2 months. A line relating height to age is constructed. How will the variability of heights behave as the age increases?

Diabetic Mice. Table 3-4 shows weight measurements for normal and diabetic mice. Use the body and kidney weights to solve Problems 2 through 9.

2. Plot kidney weight against body weight for normal parental strain A.

3. Using the plot from Problem 2, obtain a general fitted line by the black-thread method or some other method.

4. Plot kidney weight against body weight for diabetic offspring.

5. Using the plot from Problem 4, fit a line.

6. Plot kidney weight against body weight for normal parental strain A and diabetic offspring (on the same graph). Be sure to use two different marks (such as ×'s and o's) to distinguish the two groups.

7. Fit a single line to the pooled data from Problem 6.

8. Examine the residuals and report any outliers.

9. Does one line fit the two kinds of data equally well?

10. *Age preferences.* In countries that keep good records of births, the number of people at a given age usually is a rather smooth function of age. But in countries that depend more on an oral tradition, the age distribution may be more jagged. Study the Ghana population by single years of age (Table 3-5), and discuss preferences in age reporting to census takers. You may wish to graph population in thousands against age in years to bring out the effects more vividly. Are there preferences for numbers ending in 0 or 5? Do such round-number preferences seem to increase as age increases? One way to study this for numbers ending in 0 would be to compute the fraction of the 5-year group centered at a 0 ending that chooses 0 and see how this fraction varies as one goes through the ages 10, 20, 30, etc. For example, 10 is the center of ages 8 to 12, and the fraction at 10 is

$$\frac{173{,}240}{192{,}410 + 156{,}410 + 173{,}240 + 101{,}820 + 156{,}410} \approx 0.222.$$

TABLE 3-4
Weights for 10-month-old male mice from two normal strains and the diabetic offspring of crossbreeding

	Body weight (g)	*Heart (mg)*	*Liver (mg)*	*Kidney (mg)*
Normal parental strain A	34	210	2240	810
	43	223	2460	480
	35	205	1880	680
	33	225	1970	920
	34	188	1940	650
	26	149	1400	650
	30	172	1470	650
	31	201	2060	560
	31	164	1760	620
	27	188	1690	740
	28	163	1500	600
Normal parental strain B	27	118	1640	640
	30	136	1690	690
	37	156	1980	780
	38	150	1810	660
	32	140	1750	750
	36	155	1770	780
	32	157	1780	670
	32	114	1670	670
	38	144	1980	700
	42	159	2260	720
	36	149	2070	800
	44	170	2530	830
	33	131	1750	640
	38	160	2200	800
Diabetic offspring (F_1 hybrids between strain A and strain B)	42	510†	2300	1030
	44	233	2550	1240
	38	211	2070	1150
	52	264	3450	1280
	48	236	2740	1240
	46	232	2750	1100
	34	210	2080	1040
	44	211	2680	1080
	38	186	2100	870

† Assumed to be a measurement error.

Source: Data gathered by Dr. E. Elizabeth Jones in research funded by Children's Cancer Research Foundation, Boston, Mass., and Wellesley College and reproduced here with her permission.

TABLE 3-5

1960† population of Ghana by single years of age

Age in years	*Population (in thousands)*	*Age in years*	*Population (in thousands)*	*Age in years*	*Population (in thousands)*
Under 1	277	40	155	80	19
1	211	41	31	81	1
2	258	42	70	82	3
3	289	43	30	83	1
4	257	44	24	84	2
5	232	45	78	85	5
6	244	46	39	86	2
7	193	47	23	87	1
8	192	48	48	88	1
9	156	49	28	89	2
10	173	50	97	90	7
11	101	51	15	91	
12	156	52	29	92	
13	120	53	14	93	
14	130	54	22	94	
15	120	55	26	95	
16	106	56	29	96	15 (ages 91 through Unknown combined)
17	88	57	15	97	
18	131	58	23	98	
19	94	59	15	99	
20	176	60	72	100 & over	
21	106	61	9	Unknown	
22	110	62	14	Total	6726
23	82	63	9		
24	111	64	13		
25	166	65	23		
26	111	66	8		
27	88	67	7		
28	137	68	13		
29	84	69	7		
30	232	70	33		
31	52	71	3		
32	95	72	9		
33	49	73	4		
34	58	74	5		
35	120	75	12		
36	85	76	5		
37	42	77	2		
38	77	78	6		
39	48	79	3		

† Based on a 10 percent sample of census returns.

Source: Adapted from data in *Demographic Yearbook, 1962, Special Topic: Population Census Statistics*, p. 189. New York: United Nations.

Olympic Metric Mile. Table 3-6 shows to the nearest second the winning times beyond 200 seconds required in the 1500-meter run for men. Use these data to solve Problems 11 through 13.

TABLE 3-6
Winning times for the 1500-meter run for men in the Olympics

Year	*Time in seconds, minus 200*
1900	46
1904	45
1908	43
1912	37
1920	42
1924	34
1928	33
1932	31
1936	28
1948	30
1952	25
1956	21
1960	16
1964	18
1968	15
1972	16

Source: Adapted from data in *The World Almanac & Book of Facts 1980*, edited by G. E. Delury, p. 818. New York: Newspaper Enterprise Association.

11. Plot time in seconds as a function of the year.

12. Fit a line.

13. Calculate and discuss the residuals.

14. *Women principal investigators.* For grants for scientific research, the National Science Foundation has studied the fraction of their new principal investigators who are women. Their data are as follows:

Year	Percent Women
1966	6.2
1967	6.2
1968	7.0
1969	7.2
1970	7.2
1971	7.3
1972	8.1

Fit a line or curve to the data and discuss the residuals.

3-5 SUMMARY OF CHAPTER 3

In this chapter we have met the following ideas:

1. A straight line with slope m and y intercept b has as an equation

$$y = mx + b. \tag{5}$$

2. In order to be linear, a functional relation must be expressible in the form of Eq. (5).
3. Some nonlinear, exact relations can be transformed into linear relations by changing the variables.
4. Some approximate or statistical relations can be approximated by using the black-thread method to fit a line to the plotted data.
5. When we fit a line to plotted data, departures from the fitted line are called residuals.
6. Residuals aid us in analyzing and interpreting data, including the discovery of outliers.

SUMMARY PROBLEMS FOR CHAPTER 3

Forgetting. To discover how rapidly people forget, the psychologist Strong slowly read lists of 20 words to five people and asked them later to recognize the words when mixed with 20 other words. As soon as the list was read, the subjects were kept busy at other intellectual tasks, so that they could not rehearse. The investigator imposed penalties for false recognition and computed the score for retention, R, on the following basis:

$$R = 100 \times \frac{\text{number correct} - \text{number falsely recognized}}{20}.$$

Thus a subject with 14 words correct and 2 falsely recognized had a score of 60 percent.

The time interval before the subject was asked to recognize a list of words was varied from 1 minute to 1 week. The data are given below, where T is the number of minutes between the reading of the list and the testing.

T, Number of Minutes until Testing	R, Average Retention Score (Percent)
1	84
5	71
15	61
30	56
60 (1 hour)	54
120 (2 hours)	47
240 (4 hours)	45
480 (8 hours)	38
720 (12 hours)	36
1,440 (1 day)	26
2,880 (2 days)	20
5,760 (4 days)	16
10,080 (7 days)	8

Use these data to solve Problems 1 through 11.

1. Plot the data for average retention score, R, versus T.
2. Apply the black-thread method to your figure from Problem 1 to obtain an equation of the form $R = mT + b$.
3. Compute the residuals from your line in Problem 2.
4. Does there appear to be a systematic pattern in the residuals when they are plotted versus T?

5. Transform the values of T to $t = \log T$.
6. Plot R versus t on a graph.
7. Apply the black-thread method to your new figures in Problem 6 to obtain an equation of the form $R = mt + b$.
8. Compute the residuals from your line in Problem 7.
9. Has the systematic pattern from Problem 4 disappeared?
10. Use the equation fitted in Problem 7 to find the value of t, and then T, when $R = 0$.
11. The work in Problem 10 is an extrapolation from the data. Why must it be regarded with caution, or even distrust?
12. *Gravitational attraction.* An apparatus measures falling time, t, to the hundredth part of a second when an object is dropped a distance, s, in feet. The following data are gathered:

s	4	3	2	1
t	0.50	0.43	0.35	0.25

Find g by fitting a straight line through the origin with the black-thread method. (To obtain a suitable graph, plot $T = t^2$ as the horizontal axis. Recall that $s = \frac{1}{2} gt^2$.)

REFERENCES

I. J. Good (1969). A subjective evaluation of Bode's law and an 'objective' test for approximate numerical rationality. *J. American Statistical Association* 64:23–49.

D. C. Hoaglin, F. Mosteller, and J. W. Tukey (editors) (1982). *Understanding Robust and Exploratory Data Analysis.* New York: Wiley.

J. Kruskal (1978). Statistical analysis: transformations of data. In *International Encyclopedia of Statistics, Vol. 2,* edited by W. Kruskal and J. M. Tanur, pp. 1044–1056. New York: Free Press.

F. Mosteller and J. W. Tukey (1977). *Data Analysis and Regression: A Second Course in Statistics,* Chapter 4. Reading, Mass.: Addison-Wesley (contains a detailed discussion of how to transform one or more variables to straighten out curved relations).

Analysis of One-Way and Two-Way Tables

Learning Objectives

1. Breaking measurements into group effects and residuals
2. Obtaining estimates for two-way tables using the additive model with effects for rows and for columns
3. Examining residuals for evidence of departure from the additive model

4-1 COMPONENTS, RESIDUALS, AND OUTLIERS

A frequency distribution or a stem-and-leaf diagram shows the general shape of the distribution of a set of measurements. We have seen the advantages of computing a mean or a median to summarize the position of the frequency distribution. We have also found it useful to look at the variability in the data. As problems become more complex, we must ask how changes in the values of variables affect the distributions of our data. For example, the strength of a metal wire may depend on (a) the composition of the metal used, (b) how it was heat-treated, and (c) the diameter of the wire. These three variables are sources of variation in the strength of the particular wire. We want to know the effect of such variables on that strength. In addition to these sources there may also be random variation owing to imperfections in the wire arising through the process of manufacture.

This chapter deals primarily with two-way tables of measurements. We begin by developing the idea of breaking numbers into a group effect and a residual. For two-way tables, we then show how to extend this approach to group effects for both rows and columns. These component group effects go together to make up an additive model for measurements.

A SINGLE SET OF MEASUREMENTS

Before tackling such large problems, we retreat to the simple case of a single set of measurements of one variable. For this purpose we may think of each measurement as made up of two components or parts that when added give the numerical values of the measurements. For example:

$$\text{measurement} = \text{mean} + \text{residual},$$

or

$$\text{measurement} = \text{median} + \text{residual}.$$

In simple problems, the residual is the deviation from a mean, median, or other location parameter. We study the sizes of residuals, their distributions, and (in more complicated problems that we shall review later in this chapter) patterns of residuals. They tell us when we have wild measurements, when unusual effects are occurring, whether the measurements can be satisfactorily represented in simple ways or whether more complicated models are required. And they tell us how well we can represent the data by particular models.

EXAMPLE 1 Express the group of measurements

6, 9, 10, 15, 20

in terms of the mean + residual and in terms of the median + residual.

SOLUTION. The mean is 12, and we can write the measurements as 12 + residual, more specifically as

$$\begin{aligned} 6 &= 12 + (-6) \\ 9 &= 12 + (-3) \\ 10 &= 12 + (-2) \\ 15 &= 12 + 3 \\ 20 &= 12 + 8 \end{aligned}$$

or, more briefly,

measurement	=	mean	+	residual
6				−6
9				−3
10	=	12	+	−2
15				3
20				8

And, because the median is 10, we can write similarly

measurement	=	median	+	residual
6				−4
9				−1
10	=	10	+	0
15				5
20				10

EXAMPLE 2 Table 4-1 gives several groups of measurements of the same variable. Represent each measurement as the sum of three components: the grand mean, the deviation of the group mean from the grand mean, and the residual. (We call the deviation of the group mean from the grand mean the **group effect.**)

SOLUTION. The first measurement in the first group is represented as

$$\begin{array}{ccccccc} \text{measurement} & = & \text{grand mean} & + & \text{group effect} & + & \text{residual} \\ 4 & = & 21 & + & (3-21) & + & 1 \end{array}$$

Table 4-1 shows the group effects, −18, −8, and 26. These are large compared with the residuals, which are −2, −1, 0, or 1. Thus we say that the group effects are substantial and that the residuals are well behaved. The *IQR* for the raw measurements is 44, that for the residuals 2; a 20-fold reduction comes from removing the group effects.

TABLE 4-1

Groups of measurements with small variation within groups

	Raw measurements			*Residuals*		
	Group A	***Group B***	***Group C***	***Group A***	***Group B***	***Group C***
	4	12	45	1	−1	−2
	2	14	48	−1	1	1
	3	12	47	0	−1	0
	3	13	48	0	0	1
	3	14	47	0	1	0
Sum	15	65	235			
Mean	3	13	47	Grand mean $= \dfrac{15+65+235}{5\,(3)}$		
Group effect	−18	−8	26	$= \dfrac{3+13+47}{3} = \dfrac{63}{3}$		
				$= 21$		

Stem-and-leaf diagrams

Raw measurements		***Residuals***	
0	23334		
1	22344	−0	2111
2		exactly 0	000000
3		0	11111
4	57788		

$n = 15$: $Q_3 = 47$, $M = 13$, $Q_1 = 3$, $IQR = 44$

$n = 15$: $Q_3 = 1$, $M = 0$, $Q_1 = -1$, $IQR = 2$

Note that after using the component approach, we pooled the residuals from all three groups. Sometimes the within-group variabilities are not readily comparable, and we may need to use a transformation such as the logarithm to make the variabilities more comparable from one group to another.

Such orderly data as these occur more often in the physical and biological sciences than in the social sciences. For Example 2, most people would say "We don't need any statistics to see the large differences among the groups." And for the data of Example 2 that is true. Unfortunately, a good deal of data will look more like those in Example 3 and will need careful attention.

EXAMPLE 3 Suppose that we have several groups of measurements, as in Table 4-2. And suppose that the groups represent different teaching methods, and a measurement represents the average performance of a class. Represent a measurement as the sum of three components: the grand mean, the deviation of the group mean from the grand mean, and the deviation of a measurement from its group mean. Then interpret the data.

SOLUTION. The first measurement in the first group will be represented as

$$\begin{array}{ccccccc} \text{measurement} & = & \text{grand mean} & + & \text{group effect} & + & \text{residual} \\ 6 & = & 10 & + & (12-10) & + & (6-12) \end{array}$$

The residual is -6, which is large compared with the group effect of 2. Unlike those for Example 2, the group effects here are small: 2, -2, and 0. The *IQR* for the raw measurements is 7, that for the residuals is 8—there is no apparent reduction in variability that comes from removing the group effects. If all the residuals were small compared with their group effects, as in Example 2, the measurements could be well summarized by the grand mean and the group effects, but instead the residuals are large and widely spread compared to the sizes of the group effects. Thus we can regard the groups as nearly equivalent on this variable.

The situation represented by the three groups in the example just given is sometimes called a one-way layout. This merely describes the idea that there are several groups. Later we shall introduce a two-way layout, where two sets of characteristics are considered simultaneously.

TABLE 4-2
Three groups of measurements

	Raw measurements			*Residuals*		
	Group A	*Group B*	*Group C*	*Group A*	*Group B*	*Group C*
	6	4	5	−6	−4	−5
	9	6	5	−3	−2	−5
	10	8	5	−2	0	−5
	15	10	5	3	2	−5
	20	12	30	8	4	20
Sum	60	40	50			
Mean	12	8	10	Grand mean $= \frac{150}{15} = 10$		
Group effect	2	−2	0			

Stem-and-leaf diagrams

Raw measurements		*Residuals*	
3	0	2	0
2	0	1	
1	0025	0	02348
0	455556689	−0	655554322

$n = 15$: $Q_3 = 12$, $M = 8$, $Q_1 = 5$, $IQR = 7$

$n = 15$: $Q_3 = 3$, $M = -2$, $Q_1 = -5$, $IQR = 8$

Results for residuals after setting 20 aside

	Rough calculation	*Interpolated*
$n = 14$	$Q_3' = 2$	2.25
	$M' = -2.5$	−2.5
	$Q_1' = -5$	−5
	$IQR' = 7$	7.25

EXAMINING OUTLIERS OR WILD OBSERVATIONS

We note in Table 4-2 that the stem-and-leaf diagram for residuals has one observation far from the main body. Removing the effects of the group means makes the residual of 20 stand out in the total distribution. It also stood out in its own group.

To appraise wild observations, we use a rule of thumb:

> **RULE OF THUMB** If an observation is more than 1½ *IQR*s away from the nearest quartile, we set the observation aside.

The choice of 1½ is somewhat arbitrary, but research has suggested that some number in this general range is a good choice. No one number can be "best" in all circumstances.

We do not throw the observation away, but we set it aside and report it separately. Here the residual 20 is 17 units away from Q_3, which is 3. The *IQR* is 8. Because 17/8 = 2.12 exceeds 1.5, we set the 20 aside. We look at the remaining 14 residuals as they stand and recompute Q_1, M, Q_3, and the *IQR*.

For a more careful treatment, we would set the corresponding raw measurement, 30, aside and compute a table of residuals based on the table of 14 remaining raw measurements. Problems 10 through 13 at the end of this section ask for these calculations.

However, at the bottom of Table 4-2 we have merely set aside the residual 20 in the stem-and-leaf diagram and then computed quartiles, median, and interquartile range. Using ¼(14 + 1) to get the quartile rank gives ¼(14 + 1) = 3¾. For a rough calculation, we round to 4. A careful calculation would require us to interpolate. The results of the two calculations are shown at the bottom of Table 4-2. These differ only a little from one another and from the results of the original stem-and-leaf summary for residuals.

Preliminary Summary for Example 3. The data show evidence of an outlier. The *IQR* for the raw data is much the same as that for the residuals, even after setting aside the outlier. This implies that the group means are not taking out much variability (as they did in Example 2). When the measurements vary so much, we wonder whether group A's higher average score reflects really better performance or the accident of a few random errors. Later we shall find formal ways of assessing this, but for now we can note that if observation 15 or 20 from group A had been exchanged for a 5 from group C, group A would not have stood alone with the highest average. If these group scores represent classroom performance for three different methods of teaching, we shall have a clear basis for action. First, we must check up on the set-aside observation with value 30. Was it produced by a spectacular teacher, or possibly by a reporting error, or by a highly selected group of students, or what? If we understood this observation, we could probably put method C to rest. The other two methods would still be in the competition, although method A would be leading.

PROBLEMS FOR SECTION 4-1

Weather in Boston. Use the data in Fig. 2-8 to solve Problems 1 and 2.

1. For spring, which of the following temperature observations should be set aside as outliers?

15°, 17.4°, 32.5°, 79°, 82.3°

2. For autumn, which of the following temperature observations should be set aside as outliers?

4°, 5.2°, 32.1°, 47°, 73.5°, 76.6°

3. For a certain set of measurements, A, we find $Q_1 = 20.7$ and $Q_3 = 30.3$.
   a) Find the IQR of the set A.
   b) Which of the following measurements of set A should be set aside as outliers?

15.9, 23.2, 29.3, 39.9, 44.8

Electrical Resistances. Following are 10 consecutive measurements of electrical resistances:

44.5, 46.5, 41.7, 42.6, 41.7, 43.8, 41.8, 45.5, 44.5, 28.6

Use this set of measurements to solve Problems 4, 5, and 6.

4. Find Q_1, Q_3, and the IQR.

5. Continuation. According to our rule of thumb, are any of the measurements in the set outliers?

6. Compute the median and the residuals from the median.

Antelope Bands. E. T. Seton, in *The Arctic Prairies*, listed the numbers of antelopes in 26 bands seen along the Canadian Pacific Railroad in Alberta, within a stretch of 70 miles, as follows:

8, 4, 7, 18, 3, 9, 14, 1, 6, 12, 2, 8, 10,
1, 3, 4, 6, 18, 4, 25, 4, 34, 6, 5, 16, 4

Use these data to solve Problems 7, 8, and 9.

7. Make a stem-and-leaf diagram of the numbers in the 26 bands.

8. Find the median number of antelopes in a band, and find Q_1 and Q_3.

9. Continuation. Use our rule of thumb to detect outliers.

Extreme Observation. In Table 4-2 the raw measurement of 30 in group C was found to be an outlier. Remove this observation to solve Problems 10 through 13.

10. Make an analysis corresponding to that in Table 4-2.

11. Continuation. Do any of these 14 observations appear to be outliers?

12. Compare the group effects of your revised analysis with those of Table 4-2.

13. Compare the *IQR* of the raw measurements with the *IQR* of the residuals.
14. If in any group all the observations have the same value, what will the stem-and-leaf diagram of the residuals for that group look like?

Property Taxes. Use the data on property taxes in Table 2-3 from Chapter 2 to solve Problems 15 through 25.

15. Sum the property taxes for each of the four geographic areas, find the mean for each area, and find the grand mean.
16. Continuation. Find the group effect for each geographic area.
17. Continuation. Find the residuals for the Northeast.
18. Continuation. Find the residuals for the Southeast.
19. Continuation. Find the residuals for the Central region.
20. Continuation. Find the residuals for the West.

We can now represent the data as the sum of three components: grand mean, group effect, and residual.

21. Continuation. Find the *IQR* for the Northeast residuals.
22. Continuation. Find the *IQR* for the Southeast residuals.
23. Continuation. Find the *IQR* for the Central region residuals.
24. Continuation. Find the *IQR* for the West residuals.
25. Compare the *IQR* for the residuals for each of the geographic areas with the *IQR* for the raw data for each geographic area. (Your answers to Problems 19 through 22 of Section 2-3 give the *IQR* for the raw data.)

4-2 TWO-WAY TABLES OF MEASUREMENTS: INFORMAL ANALYSIS

In Section 4-1 we examined groups of data. The groups formed a one-way array. They had no special arrangement. The fact that the single measurements were vertically arranged is not relevant to this discussion. We turn now to two-way arrays. Two-way tables are tables whose entries depend on two variables or factors, a **row variable** and a **column variable.** Our aim is to describe this dependence and search out special patterns. Let us begin with an informal look at a table concerning the acidity (pH) of puffs of smoke.

EXAMPLE 4 *Average pH at various puffs.* Using a machine and analytical chemistry, scientists have measured the pH of the smoke in successive puffs of cigarettes and small cigars. The pH, they say, is an indicator of the degree

TABLE 4-3
Values for pH for cigarettes and cigars at various puffs

Average pH at	*Nonfilter cigarette*	*Filter cigarette*	*Little cigar* B	*Little cigar* C
3rd puff	6.2	6.2	6.6	6.6
5th puff	6.1	6.1	6.5	6.6
7th puff	6.1	6.0	6.5	6.6
9th puff	6.0	5.8	7.0	6.6
Last puff	6.0 (11)†	5.8 (10)	7.2 (10)	7.1(11)

† Parentheses give number of last puff.

Source: Adapted from D. Hoffman and L. Wynder (1972). Smoke of cigarettes and little cigars: an analytical comparison. *Science* 178: 1197.

of nicotine toxicity. Table 4-3 shows pH values for various puffs to one decimal. Each cell entry is an average for three cigarettes or cigars. Summarize the table. Table 4-3 differs from Tables 4-1 and 4-2 in an important way. In Table 4-3 the measurements in any row are associated with the same numbers of puffs. In Tables 4-1 and 4-2 the row measurements had no physical relation from column to column. It is well to think of each cell in a two-way array as corresponding to a population or group or universe. For example, the upper left corner cell corresponds to the universe of third puffs of nonfilter cigarettes; the entry may be the mean of several measurements from a population, as indeed it is here.

Summary. Table 4-3 is easy to summarize. The observed mean pH for any puff is between 5.8 and 7.2. The cigarettes have slightly lower values, nearer 6, the cigars higher—about 6.5 to 7, except for the last puff. The later puffs have lower pH for cigarettes, higher for cigars, especially the final puff.

All told, the table seems to say that cigarettes are alike and cigars are alike, but to be more quantitative about it, we need a more formal analysis. Furthermore, a verbal summary may be difficult to make for a large and complicated table without a formal analysis. The formal analysis offers a systematic way of examining every measurement. It also prepares us to face disagreements and altercations. Informal analysis often works satisfactorily until someone challenges the interpretation. Then a standard formal analysis may be important in helping to settle meanings and in making sure that the view expressed is not the result of a whim or a misreading. There are always disagreements as to what data mean and whether they are relevant and accurate. At the level of analysis, we need standard methods so that we can communicate precisely what we did. Then the analytic disagreements will be at the level of choice of method. Other observers will at least be able to verify that we did correctly what we claimed to do and that we know exactly what we did. In public debates, when opponents know exactly what one another did and why, then even though they may disagree, much has been gained.

We never have a single best way of analyzing data; better-informed analysts will continue to invent new and more appropriate methods. But standard methods have a great deal of value, and much time is saved in communication when they are used.

The idea of the analysis we plan can be illustrated profitably by an example.

EXAMPLE 5 *Skill related to training and endowment.* Suppose that we have a musical task to perform and that our excellence depends upon two factors. Factor 1 is the presence or absence of previous training. Factor 2 is the presence or absence of perfect pitch (some people perform as if they have built-in pitch pipes). We can imagine that the task has the property that on the average the people being tested score 100 points. On the average, previous training adds 20 points to one's performance score; absence of previous training subtracts 20 points (adds -20 points). On the average, having perfect pitch adds 30 points to one's score; not having it subtracts 30 points. Find the average scores for the four types of people.

SOLUTION. A table of outcomes is readily obtained if we assume that these scores simply add up. We got the entries in the body of Table 4-4 by direct addition. Thus

$$\text{cell entry} = \text{grand mean} + \text{row effect} + \text{column effect},$$

and the upper left-hand corner cell is obtained from

$$100 + 20 + 30 = 150 \text{ (cell entry)}.$$

We call this description of the cell entries a *model*—in this instance we have a model with additive effects. Although we obtained the cell entries by addition, we can retrieve the original effects from the entries in the table very easily. We average the elements in each row to get the row mean and then average the two row means to get the grand mean of 100 (similarly for columns). The average of column means equals the average of row means, and this is, of course, the grand mean.

TABLE 4-4
Additive performance scores based on two factors

		Perfect pitch			
		Present +30	*Absent −30*	*Row mean*	*Training effects (row)*
Training	Present +20	150	90	120	+20
	Absent −20	110	50	80	−20
	Column mean	130	70	100 = Grand mean	0
	Pitch effects (column)	+30	−30	0	

Then we subtract the grand mean from each row mean to obtain the row effects and from each column mean to obtain the column effects. These are identical with the effects and grand mean we started with, as they always are when the additive model exactly applies. We constructed the table by assuming that the grand mean, row effect, and column effect, when added, give the performance score. If the additive model does not apply exactly, we cannot reconstruct the cells from the effects we derive from the table, and we have residuals or leftovers to interpret.

The idea of the analysis we present next is to estimate the effects corresponding to the factors as if this simple additive model works. Then we examine residuals. With only one number per cell, we are using the following model:

ADDITIVE MODEL

Cell entry in row i and column j = grand mean + row i effect + column j effect + residual

In Example 5 the fit was perfect, and the residuals were all zero. We shall not see this perfection often in real data, although we do see such additions in the purchase price of a car when options are added (so much for a radio, so much for a heater, and so on). A failure of additivity occurs when, after making a list of desirable options, the salesman offers a discount if the customer buys all the options, but nothing off if the customer takes a selection.

EXAMPLE 6 A radio costs $50 extra and a heater $100, but you get a 20 percent discount if you buy both. Make an additive analysis of the price table.

SOLUTION. Table 4-5 shows the basic price table. The upper left-hand corner is 120 instead of 150 because the discount is 0.20(150) = 30, and 150 − 30 = 120. The rest of the entries are flat prices. When we try to estimate or reconstruct the cell entries by adding the grand mean and row and column effects, we find the table of estimates shown in the middle of Table 4-5. When we subtract each estimate from the corresponding entry in the original table, we get the table of residuals shown at the bottom of Table 4-5. We note the following:

1. The behavior—the discount—that led to the nonadditivity was confined to the upper left cell, but the analysis spread the residuals equally to all four cells. This analysis spreads the residuals as much as possible. It is a feature of this method of fitting. Some methods are not so extreme.
2. The sum of the sizes of the residuals (all signs taken as positive) equals the magnitude of the original change from additivity. The sense in which this is true is more complicated in tables larger than 2 by 2.

TABLE 4-5
Automobile accessory prices with options and 20% discount for package deal

		Heater			
		***Take* 100**	***Refuse* 0**	*Mean*	*Radio effects*
Radio	Take 50	120	50	85	17.50
	Refuse 0	100	0	50	−17.50
	Mean	110	25	67.50	
	Heater effects	42.50	−42.50		

Estimates

127.50	42.50
92.50	7.50

Residuals

−7.50	7.50
7.50	−7.50

PROBLEMS FOR SECTION 4-2

1. Rework Example 5 given that everyone starts with 100 points, to which effects are added. On the average, previous training adds 30 points; no previous training adds −30 points; perfect pitch adds 20 points, absence of perfect pitch adds −20 points; the effects are additive except that having both previous training and perfect pitch adds 90 points instead of 50. Find the average scores for the four types of people. Compute the row effects, column effects, cell estimates, and residuals.
2. Using the additive model, rework Example 6 given this additional change: If you buy neither the heater nor the radio, there is a charge of $20 for removing them from the automobile. Exhibit your solution as in Table 4-5.
3. Show algebraically that the average of column means equals the average of row means for 2-by-2 tables with one observation per cell.

Cooling Down Arizona. Table 4-6 contains the mean monthly temperatures for three cities in Arizona for seven different months from July through January. Use the table for Problems 4 through 10.

4. Compute the mean temperature for each of the three cities, averaging over months.
5. Compute the mean temperature for each of the months, averaging over cities.
6. Show that the average of the three column means equals the average of the seven row means.
7. Subtract the grand mean from each of the column means to get the column or city effects.
8. Subtract the grand mean from each of the row means to get the row or month effects.

TABLE 4-6
Arizona mean monthly temperatures (°F)

	Flagstaff	*Phoenix*	*Yuma*
July	65.2	90.1	94.6
Aug.	63.4	88.3	93.7
Sept.	57.0	82.7	88.3
Oct.	46.1	70.8	76.4
Nov.	35.8	58.4	64.2
Dec.	28.4	52.1	57.1
Jan.	25.3	49.7	55.3

Source: J. W. Tukey (1977). *Exploratory Data Analysis*, p. 333. Reading, Mass.: Addison-Wesley.

9. Use the additive model to find the seven Flagstaff residuals.

10. Use the additive model to find the seven Phoenix residuals.

11. Use the additive model to find the seven Yuma residuals.

12. For each row and for each column, check that the residuals add to zero.

4-3 BREAKING A TABLE INTO ITS COMPONENTS

Now that we have introduced the additive model for a two-way table, we apply it. We use it to improve our analysis and understanding of a fairly complex set of laboratory data dealing with the smoke of nonfilter and filter cigarettes and of little cigars.

EXAMPLE 7 *Formal analysis of smoke puffs.* Analyze the smoke puffs according to the additive model of Section 4-2. The data will be more manageable by hand calculation if we subtract 6 from each measurement. Doing this gives us Table 4-7, and we have carried out the numerical work of the additive analysis.

Analysis. The column effects. We note first that the cigarettes have lower pH values overall than the cigars, −0.30 and −0.40 versus 0.38 and 0.32. This rejects arguments about reduced nicotine toxicity for these little cigars as compared with cigarettes as measured in these data.

The row effects. The early and late puffs have higher pH effects than the in-between puffs. On the average, the last puff has an especially high pH. But we need to examine the joint effect of order of puff and type of smoking material for the full story.

Residuals. To continue our approach, now that we have the effects, we can estimate the cells and then obtain residuals. The estimate, based on the additive model for effects, is as follows:

ESTIMATE FROM ADDITIVE MODEL

estimate = grand mean + row *i* effect + column *j* effect

The cell estimates differ from the cell values only in neglecting the residuals. For the upper left-hand corner cell we get

$$\begin{array}{ccccccc} \text{grand mean} & + & \text{row 1 effect} & + & \text{column 1 effect} & = & \text{estimate (1,1)} \\ 0.38 & + & 0.02 & + & (-0.30) & = & 0.10 \end{array}$$

TABLE 4-7†

Values for pH, minus 6, together with additive analysis

Mean pH at	*Nonfilter cigarette*	*Filter cigarette*	*Little cigar* B	*Little cigar* C	*Total*	*Mean*	*Puff effect (row)*
3rd puff	.2	.2	.6	.6	1.6	.40	.02
5th puff	.1	.1	.5	.6	1.3	.32	−.06
7th puff	.1	.0	.5	.6	1.2	.30	−.08
9th puff	.0	−.2	1.0	.6	1.4	.35	−.03
Last puff	.0	−.2	1.2	1.1	2.1	.52	.14
Total	.4	−.1	3.8	3.5	7.6		−.01††
Mean	.08	−.02	.76	.70	.38 = Grand mean		
Type effect (column)	−.30	−.40	.38	.32	0		

	Residuals				*Total*
	.10	.20	−.18	−.12	0
	.08	.18	−.20	−.04	.02††
	.10	.10	−.18	−.02	0
	−.05	−.15	.27	−.07	0
	−.22	−.32	.30	.26	.02††
Total	.01††	.01††	.01††	.01††	.04††

† Constructed from data in Table 4-3.

†† Does not add to zero because of rounding errors.

This 0.10 does not agree perfectly with the cell value, and so we have

$$\text{residual} = \text{cell entry} - \text{estimate}$$
$$= 0.20 - 0.10 = 0.10$$

We make this calculation for every cell to find the residuals shown at the bottom of Table 4-7.

Two important benefits flow from this method of representing entries in a two-way table. The first is best understood when the numbers of rows and columns are larger than 2. For example, when we have, say, 4 rows and 5 columns, then the total number of cells is $4 \times 5 = 20$. We have to appreciate 20 numbers to carry in mind the message of the whole table. If the effects are nearly additive, then we can summarize these 20 numbers with one grand summary statistic and the numbers for the row effects and the column effects. Thus we cut the numbers we have to keep in mind from 20 to 10

(= 1 + 4 + 5). And so we get economy of understanding from this method of representation. For the general r by c table, we summarize rc numbers by $r + c + 1$, and the larger are r and c, the larger are our savings. When $r = 12$ and $c = 10$, we go from 120 numbers down to 23.

As a second benefit, the additive structure offers a baseline against which to measure the interactions between variables. Without some specific way to take account of the effects of the treatments represented by the rows and columns and by the overall level of the numbers in the table, it is difficult to say or even to imagine what *interaction* between variables means. Only by having a specific structure can we pinpoint the idea. Later we shall see that the additive structure is not the only possible structure. For example, we could have a multiplicative structure. The point is that by imposing structure we give ourselves a way to get started. Otherwise, we are lost, with many unorganized numbers on our hands and no way to appreciate the information they can provide.

Interaction. The pattern of the residuals has been emphasized by the addition of the dotted lines in the bottom part of Table 4-7. The top left quadrant has positive values; the top right has negative values. This shows that when compared with the average effects, the cigarettes have a higher pH at first, the cigars a lower one at first. For the bottom quadrants the effect reverses, but it is a continuation of the same message. With successive puffs these cigarettes have decreasing pH, and these cigars have increasing pH. This represents what we call an **interaction** or a **nonadditive effect,** between type of smoke and number of puffs. When cigars and late puffs come together, we get a much higher pH than is predictable from cigars and puffs using the additive model. This is literally another example of the whole being greater than the sum of its parts, where parts are thought of as effects. The fact that the blocks show up as they do encourages us to believe that these effects are not due only to sampling variation and measurement error. The reason is that we think of the block as having a number of residuals all agreeing on the direction of the deviation. We note that the last line has especially large residuals, and so the interactions there are especially pronounced.

To get a feel for the size of the interaction effects, get the average in each block. The resulting numbers have substantial magnitude compared to the variability of the residuals within the block as measured, say, by the *IQR* or some other measure we treat later. This average value, about 0.13 for the upper left block, is the estimated size of the interaction effect for that block.

Stem-and-leaf diagram. The stem-and-leaf diagram given in Table 4-8 does not suggest any outliers, but it does have a pit. That pit at the 0 stem is another reflection of the strong interaction in the table between type of smoke and order of puff. Near the horizontal dotted line in Table 4-7 we might expect more residuals with values near zero. However, the table gives only odd-numbered puffs, and the 8th puff appears to be the one that might

TABLE 4-8†
Stem-and-leaf diagram of smoke puff residuals

3	0
2	076
1	0800
0	8
−0	4257
−1	8285
−2	02
−3	2

† Constructed from data in Table 4-7.

produce additional small residuals in the table. Thus the pit in the stem-and-leaf diagram has a satisfactory interpretation. Finally, we might note that there was little difference between the filter and nonfilter cigarettes and little difference between the cigars.

Let us turn to some special cases.

EXAMPLE 8 *Homogeneous table.* First, suppose that we have a two-way table that has the same measurement in every cell. Do an additive analysis.

SOLUTION. This is the ultimate in uniformity. The whole story is contained in the grand mean. The row and column effects are zero, and so are the residuals. Thus the row and column effects predict the cell values perfectly. But all the effects are identical.

More generally, when the additive model holds with zero residuals, the $r \times c$ table (a table with r rows and c columns) can be represented by the r row effects, the c column effects, and the grand mean.

EXAMPLE 9 *Model with one wild observation.* In a 4-by-5 table (Table 4-9) all the measurements are zero except the measurement in row 2 and column 3. That measurement is a 1. Analyze the table.

TABLE 4-9

Model for one wild observation

Row	Column 1	2	3	4	5	Row mean	Row effect
1	0	0	0	0	0	0	−.05
2	0	0	1	0	0	.2	.15
3	0	0	0	0	0	0	−.05
4	0	0	0	0	0	0	−.05
Column mean	0	0	.25	0	0	.05 = Grand mean	
Column effect	−.05	−.05	.20	−.05	−.05		

	Residuals					Totals
	.05	.05	−.20	.05	.05	0
	−.15	−.15	.60	−.15	−.15	0
	.05	.05	−.20	.05	.05	0
	.05	.05	−.20	.05	.05	0
Totals	0	0	0	0	0	0

Analysis. The grand mean is 1/20 = 0.05. The row means are 0, except for row 2, where the mean is 0.2. The column means are 0, except for column 3, where the mean is 0.25. In the first, third, and fourth rows, the row effect is $0 - 0.05 = -0.05$. The second row effect is $0.2 - 0.05 = 0.15$. In columns 1, 2, 4, and 5 the column effect is $0 - 0.05 = -0.05$. The third column effect is $0.25 - 0.05 = 0.20$.

We are ready now to compute residuals, and we get the results shown at the bottom of Table 4-9. The residual panel of Table 4-9 illustrates that a table that is perfectly homogeneous except for one wild observation produces small residuals everywhere, with larger residuals (in absolute value) in the row and in the column containing the wild observation, and that the largest residual is at the position of the wild observation itself. Thus the fitting of the effects and the estimates tends to spread the blame for the deviation of the wild observation over all the table, but more emphatically in the row and column where the wild observation occurs. We did not get this idea so clearly in the car-heater-radio example earlier.

Because we know exactly where the wild observation is in this example, we can speak with authority on what happened. With variability everywhere, it is not so easy, and we cannot tell for sure that we have wild observations. The method of fitting tends to smooth out or suppress errors by distributing them over the whole table. If this spreading is to be reduced, we need a different sort of analysis, one that is resistant to wild measurements. We shall not treat this topic further here; the interested reader is referred to Tukey (1977, Chapters 10 and 11).

PROBLEMS FOR SECTION 4-3

1. If the row and column effects are all zero in a 2-by-2 table, can there be nonzero residuals? If not, give the proof; if so, illustrate.

2. Display a 3×4 table having column effects but having zero row effects.

3. *Adding a constant.* In the smoking example we subtracted 6 from every measurement. Show for a 2×2 table with entries

a	b
c	d

that adding a constant to every measurement does not change the effects or residuals, only the grand mean.

Solution. The grand mean is $(a + b + c + d)/4$, and the first-row mean is $(a + b)/2$. Thus the first-row effect is

$$\frac{a+b}{2} - \frac{a+b+c+d}{4} = \frac{a+b-(c+d)}{4}.$$

If we add the constant k to every cell, the first-row mean is $k + (a + b)/2$, and the grand mean is $k + (a + b + c + d)/4$. The first-row effect then is

$$k + \frac{a+b}{2} - \left(k + \frac{a+b+c+d}{4}\right) = \frac{a+b-(c+d)}{4},$$

as before. Corresponding results hold for the other row and the columns. The technique applies to $r \times c$ tables as well as 2×2 tables.

4. *Model with two wild observations.* A 4×5 table has all 0's, except for two 1's that are not in the same row or column. Compute the row and column effects and the residuals. Summarize the results in words.

5. Explain in your own words what an interaction is.

Realistic Career Choices. To study the selection of realistic career choices, a psychologist scored students as making realistic job choices if they were intellectually qualified, but not too qualified. If a student's scholastic aptitude was above the 25th percentile required for the particular occupation, but not too high above its minimum requirements, the choice was scored as realistic. The maxima were based on studies of job dissatisfaction. Table 4-10 shows for each school grade from 7 through 12 and for each scholastic quartile the percentage of students making job

TABLE 4-10

Percentages giving realistic job choices by school grade and scholastic aptitude quartile

Scholastic aptitude quartile	*School grade* 7	8	9	10	11	12	*Row total*	*Row mean*
76–99	62	66	69	64	71	79	411	68.5
51–75	52	49	57	63	58	66	345	57.5
26–50	34	33	43	56	60	60	286	47.7
01–25	2	2	35	39	46	56	180	30.0
Column total	150	150	204	222	235	261	Grand total = 1222	
Column mean	37.5	37.5	51.0	55.5	58.8	65.2	Grand mean = 50.9	

Residuals and row and column effects for percentages of realistic job choices

Scholastic aptitude quartile	*School grade* 7	8	9	10	11	12	*Row effects*
76–99	6.9	10.9	0.4	−9.1	−5.4	−3.8	17.6
51–75	7.9	4.9	−0.6	0.9	−7.4	−5.8	6.6
26–50	−0.3	−1.3	−4.8	3.7	4.4	−2.0	−3.2
01–25	−14.6	−14.6	4.9	4.4	8.1	11.7	−20.9
Column effects	−13.4	−13.4	0.1	4.6	7.9	14.3	Grand mean = 50.9

Source: J. W. Hollender (1967). Development of a realistic vocational choice. *J. Counseling Psychology* 14: 314–318. Copyright (1967) by the American Psychological Association). Reprinted by permission of the author, John W. Hollender, Ph.D., Clinical and Consulting Psychology, 465 Winn Way, Suite 160, Decatur, Georgia, 30030.

choices scored as realistic, among those who said that they had made a vocational choice. About a third had not reported a choice. Thus Table 4-10 shows that for students in grade 7 and in the top scholastic aptitude quartile, the choices of 62 percent of those choosing were scored as realistic. Use these data to solve Problems 6 through 13.

6. Verify that the column effect for grade 8 is −13.4; show your work.

7. Explain and discuss the behavior of the column effects.

8. Verify that the residual for the cell for the 51–75 quartile and grade 10 is 0.9; show your work.

TABLE 4-11

Analysis of measurements: percentages of adults with family incomes under $3000 by chronic condition and activity-limitation status, 1965–1966

Age	Persons with no chronic conditions	*Persons with 1+ chronic conditions*				Sum	Mean	Row effect
		No limitation of activity	*Limitation, but not in major activity*	*Limitation in amount or kind of major activity*†	Unable to carry on major activity			
17–44								
Observed	12	11	16	24	41	104	20.80	−12.07
Estimate	10.3	11.6	18.3	26.3	37.6	104.1		
Residual	1.7	−0.6	−2.3	−2.3	3.4	−0.1		
45–64								
Observed	12	13	21	36	51	133	26.60	−6.27
Estimate	16.1	17.4	24.1	32.1	43.4	133.1		
Residual	−4.1	−4.4	−3.1	3.9	7.6	−0.1		
65+								
Observed	43	47	54	55	57	256	51.20	18.33
Estimate	40.7	42.0	48.7	56.7	68.0	256.1		
Residual	2.3	5.0	5.3	−1.7	−11.0	−0.1		
Sum	67	71	91	115	149	493		
Mean	22.33	23.67	30.33	38.33	49.67	Grand mean = $\frac{493}{15} = 32.87$		
Column effect	−10.54	−9.20	−2.54	5.46	16.80			

† Major activity refers to ability to work, keep house, or engage in school or preschool activities.

Source: Adapted from M. J. Lefcowitz (1973). Poverty and health: a re-examination. Institute for Research on Poverty Reprint Series, Reprint 96.

TABLE 4-12

Analysis of measurements: percentage "completion" in hydrolysis

Time (min)	Millimoles of substrate per milliliter 0.005	0.004	0.003	0.002	Sum	Mean	Row effect
10							
Observed	16.2	16.7	20.0	26.5	79.4	19.85	−27.72
Estimate	11.7	15.4	21.0	31.4	79.5		
Residual	4.5	1.3	−1.0	−4.9	−0.1		
20							
Observed	29.8	30.0	32.0	46.0	137.8	34.45	−13.12
Estimate	26.3	30.0	35.6	46.0	137.9		
Residual	3.5	0.0	−3.6	0.0	−0.1		
30							
Observed	37.0	41.8	44.7	61.5	185.0	46.25	−1.32
Estimate	38.1	41.8	47.4	57.8	185.1		
Residual	−1.1	0.0	−2.7	3.7	−0.1		
40							
Observed	45.8	48.3	59.0	69.5	222.6	55.65	8.08
Estimate	47.5	51.2	56.8	67.2	222.7		
Residual	−1.7	−2.9	2.2	2.3	−0.1		
50							
Observed	51.0	59.0	65.3	70.5	245.8	61.45	13.88
Estimate	53.3	57.0	62.6	73.0	245.9		
Residual	−2.3	2.0	2.7	−2.5	−0.1		
60							
Observed	56.8	62.8	71.0	80.5	271.1	67.78	20.21
Estimate	59.6	63.3	68.9	79.3	271.1		
Residual	−2.8	−0.5	2.1	1.2	0.0		
Sum	236.6	258.6	292.0	354.5	1141.7		
Mean	39.43	43.10	48.67	59.08	Grand mean = $\frac{1141.7}{24} = 47.57$		
Column effect	−8.14	−4.47	1.10	11.51			

Source: K. M. Harmon and C. Niemann (1949). The hydrolysis of N-benzoyl-L-argininamide by crystalline trypsin. *J. Biological Chemistry* 178:747.

9. Explain and discuss the behavior of the row effects.

10. Use the data to discuss the impact of possible interactions between school grade and scholastic aptitude on realistic job choices.

11. Students in grades 10 through 12 sometimes drop out. What do you think should be the effect of dropouts on the percentage of realistic job choices?

12. Make a stem-and-leaf diagram for the residuals. Interpret the results.

TABLE 4-13

Measurements

	y			
x	*1*	*2*	*3*	*4*
1	1.6	6.3	14.1	25.1
2	3.1	12.6	28.3	50.3
3	4.7	18.8	42.4	75.4

13. Comment on the residuals for the lowest scholastic aptitude quartile in grades 7 and 8.

Age and Activity. Table 4-11 on page 110 gives the percentages of poor people in various age groups having various levels of physical activity. An additive analysis has been carried out. Use the table to solve Problems 14 through 17.

14. Discuss the effects of the age groups on the percentage with low incomes.

15. Discuss the effects of the activity groups, comparing their magnitudes with those of the age groups.

16. Discuss the interactions. Focus on the largest residual in absolute value, and explain what it means.

17. Make a stem-and-leaf diagram of the residuals, and discuss possible outliers.

Hydrolysis. Table 4-12 on page 111 gives the percentage completion in hydrolysis for a chemical reaction. Use this information to solve Problems 18 and 19.

18. The standard additive analysis has been applied to Table 4-12. Use it to develop a relation between time and millimoles of substrate and percentage completion. You may wish to graph the row effects against the time in minutes and graph the column effects against the millimoles of substrate.

19. Do you see any general interaction effects like those we found in the smoke puffs for cigars and cigarettes?

20. Table 4-13 has been specially constructed. Analyze it by taking logarithms, and discover a law of formation for the cells. (Be sure to take the logarithms of x and of y, as well as of each entry.)

21. Analyze Table 4-13 by direct measurement analysis, and discuss the relation between the residuals and the law of formation discovered from the logarithmic analysis.

4-4 SUMMARY OF CHAPTER 4

1. Using the stem-and-leaf approach, we develop components of numbers, a group effect and a residual, and then calculate residuals:

$$\text{measurement} = \text{group effect} + \text{residual}.$$

2. The distribution of these residuals offers us a way to decide which numbers to regard as outliers, namely, those more than 1.5 *IQR*s away from the nearest quartile.
3. For two-way tables, we generalize the approach used for groups to get effects both for rows and for columns. This additive approach gives us a structure that provides a baseline for computing residuals:

ADDITIVE MODEL

Cell entry in row *i* and column *j* = grand mean + row *i* effect + column *j* effect + residual

4. Residuals from the additive model for two-way tables measure with error the "interactions" between the row and column treatments. The interactions are the effects that cannot be accounted for by the individual effects of the treatments represented by the rows and columns alone. Without some specific structure, we cannot define the idea of interaction.

SUMMARY PROBLEMS FOR CHAPTER 4

Rating of Typewriters. Five typists were trained to rate typewriters—they assigned a typewriter a rating of 1, 2, 3, 4, or 5, with 5 being very satisfactory, and 1 being very unsatisfactory. Each typist rated the same five typewriters, one each of five different brands. They used them to type several different kinds of things—letters, envelopes, postcards, tables, and so on. When they were finished, their ratings produced the data shown in Table 4-14. Use these data to solve Problems 1 through 6.

1. Compute the brand means, the rater means, and the grand mean.
2. Compute the brand effects.
3. Compute the rater effects.
4. Use the additive model to compute the cell estimates.
5. Compute the residuals for each cell.
6. Examine the residuals for evidence of interaction between brands and raters. You may wish to rearrange the brands and raters in order of row means and column means.

Politics Table. Table 4-15 gives percentages of Frenchmen of various political parties saying that they discuss politics with acquaintances. Use the table to solve Problems 7, 8, and 9.

TABLE 4-14
Typewriter ratings from a low of 1 to a high of 5 given by five trained raters for five brands of typewriters

Brand	*Rater* 1	2	3	4	5	*Row totals*
1	1	1	1	1	1	5
2	4	4	2	3	3	16
3	2	3	4	4	5	18
4	1	4	3	3	4	15
5	3	4	4	4	4	19
Column totals	11	16	14	15	17	

Source: Adapted from F. Mosteller (1973). Ratings of typewriters. Set 10 in *Statistics by Example: Weighing Chances*, edited by F. Mosteller, W. H. Kruskal, R. S. Pieters, G. R. Rising, and R. F. Link, with the assistance of M. Zelinka, pp. 81–94, table on p. 82. Reading, Mass.: Addison-Wesley.

TABLE 4-15
Politics table: percentages saying they discuss politics in a French survey

Party preference	*Discuss with family*	*Discuss with friends*	*Discuss with colleagues*
Communist	68	65	69
Moderates	53	53	36
MRP	51	43	40
Radical	58	49	31
RPF	58	51	32
Socialist	53	49	42

Source: J. Stoetzel (1955). Voting behavior in France. *British Journal of Sociology*, 6:104–122, p. 119, Table XIX (d). Reprinted by permission of Routledge & Kegan Paul Ltd., publishers of the journal, and by the author.

7. Rearrange the table so that row totals decrease from top to bottom and column totals decrease from left to right. Fit the additive model to these rearranged percentages.
8. Compute residuals from the additive model.
9. Discuss what you find, both in the original table and in the residuals.

REFERENCES

G. E. P. Box, W. G. Hunter, and J. S. Hunter (1978). *Statistics for Experimenters*, Chapter 7. New York: Wiley.

W. B. Fairley (1977). Accidents on Route 2: two-way structures for data. In *Statistics and Public Policy*, edited by W. B. Fairley and F. Mosteller, pp. 23–50. Reading, Mass.: Addison-Wesley.

J. W. Tukey (1977). *Exploratory Data Analysis*, Chapters 10 and 11. Reading, Mass.: Addison-Wesley.

5

Gathering Data

Learning Objectives

1. Learning the uses of case studies or anecdotes
2. Learning the uses of sample surveys and censuses
3. Learning the various uses of observational studies.
4. Learning the uses of controlled field studies and comparative experiments

5-1 GATHERING INFORMATION

How should we gather information? Introspection and systematic reflection on previous experience can help with many problems: When we often stub our toes on the same step, we paint it white. Sometimes theory can help us, as when an electrician figures out before building a proposed new circuit what it will do. Frequently we need to check an inventory of different ways to collect data. This chapter provides such a list.

Careful study of a single case, such as an airplane crash, may be informative, but more often collections of cases inform us more reliably. Indeed, sample surveys offer a modern way of collecting cases more systematically. Most of us are familiar with public opinion surveys, but many other sorts of sample surveys are used, for example, for estimating the amount of unemployment or for computing the consumer price index, which measures the price of a standard market basket of goods in various parts of the country.

When we go beyond surveys, we usually are trying to compare the effects of different treatments or processes. The key to making valid comparisons lies in our ability to *control* related variables.

In what follows, we first give several examples of questions that might be investigated by case studies or anecdotes, by sample surveys, by observational studies, or by experiments. Then we consider in more detail what these studies are used for. And finally we illustrate how each method might be used on a given problem or similar problems.

5-2 EXAMPLES OF QUESTIONS THAT MIGHT BE STUDIED BY VARIOUS METHODS

CASE STUDIES

Airplane Crash. What caused the airplane crash that occurred in the Virgin Islands on April 27?

Success Story. What were the circumstances that enabled Thomas Alva Edison to become such a highly successful inventor?

Picnic Poisoning. What was it that made so many people ill at the company picnic?

Legislative History. What were the motivations and tensions that finally led the House and Senate to pass the Emergency School Aid Act and the president to sign it into law?

SAMPLE SURVEYS AND CENSUSES

Marital Satisfaction. How satisfied with their marital conditions (single, married, widowed, divorced, separated, and other) are American men and women? How does satisfaction relate to income level?

Quality of Cars. What fraction of cars coming off a production line have major defects? How does the occurrence of defects vary by make of car?

Unemployment. What fraction of people in the nation were unemployed in September this year? How does that fraction depend on age and geography?

Attractive Advertisements. How many informative or entertaining advertisements do people see or hear in one day in the United States?

In the sample survey questions we primarily ask either about facts and quantities as they are now or about relations between variables as they are now. For example, people regard about 20 percent of advertisements on television, on radio, or in magazines as informative or entertaining. In sample surveys we do not attempt to change the values of the variables ourselves to see what effects such a change might make, although we may relate variables. Sample surveys are a form of observational study. Let us give some examples of questions where observational studies other than sample surveys might be employed.

OBSERVATIONAL STUDIES

Although we often compare results in observational studies, we do not change the values of key variables, but compare groups in which the values already differ.

Discrimination in Graduate Admissions. Because women are not as often admitted to certain graduate schools as men, considering the relative numbers of applications, are women being discriminated against?

Driving Safety. Federal funds for improving safety conditions in city streets will run out in two months unless a firm plan for selecting and improving conditions at 10 city corners is submitted. How shall the corners be chosen?

Weather Forecasting. How frequently do the local weather forecasters err in their predictions for rain?

Astronomy. Is the rate at which the moon circles the earth speeding up or slowing down?

CONTROLLED FIELD STUDIES AND COMPARATIVE EXPERIMENTS

Note that experiments test or compare things, treatments, or processes. Some people use the term *experiment* to mean a new reform, such as a social innovation. We do not speak of social innovations as experiments, although experiments may well be used to measure the value of an innovation.

Taste Preferences. People in different parts of the country react differently to amounts of sweetness in food. If a gelatin dessert were made 10 percent sweeter, would more people in the Southeastern states buy it? How would that result compare with the effect of a 5 percent reduction in price?

Abandoning Diet. Suppose that young children with a certain genetic defect that is compensated for by a special diet during their critical years must go off the diet. Will it be better for them to go off at 4 years of age or at 5?

Traffic Flow. Does changing the posted speed limit for automobile traffic change the speed at which motorists drive? If so, does the change last?

Boyle's Law. What is the relation between pressure and volume of a gas? (Note that it is not enough to guess the law; one must verify it empirically using different gases.)

STRENGTHS AND WEAKNESSES OF THE METHODS

When we try to attack systematically a question such as one of the 16 examples given, we usually need to gather data. In this chapter we discuss the strengths and weaknesses of these methods for gathering data, and some of their other features.

5-3 THE CASE STUDY OR ANECDOTE

An airplane crash, like other rare catastrophic events, ordinarily is studied by the case-study method or anecdotal method. *Anecdote* is the technical name for such a description; it is not being used to mean a funny story. The event itself will have many special features, and these will be fully described and recorded by a team of experts. They will have checklists and calculations to make. After comparing their findings with what they know from many other studies of this kind and from much physical theory, they will try to reach a conclusion.

Once in a while an event will be so unusual that we will have nothing to compare it with, but still we try to describe it carefully. An example of such an event was the great volcanic explosion in Krakatau in 1883. It caused tidal waves that went thousands of miles, and it put so many extra particles into the atmosphere that sunsets were more spectacular for years.

These case studies in the hands of experts gradually build up bodies of knowledge, especially about those variables that seem to be more important in the event and those that seem not to matter. Every field of endeavor has its own pattern for making case studies based on experience and theory. And so we shall not attempt to explain how to do them. We must recognize their importance, however, especially as the starting point in many research investigations.

Clearly, when we want to describe a large collection of items, institutions, or individuals, the case-study approach will not do.

Some major purposes of case studies are

1. to record the circumstances surrounding a specific event or phenomenon, either as history or as a basis for further study by others later,
2. to assign causes for the event,
3. to gain knowledge useful for other occasions,
4. to determine the consequences of the event.

For our basic four examples of case studies (the airplane crash, the Edison success story, the picnic poisoning, and the legislative history of the Emergency School Aid Act), items 1, 2, and 3 are the more likely purposes. For the Krakatau explosion, the consequences are also of great importance.

5-4 SAMPLE SURVEYS AND CENSUSES

The four questions in Section 5-2 on marital satisfaction, quality of cars, unemployment, and attractive advertisements, have three main features. They involve:

1. defining the variable under study and choosing a way to measure it,
2. deciding on the target population and how to sample it,
3. actually doing the survey to find out the fraction of the population that fits the conditions described.

For each of the four questions the definitional problem must be solved, usually through preliminary data gathering—pilot studies and case studies. For example, what is a major defect in a car? Omission of brake linings? Yes. Omission of hub caps? Maybe not. Someone has to compile a list of things that are wrong with newly made cars, and then an authoritative group must arbitrarily decide which are going to be classified as major defects. Different groups might make different definitions. An automaker might not classify so many defects as major as would a consumer group. More important, the automaker might be trying to correct a specific class of defects and thus might regard a modest defect as symptomatic of the major trouble under scrutiny, and he might set aside a defect seeming major to a driver because it relates to a different part of the manufacturing system. In this illustration, the automaker is trying to get control of a problem by careful study of a part of the system.

Investigators must decide what population they want to talk about; this is called the *target population.* The unemployment study informs the president and Congress about the nation. The Bureau of Labor Statistics must decide what it means by "an unemployed person." Is the target population all adults in the country? Yes. But, people in hospitals and other institutions are excluded. And so the sampled population will differ in this respect from the target population. People who are not working must report that they are actively looking for work if they are to be counted among the unemployed.

It is important that our definitions be easy to apply and verify. At any rate, such definitional work usually is special to the purpose of the investigation; it must be redone repeatedly until we get measures that are reasonably reproducible and relevant to the matter being studied. Those working in even the most exact sciences have repeatedly had to revise their most basic definitions (for example, definitions of mass, temperature, and time). It should not surprise us, therefore, that new ventures require such shaping and reshaping. You will rarely find a definition that is completely satisfactory. If, at first blush, marital satisfaction looks hard to measure, take a good look at how to decide whether someone is unemployed, considering such possibilities as students, the retired, vacationers, those working in illegal occupations, the hospitalized, and so on.

Because developing such definitions does depend a great deal on the purpose of the study and the field of endeavor, we emphasize its importance, but we do not try to explain further how to do it here. What we explain are the basic ideas behind the rest of the data gathering and its analysis.

In the simplest situation we may take a census—look at every item or individual in the target population and record the outcome. In the United States we do this once every 10 years to see how many people we have. These data tell us how to allocate congressional representatives and how to distribute federal funds. They are also used in many other ways, such as in urban planning.

Censuses and sample surveys measure things as they are; they may tell little about what will happen if things change. By knowing how many people are above a given age, we might find the cost of a change in a law about health insurance, but we cannot learn the effect of the change on the health of the people concerned.

Given a large collection of people or items, it may be too difficult or too expensive to measure each on all the variables that interest us. We would soon despair of questioning every citizen to learn about marital satisfaction or looking carefully at every car manufactured to assess all defects. Thus, for reasons of time, cost, or the need for care, we have developed the sample survey as a compromise falling between the complete census and the informed guess. The sample survey was originally a journalistic device, but it has been refined through statistical theory and practice. In the old days, the journalist or policymaker who wanted estimates would ask a few people or look around at a few nearby items and then make a guess based on this modest convenient chunk of information. Today, by taking samples in a scientific manner, we can ordinarily make tighter inferences, usually better than the informed guess.

When it comes to finding interrelations among variables, sampling methods are nearly always essential. One might be able to make a good guess at the percentage of adult women in the United States, but one would have difficulty trying to estimate the percentage of women in the labor force who are heads of households, have children, are employed, and are between 30 and 50 years of age. Such refined estimates often are needed for considering problems of public policy, such as deciding on the types and magnitudes of day-care programs.

Let us mention also that because sampled items or individuals can be measured with greater care than in a census, the final answer may be superior to what could be obtained if a census were carried out. In the next chapter we shall discuss some of the quantitative theory of sample surveys.

When we want to compare the effects of two treatments, the sample survey has weaknesses, even if both treatments are in use in the population. The difficulty is that the assignments to the treatment groups may be related to the outcome, for reasons we want to eliminate. For example, does more schooling lead to higher income? If we merely observe what happens among members of the general population, we may be misled. Perhaps people who have had more schooling could have made higher incomes with less schooling.

The purposes of sample surveys and censuses are

1. to answer specific questions about the population related to the measurements taken at the time,
2. to provide information about interrelations among variables in the population as it stands, and to give hints about causal relations,
3. to provide a baseline for comparisons with future measurements,
4. to measure changes through a sequence of surveys or censuses.

PROBLEMS FOR SECTION 5-4

1. Why might you prefer a sample survey to a census?
2. From this book, use the sample of all pages with page numbers ending in 25, or 50, or 75, or 00, and count the numbers of tables and figures that appear on these pages. Use the results to estimate the rates of tables and figures per page.
3. Repeat Problem 2 for all pages whose page numbers end in 12, or 37, or 62, or 87, and compare the results.
4. For 100 parked cars, count the dented fenders on the sidewalk side (score as 0, 1, or 2). Record also whether the car is full size or smaller (you may wish to make finer distinctions). Relate the average number of dented fenders to the size of the car.
5. Randomly choose a page and a column from your local telephone book. Find the frequency distribution of the final digits of the phone numbers in the column. Are the 10 digits approximately equally frequent?
6. Continuation. Carry out the sample survey of last digits of Problem 5 for a different page and column, and compare the results of the two samples.
7. From a collegiate dictionary draw a sample by choosing every 50th page (50, 100, 150, etc.). Take the first defined word or phrase on the page and count how many lines are used for it (do not use a word that started on the previous page). Estimate the average number of lines used to define a word in the dictionary.
8. The *World Almanac* gives the populations of counties of states. Choose one of the numbers 1, 2, 3, 4, 5, 6, 7, 8, 9, 10 to use as a starting point in the alphabetic list of Florida counties. Then choose that and every 10th county thereafter. Find the average county population in your sample. Compare it with the true average.
9. Which are older, American senior colleges or American junior colleges? The *World Almanac* gives the years of founding for senior and junior colleges. Draw a sample of 20 from each list and find the median date of founding to help decide which group is older.
10. How old are prominent Americans? The *World Almanac* lists many, together with their birth dates. Draw a sample of 20 and compute their average age. How old is the youngest in your sample?

5-5 OBSERVATIONAL STUDIES

Sometimes we do not take a sample survey, or for some reason cannot make an experiment. We may try instead to develop our ideas from an observational study. That is, we look at data on things as they exist in nature and try to draw conclusions from them.

EXAMPLE 1 Bickel and his coauthors wanted to see whether women are discriminated against in graduate admissions at the University of California at Berkeley. They found that of the women applying for admission to graduate school, a smaller proportion were admitted than were men. They also found, on looking a level deeper, that women more often applied to departments with low admission rates (social sciences and humanities) than to those with higher admission rates (the sciences). When account was taken of what departments people applied to, it turned out that women were being admitted more frequently than men. The question then arises whether there is a bias going the other way, that is, against men.

Discussion. In this investigation a key assumption was that the talents of the men and the women applying to a given department were equal. If they were not, then the whole analysis is inconclusive. Can you think of additional items that could have been measured to help us test this assumption?

In observational studies, as in sample surveys, we do not try to change or assign the values of the variables to see what will happen. We measure things as they are. Sometimes we pretend that the observations form the equivalent of an experiment. In the driving-safety example in Section 5-2, we pretend that we know what street corners to choose on the basis of previous experience. The idea is that we think the changes will improve the community's safety. But it is only after making the change and experiencing it that the guess based on observational studies can be verified. A change at a corner can have unforeseen consequences. If we install a stoplight or 4-way stop signs, drivers may choose to take a different route, and the overall danger to pedestrians and drivers may increase, even though danger is reduced at this corner.

Observational studies are sometimes strengthened by trying to create comparable groups. Suppose investigators want to assess the effects of drinking on pedestrian accidents in the early morning hours. To get control figures, they would have to measure the percentage of such accidents in which the pedestrian was drunk, and in addition they would have to go to the sites of the accidents at those hours and find the relative frequencies of drunk and sober pedestrians.

The great observational science is astronomy. Astronomers achieve extra control by using theory to forecast what will happen, and then they check the outcome. Thus they can check competing theories in a manner something like an experiment.

Control Groups. In comparative observational studies and in comparative experiments we sometimes speak of control groups. In the admissions example, the men formed a set of control groups to which the women could be compared. In the study relating to drinking and pedestrian accidents, the people who were at the same sites where the accidents occurred at the same early morning hours formed a control group that provided an estimate of the percentage drunk.

> Ideally, control groups and experimental groups are identical, except for the treatments imposed.

A study may present the possibility of many control groups, because many treatments or many types of individuals are being studied. In the study of graduate admissions, the men applying to each department formed a control group for the women applying.

The concept of a control group suggests that an experimental group should be compared with it. Sometimes this distinction is rather arbitrary, as in the example of graduate admissions, where the women are a control group for the men as much as the men are for the women. We tend to name the groups we have special interest in, here the women, the experimental groups. Notice that the language of experimentation is being used here, even though no experiment is being performed. This is an observational study.

The classic contrast is between an experimental group that gets a new treatment, such as vitamin C for the common cold, and a control group that gets a standard treatment, say handkerchiefs and aspirin. What is important is not what we call these groups but what the comparisons of their outcomes suggest about the merits of the treatments.

The principal purposes of observational studies are (a) those of sample surveys or, more frequently, (b) those of comparative experiments. Often, sample surveys or experiments are not carried out or cannot be carried out, but groups have been or can be observed. Then the investigator will need to struggle with the evidence, giving consideration to possible biases.

PROBLEMS FOR SECTION 5-5

1. In the study of graduate admissions at Berkeley, what additional items might have been taken into account to improve (make fairer or firmer) the comparison between men and women.

2. By counting books in the biography section of a library, estimate for women authors the percentage of biographies they write about men as opposed to women, as well as the corresponding percentages for men authors. Use authors whose surnames begin with C, L, S, and W for your estimates.
3. People often say that the world is going to the dogs. They have been saying it for thousands of years; so it is likely true. If you wanted to compare the amount of crime today with that of some earlier date, what groups would you compare? Remember, you have to be able to get the data.
4. From the *World Almanac* or other source find the populations of the largest and second largest cities for five countries in Europe; take the ratio of their populations (largest to second largest). Average these ratios for the five European countries and report the result.
5. Continuation. Carry out the investigation described in Problem 4 for the largest city as compared with the third largest city. Compare the ratios in the two studies.
6. At a supermarket, compare the prices for a nationally known brand of soup with those for the brand sponsored by the store to get the average percentage differences for the same kinds of soups.
7. Use average temperatures, as given, for example, in the *World Almanac*, to compare Atlantic and Pacific coastal cities in the United States at the same latitudes. Do the Pacific cities tend to be warmer?
8. Use tables of heights for boys and girls to find out at what ages the average heights tend to be equal.
9. Large teaching hospitals tend to have higher death rates following surgical operations than do small community hospitals. If teaching hospitals have more highly trained surgeons, how can this be explained?

★ 10. Use a library to estimate the percentage of books on science written during the last 10 years by women, as compared with men. What is the corresponding percentage for books published between 1925 and 1950? Use 10 sets of 10 consecutive books from the science shelves, starting each set of 10 flush left on a different science shelf. (If a book is written by n people of both sexes, score them as $1/n$ each, so that if two women and three men coauthor a book, it is 2/5 authored by women.) Discuss the comparison.

5-6 CONTROLLED FIELD STUDIES AND COMPARATIVE EXPERIMENTS

Let us return to the last group of questions in Section 5-2. The gelatin manufacturer and the physicians attending the children with genetic disabilities cannot be satisfied with only a sample survey. An important feature of the sample survey is that it is good for measuring things as they are. And although people can sometimes use it to find out what it would take to change things, often it cannot tell what will happen when variables are deliberately changed, at least in complicated problems. Usually such efforts require a **controlled field study** or a **controlled experiment.** People spoke of Prohibition as "the great experiment." In this book we do not use the word *experiment* in the sense of something newly or tentatively introduced. In our language, Prohibition was an *innovation*. For us, an experiment is the following:

> An experiment is a carefully controlled study designed to discover what happens when one or more variables are changed in value.

We often want to know the effect of an innovation, and we may need a carefully designed experiment if we are to find out.

We shall offer a quantitative approach in Chapters 9 and 10 but for now let us consider our examples. The gelatin manufacturer may know from other considerations that Southeasterners prefer sweeter foods than, say, Northerners. But to find out whether more sweetness would make this dessert sell better, the manufacturer must compare the two levels of sweetness by trying out both versions on Southeasterners.

In our genetic-disability/diet example, we might be tempted to say "Why not keep to the diet?" But biology is complicated. Perhaps the diet is a good idea at early ages, but if continued too long it may have long-term detrimental effects that outweigh its early benefits. In the absence of theory or experience, it will be important to experiment. As children grow older, their diets are less and less under the control of others, and we may need to know how important the dietary components are while we still have control. We need to compare the outcomes when the diet is changed at different ages.

These examples suggest that we cannot confidently compare the outcomes in two parallel situations unless we observe them both, deliberately changing the value of the variable whose effects we need to know.

Sometimes we may think we can tell from a sample survey what causes something, as when we see that people in large cities suffer more from an activity like crime than people in small towns. We may be tempted to remedy this by moving people from large cities to small towns. Several scenarios are possible:

1. The small towns become large cities, and we are back where we started.

2. The crimes that city people are relieved of by the move to small towns are replaced by new transportation problems.
3. The city people are what they are, and they take the crime with them.
4. The move to the small town does relieve the crime problem and does not introduce serious new difficulties.

Unfortunately, the sample survey does not tell us if any of these scenarios is the right one. That we see all of these four outcomes as readily possible, and each easily explained and illustrated, suggests that we cannot ordinarily know without a controlled study what will happen when we make the change. We shall discuss such investigations in Chapter 17.

Let us once more drive home the distinction between measuring a population as it is and changing a variable and seeing its effect.

EXAMPLE 2 Among young adult males, a sample survey will show that, on the average, heavier men are taller. Suppose a young adult male gains 20 pounds. How does his height change?

SOLUTION. Essentially, not at all. Changing weight does not cause a change in height in this group.

The primary purposes of comparative experiments are

1. to assess the outcomes of different treatments,
2. to determine causal relations.

PROBLEMS FOR SECTION 5-6

1. Obtain two short measuring instruments marked off in different units (examples: a 1-foot ruler marked in inches and fractions such as 1/32, a 1-foot ruler marked in inches and tenths of inches, a 20-centimeter ruler marked in centimeters and millimeters). Have six different people measure the height of a desk using first one kind of ruler and then the other. Get the median result by each method and compare the results. Compare the variability for each scale.
2. Two theories of needle threading are (a) to move the tip of the thread through the unmoving eye of the needle and (b) to move the eye of the needle so that it passes around the unmoving tip of the thread. To compare these theories, you need as subjects people who are not already trained in needle threading, and you need a stopwatch and needles and thread. Time each subject three times for each

method. Compare the times for the two methods. You will need a plan for keeping one hand still, and the subjects will want good light. You will also need to define "threaded needle."

3. Repeat Problem 2 using trained subjects, that is, people who have experience in threading needles. Do they take more time with the method they have not ordinarily used than do untrained subjects?
4. Create a set of 10 nonsense three-letter syllables of the form consonant-vowel-consonant, such as *gac*, and a set of 10 meaningful three-letter words of the same form, such as *rat*. Give someone each list for exactly 1 minute, and then see how many words the subject can repeat without looking at the list. Repeat this for five people, and then determine how much better people do at remembering the meaningful words than the nonsense words.
5. Shuffle an ordinary deck of playing cards thoroughly and record the face value for the first 26. Show a subject the first 26 cards, one at a time. Ask the subject to check off on a list of 52 cards the names of the first 26. Now compare the percentage correct among 5, 6, 7, 8, 9 to that for 10, J, Q, K, A. Repeat for five subjects and summarize the data. Is there a difference in correctness of recall?
6. Chill two brands of cola to equal temperatures. Label identical sets of eight small paper cups 1, 2, 3, 4, 5, 6, 7, 8. Put each kind of cola in four different cups, randomly chosen. Ask a subject, by tasting, to assemble the cups into two groups of four of the same kind, and ask which group the subject prefers. Report and summarize the outcome for several subjects.
7. Draw a straight line horizontally across a sheet of 8½-by-11 paper. Ask a subject to use a pencil to retrace the line, not allowing any part of the hand or arm to rest on a support. Measure the largest vertical deviation from the line. Compare the largest deviation when the subject uses the "writing" hand with that when using the "other" hand. Repeat for several subjects and report the magnitude of the effect.
8. Using a telephone directory, construct a list of seven-digit numbers. Read a seven-digit number to a subject, who will then write it down from memory. For each subject, do 10 seven-digit numbers. Then compute the percentages of times subjects get the first, second, etc., digits correct. Describe the results of the experiment.
9. For Problems 1 through 8, explain which treatments or situations are to be compared.
10. What feature of each experiment in Problems 1 through 8 makes it different from a sample survey?

5-7 USING ALL METHODS

As an example, let us briefly consider a problem that might be studied by all four of the methods we have described.

EXAMPLE 3 *Additional homework.* Will giving secondary-school students of French more homework in French than they now are assigned increase their performance on a French achievement test?

Anecdotal approach. In the Newton, Massachusetts, school system, we look around to see whether some teacher decided to give more homework than others, and we describe what happened. Perhaps the students revolted and the teacher relented. Perhaps the students liked the teacher, were impressed, and did much better. Maybe the students already had too much or too little homework.

Discussion. We may discover circumstances where additional homework favors higher achievement. Whatever the finding, we shall wonder whether or not other teachers with other classes will have similar experiences. The findings may suggest a valuable controlled experiment that needs to be done.

Sample survey. We draw a sample of schools from throughout the state of Massachusetts, observe the amounts of French homework assigned, and find the performances of students on an appropriate French test. If those schools with more homework assigned have higher scores, on the average, and those with less homework have lower scores, we are inclined to suppose that more homework may lead to higher achievement.

Discussion. Note that the amounts of homework given here are what the students are used to. We have not looked into the question of what would happen if the amount were changed.

Observational study. We match up classes so that they seem similar. For example, we take classes from well-to-do neighborhoods to see how differences in amounts of homework assigned relate to performance on the test. We also look at performances in less well-to-do neighborhoods.

Discussion. We may even try to adjust the performances of students for their overall grades, but we do not change the amount of homework assigned. This approach seems less systematic than the sample survey, because anything done here can also be done in the sample survey. In some situations, however, we might not be able to do a sample survey, and we would have to rely on an observational study.

Controlled experiment. Within sets of classes from similar neighborhoods, we choose at random some classes to be given heavier French assignments. Then we compare the performances of those with the heavier assignments with the performances of those who continue with their regular amounts. A more elaborate study might measure performances before the increase in assignments. Using this before–after approach, we exercise a further measure of experimental control (as we shall discuss in Chapter 17).

Discussion. Here we have actually changed the amount of the assignment. We still have to be careful. The change was in size of assignment. We may want to know if *doing* (rather than *giving*) longer assignments changed the performance; if so, we must be careful to measure the amounts done under the two sets of assignments.

Valuable as the controlled experiment is, sometimes ethical objections and practical difficulties make such studies impossible. For example, we doubt that this country would be willing to do a controlled experiment to assess the deterrent effects of capital punishment, even if someone could see how to design it.

PROBLEMS FOR SECTION 5-7

In Problems 1 through 15, indicate whether each of the statements is true (T) or false (F).

_____ **1.** A study of a collection of cases offers less reliable information than a study of a single case.

_____ **2.** Sample surveys offer a systematic way of collecting cases.

_____ **3.** A controlled experiment would be the usual way of studying an automobile accident that took place last week.

_____ **4.** In this book, social innovations are called experiments.

_____ **5.** Every field of endeavor has its own patterns for making case studies.

_____ **6.** The great volcanic eruption in Krakatau was studied by the experimental method.

_____ **7.** In sample surveys it is not necessary to distinguish between the target population and the sampled population.

_____ **8.** A complete census is less time-consuming and less costly than a sample survey.

_____ **9.** To find interrelationships among variables, a case study usually will suffice.

_____ **10.** Measurements for a sample often can be made with more care than for a census.

_____ **11.** In observational studies we change the values of the variables to see what will happen.

_____ **12.** A sample survey is good for measuring things as they are.

_____ **13.** A controlled experiment is designed to discover what will happen when one or more variables are changed in value.

_____ **14.** The final answer in a sample survey may be superior to that obtained in a census.

_____ **15.** The science of astronomy makes considerable use of observational studies.

16. A food manufacturer wishes to market a new brand of frozen deep-dish pizza in the states of California, Oregon, and Washington. You are asked to decide how much spice should be used in the pizza. Describe

a) a case study,
b) a sample survey,
c) an observational study,
d) a controlled experiment

that would provide useful information for your decision.

17. How would you choose to study the question whether a new supersonic airplane should be allowed to land at the airport in Chicago if the concern is for the amount of noise affecting citizens living near the airport?

5-8 SUMMARY OF CHAPTER 5

Four important methods of gathering data are (1) the case study or anecdote; (2) the sample survey or census; (3) the observational study; and (4) the comparative experiment or controlled field study.

1. The case study or anecdote

a) records the event in detail,
b) assigns causes,
c) gains information useful for other occasions, and
d) determines the consequences.

2. The sample survey or census

a) describes the population at the time,
b) provides information about interrelations among variables,
c) gives a baseline for future measurements, and
d) measures changes.

3. Observational studies usually are intended to do in less systematic ways

a) the jobs of sample surveys, or
b) the jobs of comparative experiments.

4. Controlled field studies and comparative experiments

a) assess the outcomes of different treatments, and
b) determine causal relations.

Ideally, control groups and experimental groups are groups that are identical except for the treatment given. The comparison of their outcomes tells us how well the treatments perform.

SUMMARY PROBLEMS FOR CHAPTER 5

Problems 1 through 10 deal with the following situation: Your purpose is to find out how to make the fast checkout line at a supermarket work best by choosing the maximum number of items customers are allowed to have when they enter.

1. How might a case study of what goes on at one store using one maximum number tell you something about this question?
2. How might a sample survey of managers of supermarkets help you?
3. Why might you want a sample survey of customers?
4. What might you learn from an observational study of checkout lines with different maxima?
5. How might you design an experiment to find the best number?
6. In Problem 5, you need to decide what outcome you want to maximize. What are some possible variables? (Consider this from the point of view of the market and that of the customer.)
7. What is the difficulty with an experiment that uses a different maximum number of items each day of the week, say 5 on Monday, 6 on Tuesday, 7 on Wednesday, and so on?
8. Why might you wish to study several separate markets, even if you are trying to decide on a plan for one?
9. Suppose a sample survey of markets shows that the most used maximum number is 10 and that the number varies from 6 through 12. How does this help in designing an experiment?
10. A case study of one store shows that it never uses a fast checkout line. What more would you like to know from the case study?

REFERENCES

P. J. Bickel, E. A. Hammel, and J. O'Connell (1977). Sex bias in graduate admissions: data from Berkeley. In *Statistics and Public Policy*, edited by W. B. Fairley and F. Mosteller, pp. 113–130. Reading, Mass.: Addison-Wesley.

R. Ferber, Chair; P. Sheatsley; A. Turner; and J. Waksberg. Subcommittee of the Section on Survey Research Methods (1980). *What Is a Survey?* Washington, D.C.: American Statistical Association.

D. C. Hoaglin, R. J. Light, B. McPeek, F. Mosteller, and M. A. Stoto (1982). *Data for Decisions: Information Strategies for Policymakers*. Cambridge, Mass.: Abt Books.

6

Sampling of Attributes

Learning Objectives

1. Sampling from populations of two kinds of elements
2. Estimating population proportions from sample proportions
3. Variances for sampling with and without replacement
4. Probability distributions for sample estimates of proportions
5. The effect of the relation between the sample size and the population size on the variance of $\overline{p}$

6-1 SAMPLES AND THEIR DISTRIBUTIONS

In gathering data, we need to know how close the estimates from a sample are likely to be to the corresponding properties of the population. To understand this, we need to know how the variability of sample estimates relates both to properties of the population and to sample size. We begin with a special case much used in sample surveys: data that have only two outcomes. Our findings will help us later in considering the analysis of experimental data and of observational studies, but it is convenient and instructive in first reading to think of this work as dealing with sample surveys. We call such characteristics or qualities *attributes.*

In many studies, important variables can take only one of two values (male or female, sick or well, defective or nondefective) for each individual or item in the collection being studied.

> The collection under study is called **the population.**

Examples of populations are: all full-time students at the University of Texas; all licensed airplane pilots (male and female); all cars produced in April 1985 in Detroit; all U.S. hospitals with more than 100 beds; all peach trees in the Johnson peach orchards in Georgia; or even the Williams family—father (F), mother (M), their young son (S), and their daughter (D)—living at 30 Pierce Road in Belmont, Massachusetts.

If data have several categories, we often can reduce them to two: the category of interest to us and a pool of all the rest of the categories. Thus for a response to a survey question, the categories may be Strongly Agree, Agree, Disagree, and Strongly Disagree. For some analytic purposes we might use only the categories Strongly Agree and the rest (Do Not Strongly Agree, or, equivalently, any of Agree, Disagree, and Strongly Disagree).

EXAMPLE *Sampling from the Williams family.* Let us study what happens if we draw a sample of people from the Williams family. The father (F) and mother (M) are voters, and the son (S) and daughter (D) are too young to vote. Let us divide the family into voters and nonvoters, and let us choose samples of various sizes and use their outcomes to estimate the proportion of voters. For this small population, we know that 0.5 is the proportion of voters. This number will be called the **true value** for the population. In most problems we do not know the true value. How then will we learn anything by studying a population where we do? By looking at all possible cases, we see what happens when we sample. By studying variation in a known situation, we learn about variation more generally. The population size is usually denoted by N, and for the Williams family $N = 4$. The population proportion or fraction having the attribute that we study is designated by p. For the Williams family, the fraction of voters is $p = 2/4 = 1/2$.

SAMPLES

Samples of Size 1. If we choose one person from the population (the Williams family) and regard that sampled person's voting status as an estimate of the fraction of voters in the population, then in half of the 4 possible samples we get 0 voters as our estimate; in the other half we get 100 percent, or a proportion of 1, as the estimate. Thus if we draw F or M, the estimate of the proportion is 1, and if we draw S or D, the estimate is 0. In general, let us call the estimate of p based on the sample, $\bar{p}$ (we read this symbol as p-bar). We can exhibit this information in a table:

Distribution of $\bar{p}$ for Samples of Size 1: Voter Proportions

Estimate, $\bar{p}$:	0	1
Frequency:	$1/2$	$1/2$

Although the estimate is always as far away as it can possibly be from the true value, $p = 1/2$, the average of the estimates is $1/2$, because half the time it is 0, and half the time it is 1. Next, let us take a larger sample.

Samples of Size 2. Disregarding order within samples, we can draw 6 different samples of size 2 from the Williams family: F-M, F-S, F-D, M-S, M-D, S-D. And because F and M are the voters, these samples produce the estimates 1, $1/2$, $1/2$, $1/2$, $1/2$, and 0, respectively, which we can summarize as follows:

Distribution of $\bar{p}$ for Samples of Size 2: Voter Proportions

Estimate, $\bar{p}$:	0	$1/2$	1
Frequency:	$1/6$	$4/6$	$1/6$

This result is much more satisfying than that for samples of size 1. Now the estimate is as far from the true value as it can be in only one-third of the samples, and we get the correct answer two-thirds of the time. Although in this example the term *correct* applies exactly, the essential idea is that several of the samples give results close to the true value. We should not overemphasize the exactness of the result, because that idea will not ordinarily hold. Closeness is what we can hope for and try to arrange.

Samples of Size 3. The 4 samples of size 3 are F-M-S, F-M-D, F-S-D, and M-S-D, with proportions of voters being ⅔, ⅔, ⅓, and ⅓, respectively. We can represent this as follows:

Distribution of $\overline{p}$ for Samples of Size 3: Voter Proportions

Estimate, $\overline{p}$:	⅓	⅔
Frequency:	½	½

Now our estimate is never correct, but it is never more than ⅙ (= ½ − ⅓ = ⅔ − ½) away from the true value, and never closer either. Samples of size 2 and size 3 are improvements over samples of size 1, if we want to estimate p. It would be good to have a measure of closeness so that we could see whether samples of size 3 give "closer" estimates, on the average, than samples of size 2.

Samples of Size 4: A Census. For completeness, let us notice that the only sample of size 4 is F-M-S-D, averaging ½ voters, represented as follows:

Distribution of $\overline{p}$ for Samples of Size 4: Census

Estimate, $\overline{p}$:	½
Frequency:	1

(Note: $\overline{p} = p$ when the sample is the whole population.)

Mean of the Estimates: Unbiased Estimate. Let us note that for every sample size for the Williams family, the mean of the estimates from the samples comes out to be ½. For example, in samples of size 1 we saw that two samples gave 0 and two gave 1, totaling 2 ($2 \times 0 + 2 \times 1 = 2$). When 2 is divided by the number of samples, we get $\frac{2}{4} = \frac{1}{2}$. Similarly, for the six samples of size 2, one gives 0, four give ½, and one gives 1, totaling 3 ($1 \times 0 + 4 \times \frac{1}{2} + 1 \times 1 = 3$). Dividing by 6, the number of samples, gives ½. You are asked to verify that this averaging works out correctly for the other sample sizes. This idea generalizes. Every sample yields an estimate of the true proportion, and the mean of these estimates is the true proportion. More formally:

UNBIASEDNESS

In sampling from a finite population consisting of two kinds of elements, if $\bar{p}$ is the sample estimate of the proportion of type *A* elements in the population, then for a given sample size, *the mean of the sample estimates from all the samples is the true proportion* p of type *A* elements in the population. Because of this property, we say that $\bar{p}$ is an **unbiased estimate** of p.

PROBLEMS FOR SECTION 6-1

1. Verify that samples of size 3 from the Williams family give an unbiased estimate of the proportion of voters.
2. Samples of size 1 are drawn from the Williams family. Use the composition of the samples to estimate the proportion of adult males in the population (the Williams family), just as we did for voters. Verify that the estimate $\bar{p}$ is an unbiased estimate of the true proportion, p, which is ¼.
3. Continuation. Carry out the same analysis for samples of size 2 for the Williams family.
4. Continuation. Carry out the analysis for samples of size 4.
5. The Gonzales family in Kalamazoo, Michigan, has three members: father (F), mother (M), and daughter (D). The true proportion of females in this population is ⅔. Verify that for samples of size 1 and for samples of size 2, the mean of the sample estimates is ⅔.
6. A bowl contains five different pieces of fruit: an apple (A), a grapefruit (G), a lemon (L), a pear (P), and a tangerine (T). The true proportion of citrus fruits in the bowl is 0.6. Verify that the mean of the sample estimates of the proportion of citrus fruits for samples of size 3 is also 0.6.
7. Continuation. Carry out the fruit bowl analysis for samples of size 4.

6-2 MEASURES OF VARIATION

We need a measure of how much sample estimates vary around the true value in the population. Such a measure tells us the precision of our estimating mechanism.

A MEASURE OF VARIATION: THE MEAN ABSOLUTE DEVIATION

For samples of size 1, drawn from the Williams family, the estimate of the proportion of voters for each sample deviates by ½ ($1 - \frac{1}{2} = \frac{1}{2}$ and $0 - \frac{1}{2} = -\frac{1}{2}$) from the true value, and so the mean absolute deviation is ½. (The term *absolute* means that we disregard the sign. We often use vertical bars to denote the absolute value such as in $|-3| = |3| = 3$, for example.)

For the six samples of size 2, two absolute deviations are ½, and the rest 0, and so the mean absolute deviation is

$$\frac{2 \times \frac{1}{2} + 4 \times 0}{6} = \frac{1}{6}.$$

For samples of size 3, the mean absolute deviation is ⅙, and for samples of size 4, it is 0. Figure 6-1 graphs these results. Thus, for this measure of variation, the variability declines with increasing sample size, though not as steadily as we might like, because samples of size 2 and size 3 gave identical mean absolute deviations.

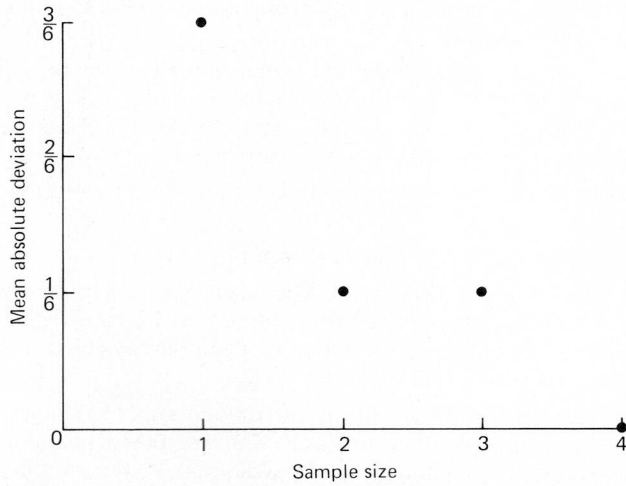

Figure 6-1 Mean absolute deviation plotted against sample size in the Williams family example.

We use another measure of variation more frequently than we use the mean absolute deviation. This measure is called the **variance**.

VARIANCE

The variance also involves the deviation from the true value, but instead of getting rid of the sign by way of absolute values, *it squares the deviations and then averages them.* For our example of the Williams family, samples of size 1 give deviations of $1 - \frac{1}{2} = \frac{1}{2}$ and $0 - \frac{1}{2} = -\frac{1}{2}$. Thus the squared deviations are $(\frac{1}{2})^2 = \frac{1}{4}$ and $(-\frac{1}{2})^2 = \frac{1}{4}$. The mean square deviation or variance is

$$\frac{2 \times \frac{1}{4} + 2 \times \frac{1}{4}}{4} = \frac{1}{4}.$$

For the set of sample sizes used with the Williams family, we get the results shown in Table 6-1. In Fig. 6-2 we have plotted the sizes of the variances against the sample sizes. We note that this measure of variation in this problem gives us a steady reduction in variation as sample size increases. This is an attractive property of the variance, and it has others. We use the variance often, and so it is good to know exactly what the variance is and how to compute it.

TABLE 6-1

Variances of sample estimates of voter proportions for the four sample sizes in the Williams family

Samples of size	*Average squared deviation*
1	$\frac{1}{4}$
2	$\frac{1 \times \frac{1}{4} + 4 \times 0 + 1 \times \frac{1}{4}}{6} = \frac{1}{12}$
3	$\frac{1}{36}$
4	0

When we refer to a variance, we may use the abbreviation "Var X" for the variance of the variable X. Alternatively, we use the symbol σ^2, where σ is the lower-case Greek letter sigma.

> An equivalent quantity called the **standard deviation** is the square root of the variance.

The standard deviation is measured in the original units, whereas the variance is measured in units squared. We shall see more of the standard deviation later. We often symbolize the standard deviation by σ.

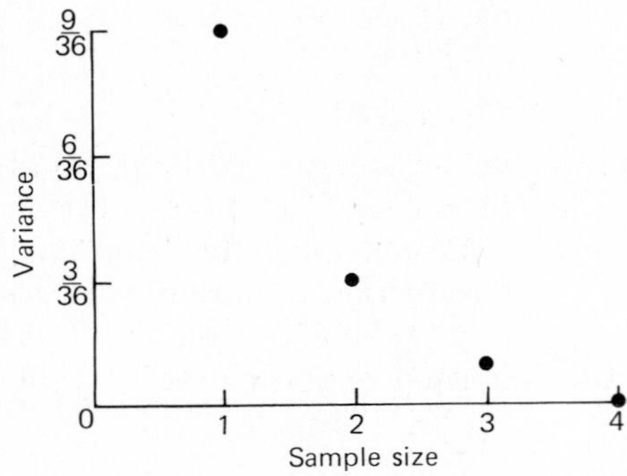

Figure 6-2 Variance plotted against sample size for samples from the Williams family.

Summary of the Williams Family Example. This example has illustrated several ideas:

1. A *population,* the Williams family.
2. An *attribute* or *characteristic* being measured (voter or nonvoter) associated with each member of the population.
3. A *true value,* p, the fraction of voters in the population.
4. An *observed value,* $\bar{p}$, of the estimate of the true value from a given sample: the fraction of voters in the sample.
5. The *variation* in the value of the estimate from one sample to another for samples of the same size.
6. The *distribution* of the sample estimates from samples of a given size; the distribution is the whole table of values of $\bar{p}$ together with their associated relative frequencies.
7. The *mean estimate* for all samples of a given size, which turns out to be the true value for the population in this example and in others—unbiasedness.

8. *Measures of variation:* the *mean absolute deviation,* the *variance* (average squared deviation from the true value), and the *standard deviation* (square root of the variance).
9. The *reduction in variance* with increasing sample size.

This is quite a bundle of ideas to get from such a small example, and yet the ideas illustrated here are basic to the theory of sampling and sampling variability, not only in sample surveys but also elsewhere in statistics. Indeed, the control and measurement of variation are major jobs of statistics in both the design and analysis of investigations.

By now, the reader will have a set of unanswered questions, such as these:

Query: *Sampling with replacement.* What if, in drawing a sample, the individual drawn is replaced before the next draw? This is called **sampling with replacement.** In samples of size 2 from the Williams family, F-F, M-M, D-D, and S-S are the additional possible samples.

Answer: The same ideas continue to apply. We would simply have more possible samples, actually 16 (because we have 4 ways to draw the first individual, and for each of these, 4 ways to draw the second, and $4 \times 4 = 16$). We distinguish the order F-M from M-F now, but that did not matter when we drew without replacement.

Query: *Large populations.* What if we have a much larger population?

Answer: We have more possible samples. For large enough samples, counting on our fingers gets out of hand, and we begin to want formulas or tables.

Query: *Large samples.* What if we have larger samples as well as larger populations?

Answer: Same as the preceding answer.

Query: *General p.* What if we have a true value other than ½?

Answer: Except for the symmetry in the distribution, all the ideas discussed earlier still work.

EXAMPLE Suppose that the Williams family has 3 voters and 1 nonvoter. Samples of size 2 are drawn without replacement.

SOLUTION. Six samples are possible, three with 2 voters and three with 1 voter.

Samples of Size 2: Voter Proportions		
Proportion:	½	1
Frequency:	½	½

The average of the sample proportions is

$$\frac{1}{2} \times \frac{1}{2} + \frac{1}{2} \times 1 = \frac{3}{4},$$

which agrees exactly with the true proportion of voters, as expected from our principle of unbiasedness in Section 6-1.

The mean absolute deviation is

$$\frac{1}{2}\left|\frac{1}{2} - \frac{3}{4}\right| + \frac{1}{2}\left|1 - \frac{3}{4}\right| = \frac{1}{4}.$$

The variance is

$$\frac{1}{2}\left(\frac{1}{2} - \frac{3}{4}\right)^2 + \frac{1}{2}\left(1 - \frac{3}{4}\right)^2 = \frac{1}{16}.$$

The standard deviation is

$$\sqrt{\frac{1}{16}} = \frac{1}{4}.$$

Query: *Sample size.* If we had a population of a million, how large a sample size would we need to estimate the true p?

Answer: The question cannot be answered in its present form.

Query: *Accuracy in a large sample from a large population.* Suppose the true value is "near" 0.5 (say between 0.2 and 0.8) and the population is large (say a million). What accuracy can we expect from a sample of 1000?

Answer: In about 95 percent of our samples the estimate will be closer than 0.03 to the true value!

Query: *Help for the skeptic.* How can ¹⁄₁₀₀₀ of a population tell me much about it?

Answer: It is the careful design of the investigation that makes it happen, and the knowledge comes from making the design match some mathematical ideas that we used earlier, such as allowing every sample of a given size an equal chance of being selected. Just looking around for any old sample of 1000 will not give the assurance described, because it may be easier to find the old folks at home than the young people; and because age may be related to the attribute (in our example, voting), we could get a **biased** estimate.

We study these more complicated situations in the next sections.

PROBLEMS FOR SECTION 6-2

1. Verify that the average squared deviation of sample estimates of voter proportions for samples of size 3 from the Williams family is $\frac{1}{36}$.

2. Compute the mean of the sample estimates for samples of size 2 from the Williams family, when sampling is done with replacement.

3–6. For the bowl-of-fruit example (Problem 6 following Section 6-1), calculate the variances for sample estimates of the proportions of citrus fruits for samples of size 1, 2, 3, and 4.

7. Using the values from Problems 3 through 6, plot the variances against the sample sizes, as in Fig. 6-2. Does the variance increase or decrease with sample size?

8. In sampling without replacement, we could ignore the possible order of drawing the items because every possible composition of sample elements could occur in the same number of orders. In samples of two from the Williams family, we could draw any pair in two orders. In samples of two drawn with replacement, not every composition of sample has the same number of possible orders. Explain and illustrate.

9. Compute the distribution of $\bar{p}$ and its mean absolute deviation for samples of size 3 drawn from the voter population consisting of 3 voters and 1 nonvoter.

10. Samples of size 2 are drawn from a population with 3 voters and 2 nonvoters. Find the distribution of $\bar{p}$, the proportion of voters in the sample, and verify that $\bar{p}$ is an unbiased estimate of p. Then compute the variance and standard deviation of $\bar{p}$.

6-3 SAMPLING WITHOUT REPLACEMENT FROM A FINITE POPULATION WITH TWO CATEGORIES

In our samples from the Williams family, we drew without replacement, because that seems a natural method of sampling. For one thing, the sample must then contain no duplicate individuals, and we seem to get more information about the population than we would by drawing with replacement. Later we shall give this idea a more quantitative basis. We shall next study ideas of bias and variability in samples and populations of any size.

First, we need to establish some additional notation. Let us illustrate again with the Williams family.

EXAMPLE *Notation.* For the Williams family, the size of the population is

$$\text{population size:} \quad N = 4.$$

The numbers of voters and nonvoters are

$$\text{voters in population:} \quad X = 2$$
$$\text{nonvoters in population:} \quad N - X = 2.$$

The true proportion of voters is

$$\text{population proportion of voters:} \quad p = \frac{X}{N} = \frac{2}{4} = \frac{1}{2}.$$

Let us introduce n for sample size and x for the number of voters in a sample, and let us suppose that we choose

$$\text{sample size:} \quad n = 3$$
$$\text{number of voters in sample:} \quad x = 1$$
$$\text{number of nonvoters in sample:} \quad n - x = 2\,.$$

Then the sample proportion of voters, which is also the sample estimate of the population proportion of voters, is

$$\text{sample estimate of population proportion of voters:} \quad \bar{p} = \frac{x}{n} = \frac{1}{3}.$$

We shall next apply our notation more generally, as follows: Let us suppose, as before, that in a population of N elements, each has exactly one of two attributes, A or not-A, and there are X A's and $N - X$ not-A's. Consider further all possible samples of size n drawn from this population. In the **population,** the true proportion of A's is p. That is,

$$p = \frac{\text{number of } A\text{'s in the population}}{\text{number of elements in the population}} = \frac{X}{N}.$$

In a **sample** of size n, let the number of A's be called x. Then the observed proportion of A's in the sample is

$$\bar{p} = \frac{\text{number of } A\text{'s in the sample}}{\text{number of elements in the sample}} = \frac{x}{n}.$$

We use $\bar{p}$ to estimate p. This is our usual everyday estimate (it's like a batting average).

We have already indicated without proof that if we compute $\bar{p}$ for every possible sample of size n and take the mean value of all the $\bar{p}$'s, we get the true value p. Sometimes we use the term **expected value** of $\bar{p}$, meaning the average over all samples of a given size. Let us state this important fact more formally:

RESULT 1: UNBIASEDNESS

In a population of size N containing the proportion p of A's, if $\bar{p} = x/n$ is the proportion of A's in a sample, the mean value (or expected value) of $\bar{p}$ over all possible samples of size n is p.

Because we use $\bar{p}$ to estimate p, and the $\bar{p}$'s differ from sample to sample, it is pleasant that on the average we get the true value. We illustrated this repeatedly in Section 6-1. We called this property **unbiasedness**.

If $\bar{p}$ averages to something more than p, we say that it has a positive bias; if it averages to something less, we say that it has a negative bias. When the bias is zero, as it is here, we say that $\bar{p}$ is an **unbiased estimate** of p.

Naturally, a particular sample value of $\bar{p}$ may be higher or lower than or equal to p. But note:

Bias is not a property of an estimate from one sample. It is a property of an estimate averaged over all samples.

EXAMPLE *Biased estimate.* When counting the numbers of defective tubes in samples of 50 color television display tubes drawn randomly from manufactured lots of 1000 tubes, an inspector always counts one defective tube too many because, unknown to him, his counter has a 1 stuck in it. What is the bias of his estimate of the proportion of defectives?

SOLUTION. He always counts $x + 1$ and estimates p by

$$\bar{p}^* = \frac{x+1}{n} = \frac{x}{n} + \frac{1}{n}.$$

Because $\bar{p} = x/n$, we can rewrite

$$\bar{p}^* = \bar{p} + \frac{1}{n}.$$

The average value of $\bar{p}$ is p, and because $1/n$ is a constant, its average value is $1/n$, or $1/50$ in our example. That is, $1/50$ is always $1/50$. Therefore, the bias of $\bar{p}^*$ is $1/50 = 0.02$.

EXAMPLE In the preceding example, suppose that the error in counting is corrected and the number of defective tubes in a single sample of 50 is 2, so that the estimate is 4 percent defective ($\bar{p} = 0.04$). Later, for some reason, every item in the lot of 1000 is examined, and 3 percent are found to be defective. Is this a biased sample?

SOLUTION. No. Bias is not a property of a single sample; it is a property of a method of estimation or a method of sampling. Bias is a long-run idea, not a one-sample concept.

A Statistic Estimates a Parameter. When we think of $\bar{p}$, we should think of it not simply as having a value like 0.04 in our example but as a formula for computing observed proportions. Such a formula is called a **statistic.** Examples of statistics that we have discussed thus far are the observed proportion and the sample mean. The corresponding population values (the true proportion and the population mean) are called **population parameters,** or just **parameters.** Typically, statistics are used to estimate parameters. For example, the observed proportion is used to estimate the true proportion.

PROBLEMS FOR SECTION 6-3

1. Distinguish between n and N.
2. Explain the difference between $\bar{p}$ as a number and $\bar{p}$ as a statistic.
3. Why can't the outcome of a single sample be regarded as biased when its $\bar{p}$ does not equal p?
4. What parameters do we have for measuring variability?

Use the following information for Problems 5 through 9: In a population of 12 items, 2 are defective. If random samples of 2 are drawn, the distribution of $\bar{p}$, the proportion of defectives in the sample, is

$\bar{p}$:	0	$\frac{1}{2}$	1
Frequency:	$\frac{45}{66}$	$\frac{20}{66}$	$\frac{1}{66}$.

5. In this problem, what are the values of N, n, and p?

6. Verify that the expected value of $\bar{p}$, using the distribution, is identical with p.

7. Compute the mean absolute deviation of $\bar{p}$ for this distribution.

8. Compute the variance of $\bar{p}$ for this distribution and its standard deviation.

9. Check that the frequency distribution is correct for this sampling. Show your work.

6-4 VARIABILITY OF $\bar{p}$ IN SAMPLING WITHOUT REPLACEMENT

Although it is a comfort to know that $\bar{p}$ averages to p, it would be even more of a comfort to know that $\bar{p}$ is often close to p. We have seen some distributions of $\bar{p}$ for the Williams family, and we have computed the variance of $\bar{p}$ for samples of various sizes in our examples. As our next step in understanding how close an estimate $\bar{p}$ is likely to be to the true value p, we give a general form for the variance of $\bar{p}$ in the two-category sampling problem. We call two-category problems binomial (meaning having two names).

Variance of $\bar{p}$ in Sampling without Replacement. Without proof, we give the variance of the estimate $\bar{p}$ as follows:

RESULT 2: VARIANCE WITHOUT REPLACEMENT

$$\text{variance of } \bar{p} = \text{Var}\,\bar{p}$$

$$= \frac{p(1-p)}{n}\left(\frac{N-n}{N-1}\right)$$

$$= \left(\frac{\text{population variance}}{\text{sample size}}\right)\left(\begin{array}{c}\text{improvement factor}\\ \text{for sampling}\\ \text{without replacement}\end{array}\right) \qquad (1)$$

Before attempting to understand formula (1) in detail, let us check it out on a problem that we have already solved.

EXAMPLE Compute the variance of $\bar{p}$ for the estimated proportion of voters for samples of size 2 for the Williams family using formula (1), and compare the result with the result we computed in Section 6-2.

SOLUTION. For the Williams family, $N = 4$, $n = 2$, and $p = \frac{1}{2}$, and so formula (1) gives

$$\text{Var}\,\bar{p} = \frac{\frac{1}{2}(1 - \frac{1}{2})}{2}\frac{(4 - 2)}{(4 - 1)} = \frac{1}{12}.$$

This agrees with our direct calculation in Section 6-2.

Appreciating Formula (1). We turn now to a more careful examination of formula (1). In its algebraic form it is useful for the specific problem of sampling without replacement from a population of two kinds of elements. In its verbal form, it is a formula of even more generality, as we shall later see. Here we concentrate on the algebraic form.

Let us verify first that the variance of the population is $p(1 - p)$. The total population has Np A's (voters) and $N(1 - p)$ non-A's (nonvoters). We look at each person separately, as if each were a sample of size 1. When a sample consists of one A, the squared deviation is $(1 - p)^2$. When the sample consists of one non-A (that is, zero A's), the squared deviation is $(0 - p)^2 = p^2$. Weighting each of these squares by the number of them, we get the total sum of squared deviations:

$$Np(1 - p)^2 + N(1 - p)p^2.$$

We factor out $Np(1 - p)$ and get

$$Np(1 - p)(1 - p + p) = Np(1 - p).$$

This quantity is the total sum of squared deviations for samples of size 1, and there are N such samples. Thus the average squared deviation, or variance of estimates from samples of size 1, is

$$\frac{Np(1 - p)}{N} = p(1 - p).$$

> Variance of a binomial observation:
>
> $$p(1 - p).$$

If we think of each item as a sample of size 1, then we have a $\bar{p}$ for each item that is either 1 or 0, and the variance $p(1 - p)$ is the variance of estimates for samples of size 1. In general, not just for this example, we have the following:

> The population variance is the variance of the observations for samples of size 1.

EXAMPLE $N = 5$. In a population of 3 A's and 2 not-A's, compute the variance of the number of A's for samples of size 1 and the population variance.

SOLUTION. Here, $p = 3/5$. The population variance is, by direct calculation,

$$\frac{(1-p)^2 + (1-p)^2 + (1-p)^2 + (0-p)^2 + (0-p)^2}{5}$$

$$= \frac{3(1-p)^2 + 2p^2}{5} = \frac{3(2/5)^2 + 2(3/5)^2}{5}$$

$$= \frac{2(3)}{5^3}(2+3) = \frac{2(3)}{5^2} = \frac{3}{5}\left(\frac{2}{5}\right).$$

Now, using the formula for the sampling variance for samples of size $n = 1$, $N = 5$, $p = 3/5$, we get

$$\text{Var } \bar{p} = \frac{(3/5)(2/5)}{1}\frac{(5-1)}{5-1} = \frac{3}{5}\left(\frac{2}{5}\right) \quad \text{or} \quad p(1-p).$$

Thus the two results are identical.

Now let us look at the n in the denominator of the factor $p(1 - p)/n$ in formula (1). The larger the sample size, the smaller the factor $p(1 - p)/n$, and therefore the smaller the variance of $\bar{p}$.

Finally, let us take a look at the improvement factor $(N - n)/(N - 1)$ in formula (1). Several matters stand out:

1. If the sample size is N (that is, if we draw a sample consisting of the whole population), then the factor $(N - n)/(N - 1)$ is zero, and so the variance is zero.

2. If the sample size is 1, the factor $(N - n)/(N - 1)$ reduces to 1, and then $\text{Var}\ \bar{p} = p(1 - p)$. This is as it should be, because we have proved formula (1) for samples of size 1.
3. As the sample size n grows from 1 to N, the improvement factor goes down from 1 to 0, and so two pressures send the variance of $\bar{p}$ to zero: the growing n in the denominator and the diminishing $N - n$ in the numerator. Both effects reduce the value of $\text{Var}\ \bar{p}$ toward zero.
4. In most practical survey problems, N is large compared with 1, ordinarily at least in the hundreds. If we neglect the 1 of $N - 1$, then the improvement factor (or correction for finite population) is approximately

$$\frac{N - n}{N} = 1 - \frac{n}{N} = 1 - f,$$

where f is n/N, the sampling fraction.

SMALL GAINS FROM THE FINITENESS OF THE POPULATION

In speaking earlier of the effect of the sample size on the variance, we pointed out that even though the improvement factor for finite sampling tends to 1 when the population becomes large compared with the sample size, we still have the n in the denominator to help us. When someone asks how large the sample size needs to be, we must say that it makes little difference whether the population is infinite, 200 million, 2 million, 200,000, or even 20,000. If we are taking a random sample of 1000 from any of these populations, the improvement factor for the variance does not reduce much from 1 until we get down to 20,000, when we pick up 5 percent ($n/N = 1000/20{,}000 = 0.05 = 5$ percent).

Next we ask what size sample from a finite population gives the same precision as a sample from an infinite population. If n_∞ is the sample size to be drawn from an infinite population and n_f is the sample size to be drawn from a finite population with the same p, we want to equate the variances for the sampling from the infinite population and that from the finite population. For the infinite population,

$$\sigma_\infty^2 = \frac{p(1 - p)}{n_\infty}$$

(∞ is the symbol for infinity). For the finite population,

$$\sigma_{\text{finite}}^2 = \frac{p(1 - p)}{n_f}\left(\frac{N - n_f}{N - 1}\right).$$

We need to equate these and solve for n_f in terms of N and n_∞. When we do, we get the following:

> Sample sizes required for equal variances:
>
> $$n_f = n_\infty\left(\frac{N}{N + n_\infty - 1}\right) \approx n_\infty\left(1 - \frac{n_\infty}{N}\right)$$

EXAMPLE To elaborate, let us consider the following populations:

1. hypothetical infinite population
2. voting population of United States (120,000,000)
3. voting population of Illinois (6,500,000)
4. voting population of Oklahoma (1,500,000)
5. voting population of Albany, Georgia (32,000)

If we take a sample of size 3000 from the hypothetical infinite population, what sample sizes will we need in order to get the equivalent precision (variability) in the other populations?

SOLUTION. We will need the following:

2. United States: 3000
3. Illinois: 2999
4. Oklahoma: 2994
5. Albany, Georgia: 2743

Thus we see that we need about the same sample size for Oklahoma as for Illinois or for the nation as a whole. Only when we get down to a population the size of Albany, Georgia, do we pick up much savings (approximately 8.6 percent).

PROBLEMS FOR SECTION 6-4

1. Use formula (1) for the bowl-of-fruit example (Problem 6 following Section 6-1) to get the variances of estimates of p from samples of size 3 and from samples of size 4 for sampling without replacement. Are your answers the same as for Problems 5 and 6 following Section 6-2?

2. If $p = 1/4$ and $N = 100$, what is the variance of the sample proportion for samples of size 10?

3. If $p = 1/4$ and $N = 50$, what is the variance of the sample proportion for samples of size 10?

4. If $p = 1/3$ and $N = 100$, what is the variance of the sample proportion for samples of size 10? What about samples of size 20?

5. Suppose that two towns have sizes $N_1 = 100$ and $N_2 = 200$. The proportion of voters in each is $p = 0.75$. You plan to take a sample of size $n_1 = 25$ from the first town. How large a sample, n_2, will you need to take from the second town so that the variances of the sample estimates of the proportions of voters in both samples will be equal?

6. Use the exact formula to verify that a sample of size 2743 from a population of size 32,000 has precision equivalent to that of a sample of size 3000 from an infinite population, assuming that both populations have the same p.

7. Continuation. For the same population, what sample size will yield equivalent precision to a sample of size 1500 from an infinite population?

★ 8. Carry out the algebra required to show that

$$n_f = n_\infty \left(\frac{N}{N + n_\infty - 1}\right).$$

★ 9. By dividing numerator and denominator of $N/(N + n_\infty - 1)$ by N, holding n_∞ fixed, and letting N grow large, verify the approximation

$$n_f = n_\infty \left(1 - \frac{n_\infty}{N}\right).$$

★ 10. If the same sample size n is used for both an infinite binomial population and one of size N, find the ratio $\sigma^2_{\text{finite}}/\sigma^2_\infty$.

6-5 INFINITE POPULATIONS AND SAMPLING WITH REPLACEMENT

When we sample one element at a time with replacement, the population always has the same composition at each draw. If we did not keep track of the name of the element being sampled, it would be as if we had an infinite population whose relative composition is unchanged by the withdrawal of one element, or even a finite number of elements. When we toss a coin and get a head, we have not depleted the store of heads on the coin. At each toss we still have one head and one tail equally likely on a fair coin. Essentially, we are replacing the head in the population of possible outcomes after the toss. And thus we can think of the population as infinite.

This last point has implications. In sampling with replacement, we can think of the population as being infinite in size, because we can draw the same item as many times as we like without depleting the population. Thus N is inevitably large compared with both n and 1. Let us rewrite

$$\frac{N - n}{N - 1}$$

by factoring N out of both numerator and denominator. We get

$$\frac{N - n}{N - 1} = \frac{N\left(1 - \frac{n}{N}\right)}{N\left(1 - \frac{1}{N}\right)} = \frac{1 - \frac{n}{N}}{1 - \frac{1}{N}}.$$

Now, holding the sample size n fixed as the population size N grows large, $1/N$ and n/N tend to zero, and the ratio

$$\frac{1 - \frac{n}{N}}{1 - \frac{1}{N}}$$

tends to 1. Thus, if N is very large compared with n, the improvement factor makes little difference to the variance of $\bar{p}$. For example, if we draw a sample of $n = 1000$ from a population of $N = 1{,}000{,}000$, $f = n/N = 0.001$, and the correction factor is 0.999 instead of 1.000, which is scarcely worth computing. Although we lose nearly all the force of the correction factor, the n in the denominator of formula (1) is just as forceful as ever in reducing the variance. Doubling the sample size will halve the variance, even for large N. In summary, we know that the variance of our sample estimates is no larger than this:

> Sampling with replacement or from an infinite population:
>
> $$\text{Var}\,\bar{p} = \frac{p(1 - p)}{n} \qquad (2)$$

This holds no matter how large N is compared with n. In many practical problems this is the appropriate formula, whether because N is much larger than n or because the sampling is done with replacement or from an essentially infinite population.

EXAMPLE *Sampling with replacement.* Let us study estimates of the voter proportion from samples of 2 drawn with replacement from the Williams family and thus check that formula (2) is a special instance.

SOLUTION A. *Formula.* First, formula (2) says that with replacement

$$\text{Var } \bar{p} = \frac{p(1 - p)}{n}.$$

For $p = ½$ and $n = 2$, formula (2) gives

$$\text{Var } \bar{p} = \frac{½(½)}{2} = \frac{1}{8}.$$

SOLUTION B. *Enumeration.* If we have F-F, we have two voters and $\bar{p} = 1$; for F-D we have one voter and $\bar{p} = ½$; for D-D, we have no voters and $\bar{p} = 0$. The 16 samples of size 2 and the corresponding estimates of fraction of voters are shown in Table 6-2 in parallel panels. Thus the distribution of $\bar{p}$'s is

Estimate, $\bar{p}$:	0	$\frac{1}{2}$	1
Frequency:	$\frac{4}{16}$	$\frac{8}{16}$	$\frac{4}{16}$

The average squared deviation is

$$\text{Var } \bar{p} = \frac{4}{16}\left(0 - \frac{1}{2}\right)^2 + \frac{8}{16}\left(\frac{1}{2} - \frac{1}{2}\right)^2 + \frac{4}{16}\left(1 - \frac{1}{2}\right)^2 = \frac{2}{16} = \frac{1}{8},$$

and this agrees with the result from formula (2). This is reassuring, for we did not derive the original formula for Var $\bar{p}$, except for samples of size 1. In samples of size 1, sampling with replacement and sampling without replace-

TABLE 6-2

The 16 samples of size 2 from the Williams family and their corresponding estimates of $\bar{p}$

Samples				*Estimate,* $\bar{p}$			
F-F	F-M	F-S	F-D	1	1	½	½
M-F	M-M	M-S	M-D	1	1	½	½
S-F	S-M	S-S	S-D	½	½	0	0
D-F	D-M	D-S	D-D	½	½	0	0

ment are identical and of course give the same variance. The result ⅛ obtained here is larger than the $\frac{1}{12}$ obtained in sampling without replacement in Table 6-1.

EXAMPLE *Quality control.* In controlling the quality of manufactured products, industrial engineers may draw random samples from the production line or from a completed lot of material to estimate the proportion of defectives, p, being manufactured.

Part a: Suppose that a sample of 1000 is drawn from a production line. Estimate the standard deviation of the estimate of p if the observed defective rate is 2 percent.

SOLUTION. Our attitude toward the production line is that it is producing items with proportion defective p. The population we are sampling from is open-ended, and so we regard it as infinite. Because we do not know the true value of p, we use $\bar{p}$ to estimate it in the formula for the standard deviation. Therefore,

$$\sigma_{\bar{p}} = \sqrt{\frac{p(1-p)}{n}} \approx \sqrt{\frac{\bar{p}(1-\bar{p})}{n}} = \sqrt{\frac{.02(.98)}{1000}} = .0044.$$

Part b: Estimate the standard deviation of the estimate of p for the lot if the sample of 1000 is drawn from a lot of 4000 and the observed defective rate is 2 percent.

SOLUTION. If the sample is drawn from a lot of 4000, then we use

$$\sigma_{\bar{p}} \approx \sqrt{\frac{\bar{p}(1-\bar{p})}{n}\frac{(N-n)}{N-1}}$$
$$= \sqrt{\frac{.02(.98)(3000)}{1000(3999)}}$$
$$= .0038.$$

And so we do get a modest reduction in the estimated standard deviation of $\bar{p}$ for the lot as compared with $\bar{p}$ for the production line.

EXAMPLE *Public opinion polling.* Public opinion polling agencies frequently use samples of about 800, 1600, and 3200. Although they use methods somewhat different from random sampling, the standard deviations are still close to those we have been using. Find the estimated standard deviation of the observed proportion of a national sampling if the observed proportion responding "yes" to the question is 0.8.

SOLUTION. The estimated standard deviations are given by the formula

$$\sigma_{\bar{p}} \approx \sqrt{\frac{\bar{p}(1-\bar{p})}{n}},$$

because samples of these sizes are negligible compared with the national population. Carrying out the arithmetic, we get the following:

Sample size:	800	1600	3200
Estimate of $\sigma_{\bar{p}}$:	.014	.010	.007

Note that when we quadruple the sample size from 800 to 3200, the estimate of $\sigma_{\bar{p}}$ is cut in half.

Putting the Variance to Work. Cheering as it is to know that larger samples lead to smaller variances, how can we use this information to tell us how close the sample estimate $\bar{p}$ is likely to be to the true value p? The answer depends on an almost incredible theorem that gives a good approximation in many circumstances. We study this in the next chapter.

PROBLEMS FOR SECTION 6-5

1. If $p = 0.25$ and $n = 50$, what is the variance of $\bar{p}$ for sampling with replacement?
2. If $p = 0.75$ and $n = 50$, what is the variance of $\bar{p}$ for sampling with replacement?
3. Why should the answers for Problems 1 and 2 be the same?
4. A geologist, wishing to estimate the proportion of quartzite pebbles in a stream, takes a sample of size 30. If the true proportion is $p = 0.7$, what is the variance of $\bar{p}$? What if $p = 0.6$?
5. Continuation. If the geologist wants to cut the variance in half, how big a sample should she draw?
6. For a sample of size 10 drawn with replacement, the variance of $\bar{p}$ is 0.016. What values might the true proportion p have?

7. Explain in your own words why finite samples of size n drawn with replacement from a finite population with proportion p of items of type A generate the same variance for $\bar{p}$ as does sampling without replacement from an infinite population with the same composition and sample size.
8. In sampling with replacement, samples of size n have been used to estimate $\bar{p}$. What sample size is needed to reduce the variance by half? To reduce the standard deviation by half?
9. If p is near zero, then the variance of $\bar{p}$ for sampling with replacement is approximately p/n. Explain why.
10. Continuation. If p is near 1, approximate the variance of $\bar{p}$.
11. *Class project: parking meters.* If your town or city has parking meters, it should be an instructive study to find out how they behave when fed a given amount of money.

    a) What fraction of meters accept money?
    b) What fraction activate their dials on being given money?
    c) For those that accept money, compare the parking time the machines deliver before showing a violation with what they are supposed to deliver.
    d) Estimate the standard deviations of the estimates in parts a and b (use the observed proportion instead of the true proportion p in the formula).

You will be more likely to study the shorter times rather than the longer ones because an observation costs less and because it doesn't take as long to measure the times.

To do the study properly, you will want to map the parking meters in a well-defined region and then draw a random sample of them. Simple random sampling is satisfactory, but you may want to draw a meter at random and then measure that meter and perhaps the meters nearest it (perhaps two). It is important that chosen meters be predesignated and not be subject to the whim of the field worker. "The nearest two" is not well defined, but it needs to be.

(Good methods of drawing random samples using random numbers are described in Chapter 7.)

You will need to decide what to do about meters already in use. One possibility for meters that have remaining time showing on their dials is to start a measurement at whatever the pointer indicates and then time it to the end. A distraction will occur when a parker comes to put additional money into a meter before it runs out. It will be best not to intervene in that process, even though nonintervention will cost you a measurement. If you are using a watch, and the parking-meter dial has an easily visible pointer, you may nevertheless be able to get a comparison between elapsed times as shown on the dial and as shown on your watch.

A carefully designed and well-executed study of this kind can create a good deal of popular interest. People like to know if such equipment is giving them what it promises.

After formulating an initial plan, be sure to carry out a pilot study. A first foray into the field will uncover new problems that must be taken into account in the design.

6-6 FREQUENCIES, PROPORTIONS, AND PROBABILITIES

In this chapter we have been looking at all possible samples of a given size from a population, and we have calculated how often an observed proportion $\bar{p}$ takes on various values. For example, if $\bar{p}$ gives the proportion of voters in a sample of size 3 from the Williams family, we have the following:

Estimate, $\bar{p}$:	⅓	⅔
Frequency:	½	½

The frequencies for the values of $\bar{p}$ are often called **probabilities.** We tend to use the terms interchangeably. When we refer to the collection of probabilities associated with the possible values of $\bar{p}$, we speak of the **probability distribution** of $\bar{p}$. For probability distributions, the frequencies or probabilities for the possible values add to 1.

EXAMPLE $N = 10$ and $p = 0.4$. We consider drawing samples of size 5 without replacement from this population, and we get, to three decimals, the following probability distribution for $\bar{p}$ (by methods not given here):

Estimate, $\bar{p}$:	0	.2	.4	.6	.8	1.0
Probability:	.024	.238	.476	.238	.024	.000

In this example, the probability distribution of $\bar{p}$ is the collection of the 6 different probabilities associated with the 6 possible values of $\bar{p}$.

Mathematicians have developed a formal set of rules for dealing with probabilities and probability distributions. We discuss these in Chapter 8.

6-7 SUMMARY OF CHAPTER 6

For two-category data, we have the following notation:

N: population size;

n: sample size;

p: population proportion of type A;
$\overline{p}$: sample proportion of type A.

1. $\overline{p}$ has a probability distribution for random samples of size n.
2. $\overline{p}$ is an unbiased estimate of p.
3. If we score 1 for an observation of type A and 0 for type non-A, then the variance for a single observation is

$$p(1 - p).$$

4. The variance for $\overline{p}$ for sampling without replacement is

$$\operatorname{Var} \overline{p} = \frac{p(1 - p)}{n}\left(\frac{N - n}{N - 1}\right). \tag{1}$$

5. The standard deviation is $\sqrt{\text{variance}}$.
6. If sampling is with replacement, the variance of $\overline{p}$ is

$$\operatorname{Var} \overline{p} = \frac{p(1 - p)}{n}. \tag{2}$$

7. Sampling with replacement is equivalent to sampling from an infinite population with the same proportional composition.
8. If a finite population and an infinite population have the same composition, then the sample size n_f from the finite population drawn without replacement required for the same variance given by a sample of size n_∞ from the infinite population is

$$n_f = n_\infty\left(\frac{N}{N + n_\infty - 1}\right) \approx n_\infty\left(1 - \frac{n_\infty}{N}\right).$$

9. The main message is that from the point of view of variance of estimates, population size does not matter much in most practical work. It can be regarded as infinite. But the sample size does matter.

SUMMARY PROBLEMS FOR CHAPTER 6

1. For the probability distribution of $\overline{p}$ given in Section 6-6, with $N = 10$, $p = 0.4$, and $n = 5$, find the mean, variance, and standard deviation, using the formula for the variance.
2. Continuation. Compute the mean and variance directly from the frequencies of $\overline{p}$ given in the distribution in Section 6-6.

3. If $N = 10$, $p = 0.4$, and $n = 5$ and samples are drawn with replacement, then to three decimals the probability distribution of $\overline{p}$ is as follows:

Estimate, $\overline{p}$:	0	.2	.4	.6	.8	1.0
Probability:	.078	.259	.346	.230	.077	.010

Use the formulas to find the mean and standard deviation and variance of $\overline{p}$.

4. Continuation. Compute the mean and variance of $\overline{p}$ directly from the frequencies, and compare the results with those in Problem 3.

5. How would changing N from 10 to 20 or to infinity in Problems 3 and 4 change the mean and variance?

6. In sampling without replacement from a senior class of size $N = 1000$, a sample of size $n = 100$ is drawn. The fraction of men in the class is actually $p = 0.6$. Find the mean and variance and standard deviation of $\overline{p}$.

7. In Problem 6, what is the value of the improvement factor?

8. Graph $\sqrt{p(1 - p)}$ against p to see how the standard deviation of $\overline{p}$ changes with p for a fixed sample size.

9. Continuation. Sometimes the graph in Problem 8 is referred to as "flat in the middle," and the value $\sqrt{p(1 - p)}$ is taken as ½ for all p from 0.2 to 0.8. What is the maximum percentage error when this is done?

★ **10.** If there are N items in the population, then there are N^2 samples of size 2 drawn with replacement and $N^2 - N$ samples of size 2 drawn without replacement. Explain why.

★ **11.** We found

$$n_f = n_\infty \left(\frac{N}{N + n_\infty - 1} \right).$$

Find n_∞ in terms of N and n_f.

★ **12.** Two populations of size N and M have the same proportion p of voters. If, without replacement, a sample of n is drawn from the first and m from the second, find a formula relating n to m so that the variances of $\overline{p}$ are the same for the two samples.

REFERENCES

W. G. Cochran (1977). *Sampling Techniques*, third edition, Chapters 1 through 3. Wiley, New York.

W. E. Deming and A. Stuart (1978). Sample surveys. In *International Encyclopedia of Statistics, Vol. 2,* edited by W. H. Kruskal and J. M. Tanur, pp. 867–889. Free Press, New York.

R. Ferber, P. Sheatsley, A. Turner, and J. Waksberg (1980). *What Is a Survey?* American Statistical Association, Washington, D.C.

F. Mosteller, R. E. K. Rourke, and G. B. Thomas, Jr. (1970). *Probability with Statistical Applications,* second edition, Chapters 1, 3, 4, and 5. Addison-Wesley, Reading, Mass.

M. J. Slonim (1960). *Sampling in a Nutshell.* Simon & Schuster, New York.

7

Large Samples, the Normal Approximation, and Drawing Random Samples

Learning Objectives

1. Working with large samples
2. Using the standard normal distribution to approximate binomial and related probabilities
3. Using random numbers to draw samples from finite populations

7-1 PRACTICAL COMPUTATION OF PROBABILITIES AND METHODS OF DRAWING RANDOM SAMPLES

When samples are large, the procedures we used for small-sample enumeration in Chapter 6 become impractical. We need tables and approximations. Binomial tables are available for computing the probability that $\bar{p}$ is within a given distance of p. For our work, what is more important is that a table of the standard normal distribution can give good approximations, not only in binomial problems but also in problems we shall study in later chapters.

The process of drawing random samples can be aided by the use of random numbers. In Section 7-4 we explain two methods for drawing such samples.

7-2 LARGE SAMPLES FROM LARGE POPULATIONS WITH TWO KINDS OF ELEMENTS: APPROXIMATE DISTRIBUTION OF $\bar{p}$ WHEN p IS NOT NEAR 0 OR 1

To find out how often we can expect an observed proportion, $\bar{p}$, to be within a given distance of the true value, we need a little more equipment, and we need some examples with larger samples than those we have dealt with in the Williams family. We shall regard the population as infinite throughout this section.

We are involved here with the binomial distribution. The distribution gives the probability of getting a given number of successes in n trials when p is the probability of success on each trial and the outcome of one trial has no influence on the outcomes of others. For example, when drawing members with replacement from the Williams family, "success" corresponded to drawing a voter, and the probability of success was ½.

EXAMPLE 1 $n = 25$, $p = ½$. For an infinite population with the fraction of A's equal to ½, find the proportion of samples of size 25 that give each possible estimate $\bar{p}$.

SOLUTION. Many sources give substantial sets of binomial tables for solving this problem precisely, and many books give a formula to help answer such a question. By using a table, we can get some understanding of the distribution of the $\bar{p}$'s and of the approximation that we use for this distribution. Table 7-1 shows to three decimals the probabilities of the possible values of $\bar{p}$ for $n = 25$, $p = ½$. The symmetry of Table 7-1 shows that for this distribution the proportion of samples that get x A's is identical with

TABLE 7-1

Distribution of sample estimates $\bar{p}$ for samples of size $n = 25$ for a binomial with $p = \frac{1}{2}$†

Number of A's in sample x	*Estimate of p* $\bar{p}$	*Probability of estimate*	*Number of A's in sample* x	*Estimate of p* $\bar{p}$	*Probability of estimate*
0	.00	0+††	13	.52	.155
1	.04	0+	14	.56	.133
2	.08	0+	15	.60	.097
3	.12	0+	16	.64	.061
4	.16	0+	17	.68	.032
5	.20	.002	18	.72	.014
6	.24	.005	19	.76	.005
7	.28	.014	20	.80	.002
8	.32	.032	21	.84	0+
9	.36	.061	22	.88	0+
10	.40	.097	23	.92	0+
11	.44	.133	24	.96	0+
12	.48	.155	25	1.00	0+

† Note that the same probability applies to x and $n - x$ and to $\bar{p}$ and $1 - \bar{p}$.

†† For entries 0+, the probability is less than .0005 but greater than 0.

that for $n - x$ *A*'s. For example, when $x = 10$, the probability is 0.097, just as it is for $n - 10 = 15$. This result is peculiar to distributions with $p = \frac{1}{2}$. When p is not $\frac{1}{2}$, we do not get exact symmetry in the $\bar{p}$ distribution, although sometimes we get approximate symmetry.

The events with the largest probability are $x = 12$ and $x = 13$, and each occurs in 0.155 or 15.5 percent of the samples of size 25.

The sum of the probabilities should be 1.000, but because of rounding in the table, they add to only 0.998. If we compute each probability exactly, the sum adds to 1. The probabilities for $x = 0, 1, 2, 3, 4, 21, 22, 23, 24$, and 25 each are less than 0.0005.

If we sum the probabilities from $x = 0$ to $x = 8$ ($x \leq 8$), we find the sum to be about 0.054 (allowing 0.001 for the total of the five 0+'s). Thus about 5 percent of the samples have 8 or fewer *A*'s, and about 5 percent have 17 or more *A*'s. By subtraction, this means that about 90 percent of the samples have *x*'s from 9 through 16, and therefore 90 percent of the $\bar{p}$'s lie between

$9/25 = 0.36$ and $16/25 = 0.64$. Thus, in this problem, in only about 10 percent of the samples does $\bar{p}$ depart from p by more than 0.14.

We want to relate the distances of $\bar{p}$ from p to the probabilities in the table. One measure of variation we introduced in Section 6-2 and used again in Section 6-4 was the variance, a quantity based on the squares of distances. To get a distance measure of variation, we took the square root of the variance, which is the standard deviation:

$$\text{standard deviation} = \sqrt{\text{variance}}$$

In our problem we can compute the standard deviation, σ, of our estimate $\bar{p}$ as follows:

$$\sigma = \sqrt{\text{variance}} = \sqrt{\frac{p(1-p)}{n}} = \sqrt{\frac{\frac{1}{2}(\frac{1}{2})}{25}} = \frac{1}{10}.$$

Thus the standard deviation for the distribution of the estimate $\bar{p}$ is 0.1.

What proportion of the samples give values of $\bar{p}$ within 0.1 of p in our problem? Those values of $\bar{p}$ run from 0.4 to 0.6. From Table 7-1, we need

$$2(.155 + .133 + .097) = 2(.385) = .770.$$

So in this problem, we find 77 percent of the samples giving values within 1 standard deviation (that is, within 1σ) of the true value. What proportion lie within 2 standard deviations of p? We get 0.956 or 95.6 percent by summing from $\bar{p} = 0.3$ to $\bar{p} = 0.7$, that is, for $8 \le x \le 17$. We note here that we cannot have $\bar{p} = 0.3$, only 0.28 or 0.32.

Figure 7-1 shows a graph representing the probabilities of the various values of $\bar{p}$. Note that its general shape is that of a bell: The heights of the ordinates represent the probabilities with which given $\bar{p}$ values occur. Each bar is centered over a possible value of $\bar{p}$. We have used bars rather than dots immediately above the possible values of $\bar{p}$ because we are planning to approximate the probabilities by the areas under a smooth curve. Thus we plan later to act as if the probability at a given value of $\bar{p}$ were spread smoothly over an interval. For example, we will pretend that $\bar{p} = 0.52$ runs from 0.50 to 0.54, which are the midpoints between $\bar{p} = 0.48$ and $\bar{p} = 0.52$ and between $\bar{p} = 0.52$ and $\bar{p} = 0.56$.

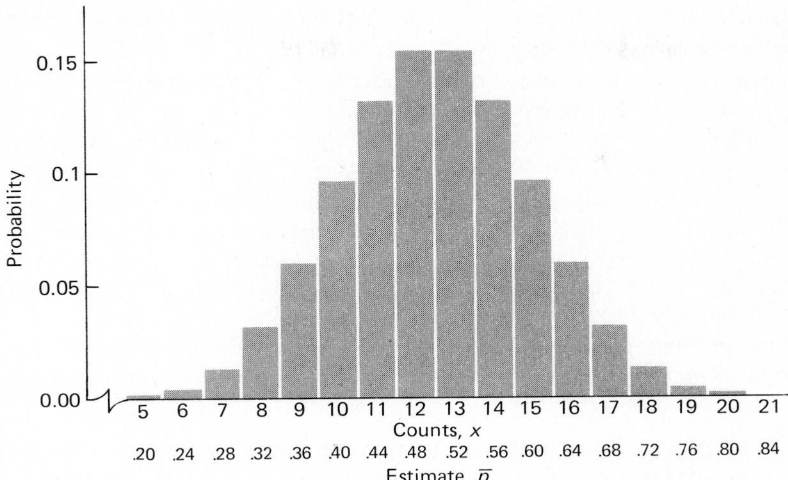

Figure 7-1 Histogram showing the probabilities associated with each outcome, expressed as either x, the count, or $\bar{p}$, the estimate, for samples of size $n = 25$ drawn from a large population. Note that although probabilities are concentrated at the integral values of x, or at $\bar{p}$'s that are multiples of 0.04, the histogram has spread these probabilities over an interval. This approach improves the approximation, as described in the discussion of the refinement.

PROBLEMS FOR SECTION 7-2

All the problems in this set refer to the binomial distribution with samples of size $n = 25$ and $p = \frac{1}{2}$. Use Table 7-1 to solve the following problems.

1. What proportion of the samples yield values of x equal to or greater than 15?
2. What proportion of the samples give values of $\bar{p}$ within 0.06 of p?
3. What proportion of the samples give values of $\bar{p}$ within 0.3 of p? What proportion is within 0.32?
4. What proportion of the samples give values of $\bar{p}$ that lie within $\frac{1}{2}$ of a standard deviation of p? What proportion lie within 3 standard deviations?
5. Which is more likely, that x equals one of 5, 6, 7, or 8 or that it equals 9?
6. Find the smallest set of x's that contains at least 75 percent of the total probability.
7. Event A is $x = 11, 12, 13, 14$, and event B is $x \leq 12$. You get to choose either event A or event B, and you get a prize if your event occurs when the binomial experiment is carried out. Which event do you choose?
8. As inspector of weights and measures, you weigh 25 packages, each of which is supposed to weigh a pound. You expect that about half will weigh slightly more

than a pound, and half slightly less. You find that 20 of the 25 packages weigh less than 1 pound. What is your conclusion?

9. A machine tool is supposed to be creating blocks of metal 1.000 inch thick. However, the tool wears and occasionally needs resetting. A sample of 25 recently made blocks is measured, and 4 are less than 1.000 inch in thickness, the rest being larger. Do you think it is time to reset the machine? Why or why not?

10. A congressman receives a large amount of mail concerning a controversial issue. Among the first 25 pieces of mail, 15 of his constituents favor the congressman's stand on the issue. What is the chance that the sample will favor his position at least this much if the total mail he is going to get runs half favorable and half unfavorable to his position?

7-3 FITTING THE STANDARD NORMAL DISTRIBUTION TO A BINOMIAL DISTRIBUTION

Even when we deal with discrete distributions like the binomial distributions, continuous distributions can give useful approximations. The most important family of continuous distributions is composed of normal distributions, one of which is graphed in Fig. 7-2. For these distributions, probabilities are represented as areas between the curve and the horizontal axis. For example, the probability of a number between 0 and 1 being drawn from the normal distribution in Fig. 7-2 is about 0.34, a number we can get from Table 7-2 or Table A-1 in Appendix III in the back of the book.

Although we do not explain how to calculate it, the mean of the normal distribution in Fig. 7-2 is 0, and the standard deviation is 1. The distribution in Fig. 7-2 is called the **standard normal distribution.**

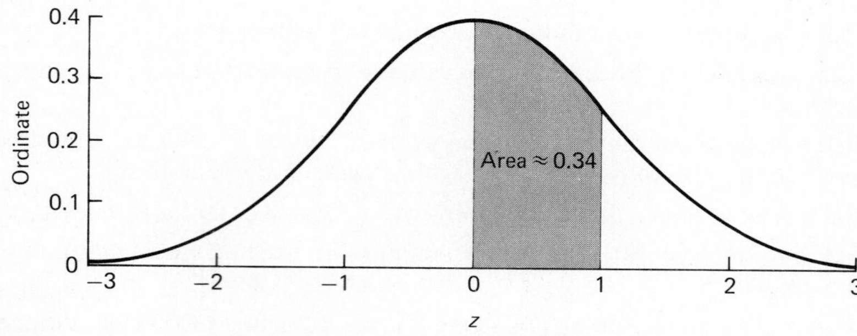

Figure 7-2 Ordinates of the standard normal distribution. The probability of z falling into an interval is represented by the area under the curve and over the interval. For example, the probability that z is between 0 and 1 is about 0.34.

TABLE 7-2†
Standard normal distribution

z	.0	.1	.2	.3	.4	.5	.6	.7	.8	.9
0	.000	.040	.079	.118	.155	.191	.226	.258	.288	.316
1	.341	.364	.385	.403	.419	.433	.445	.455	.464	.471
2	.477	.482	.486	.489	.492	.494	.495	.497	.497	.498
3	.499	.499	.499	.4995						

† The area under the standard normal curve from 0 to z, shown shaded, is P:

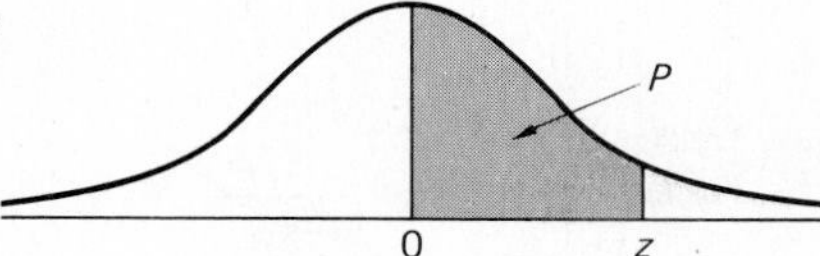

Examples:
Probability between z = 0 and z = 1.3 is .403.
Probability between z = −1.3 and z = 0 is .403.
Probability between z = −1.3 and z = +1.3 is .403 + .403 = .806.
Probability of z ≥ 1.3 is .500 − .403 = .097.
Probability of z ≥ −1.3 is .500 + .403 = .903.

When we use it to approximate a binomial, we adjust the scale of the binomial $\overline{p}$ so that it has mean 0 and standard deviation 1. We do this by computing standard scores for the values of $\overline{p}$ of interest. In general, not just for the binomial, the **standard score** for a variable x with mean μ and standard deviation σ is

$$z = \frac{x - \mu}{\sigma}$$

For the binomial $\overline{p}$, the standard score is

$$z = \frac{\overline{p} - p}{\sqrt{\dfrac{p(1 - p)}{n}}}.$$

Note that these z scores have 0 means and standard deviations 1. So the change to z scores equates the means and standard deviations of the normal and the binomials.

EXAMPLE 2 Binomial $n = 4$ and $p = \frac{1}{2}$. Figure 7-3 shows a plot of the histogram for the binomial with a sample of size 4 with probability of success p equal to $\frac{1}{2}$. It shows two horizontal scales, one for $\bar{p}$ and one for z. Check the value of the z scale when $\bar{p} = 0.75$.

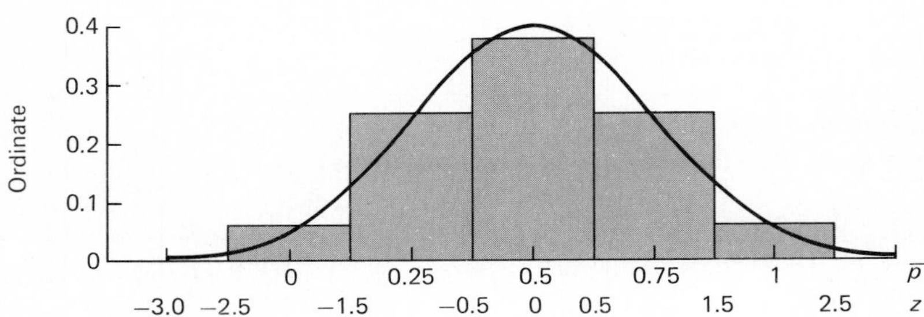

Figure 7-3 Standard normal distribution and histogram for the binomial distribution with $n = 4$ and $p = \frac{1}{2}$.

SOLUTION. We have $p = \frac{1}{2}$, and so the standard deviation of $\bar{p}$ is

$$\sigma = \sqrt{\frac{\frac{1}{2}(\frac{1}{2})}{4}} = \frac{1}{4} = .25.$$

Thus the standard score for $\bar{p} = 0.75$ is

$$z = \frac{\bar{p} - p}{\sigma} = \frac{.75 - .50}{.25} = 1.$$

We note visually that the areas in Fig. 7-3 corresponding to the bars of the histogram are approximated by the areas under the normal curve falling between the verticals bounding the bars.

EXAMPLE 3 In Fig. 7-3, compare the area under the standard normal distribution that is associated with the bar for $\bar{p} = 0.75$ with the correct probability.

SOLUTION. The probability of exactly 3 successes out of 4 trials when $p = \frac{1}{2}$ is 0.25. Table 7-2 gives the area under the normal curve between 0 and

z. The histogram bar for $\bar{p} = 0.75$ runs from $z = 0.5$ up to $z = 1.5$. We need the difference in the areas, and we get the following:

	Value from Table 7-2	
area from $z = 0$ to $z = 1.5$:	.433	
area from $z = 0$ to $z = .5$:	.191	
area from $z = .5$ to $z = 1.5$:	.242	(difference)

This value 0.242 approximates the correct value 0.250, which can be found from Table A-5 in Appendix III in the back of the book. Table A-5 is a special binomial table for $p = ½$.

The example just carried out gives an especially close result because the binomial is symmetric for $p = ½$, as is the normal. The next example deals with an asymmetric binomial.

EXAMPLE 4 Binomial $n = 12$ and $p = ¼$. Compare the binomial probability of 3 or more successes with that given by the normal approximation.

SOLUTION. From binomial tables we can find that the probability of 3 or more successes is 0.609. Figure 7-4 shows a graph of the histogram using the z score along with a superimposed graph of the standard normal curve. The

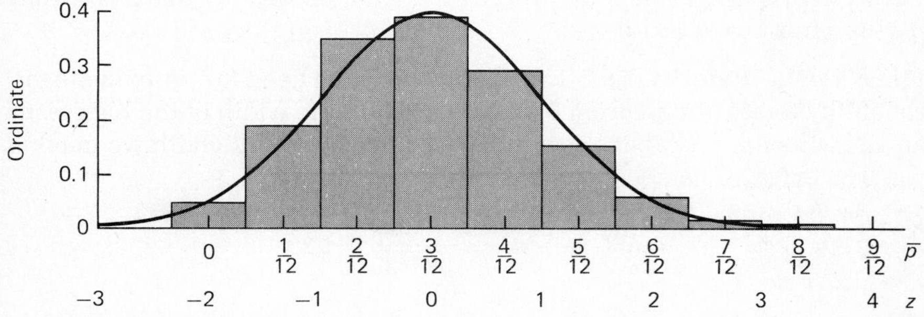

Figure 7-4 Standard normal distribution and histogram for the binomial distribution with $n = 12$ and $p = ¼$.

normal curve to the right of $z = 0$ has area 0.5. To this we need to add the area corresponding to the left half of the bar for $\overline{p} = 3/12$. The lower limit of that bar falls midway between $\overline{p} = 2/12$ and $\overline{p} = 3/12$, or at $\overline{p} = 2.5/12$. This gives a z score of

$$z = \frac{\overline{p} - p}{\sigma} = \frac{2.5/12 - 3/12}{\sqrt{(1/4)(3/4)/12}} = -\frac{1}{3} \approx -.333.$$

Because the normal curve is symmetric, we read the area between $z = 0$ and $z = +0.333$, and we get, after interpolation, the probability 0.130. This must be added to the 0.500 we got from the right half of the curve to give a final approximation of 0.630. This 0.630 can be compared with the correct value 0.609.

Note. The discrete distribution has finite limits, but the normal has infinite limits. Consequently, we need to handle the tail of the normal to the right of the rightmost bar in a special way. We add the probability under the normal curve to the right of the rightmost bar to that of the rightmost bar, and similarly for the extreme left.

ANOTHER NORMAL TABLE

Sometimes it is especially convenient to have a standard normal table that gives probabilities for intervals from $-z$ to z. Table 7-3 does this.

EXAMPLE 5 Use Table 7-3 to approximate the probability that $\overline{p}$ is within 0.1 of ½ when $p = 1/2$ and $n = 25$.

SOLUTION. Refer to Fig. 7-1. In this problem we need to approximate the probability for $\overline{p}$ running from 0.4 to 0.6 and half the width of the histogram bars at each end. Each bar has width 0.04, and so the full width we need is from 0.38 to 0.62. The z scores then are

$$z = \frac{\overline{p} - p}{\sigma} = \frac{.62 - .50}{\sqrt{(1/2)(1/2)/25}} = 1.2,$$

and

$$z = \frac{.38 - .50}{\sqrt{(1/2)(1/2)/25}} = -1.2.$$

Reading Table 7-3 from -1.2 to 1.2 gives 0.770, which happens to be correct to three decimals, as we learned in Section 7-2 from a table of the binomial distribution.

TABLE 7-3

Standard normal distribution†

z	.0	.1	.2	.3	.4	.5	.6	.7	.8	.9
0	.000	.080	.159	.236	.311	.383	.451	.516	.576	.632
1	.683	.729	.770	.806	.838	.866	.890	.911	.928	.943
2	.954	.964	.972	.979	.984	.988	.991	.993	.995	.996
3	.997	.998	.999	.999	.999	.9995				

† The probability that a measurement from the standard normal distribution falls between $-z$ and z:

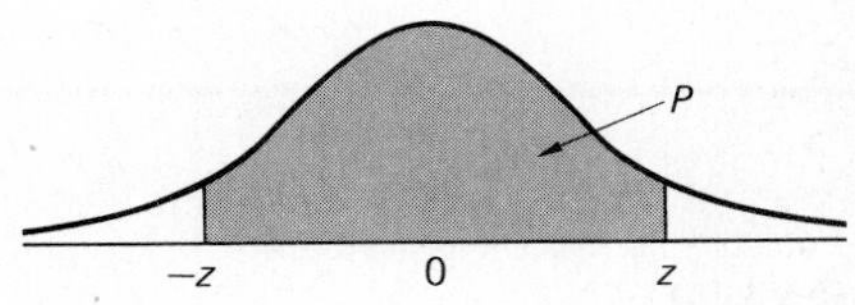

Examples:
Probability between $z = -1.3$ and $z = +1.3$ is .806.
Probability between $z = 0$ and $z = 1.3$ is $.806/2 = .403$.
Probability between $z = -1.3$ and $z = 0$ is $.806/2 = .403$.
Probability of $z \geq 1.3$ is $(1 - .806)/2 = .097$.

PROBLEMS FOR SECTION 7-3

1. How is probability represented by the standard normal curve?
2. To get the scale for $\bar{p}$ to match that of the standard normal curve, how do we adjust it?
3. The z score for the binomials and the z score for the standard normal have two common properties. What are they?
4. For the standard normal curve, use Table 7-2 to find the probability between

   a) $z = 0$ and $z = .6$
   b) $z = 0$ and $z = 1.65$
   c) $z = 0$ and $z = 1.96$
   d) $z = 0$ and $z = -.2$

5. For the standard normal curve, find the probability between
   a) $z = -1$ and $z = +1$
   b) $z = -2$ and $z = +2$
   c) $z = -.67$ and $z = +.67$
6. For the standard normal curve, find the probability between
   a) $z = -.5$ and $z = 1.5$
   b) $z = -1.65$ and $z = 1.96$
   c) $z = 1$ and $z = 2$
   d) $z = -1$ and $z = -2$
7. For the binomial with $n = 4$ and $p = \frac{1}{2}$, the probability is 0.375 that $\bar{p} = 0.5$. Use the z score and Table 7-2 to get the normal approximation to this probability.
8. For the binomial with $n = 4$ and $p = \frac{1}{2}$, use Fig. 7-3 to get the normal approximation that $\bar{p}$ is equal to or greater than 0.5.
9. For the binomial with $n = 12$ and $p = \frac{1}{4}$ in Fig. 7-4, obtain the normal approximation to the probability that $\bar{p} \geq \frac{5}{12}$.
10. Continuation. Find the normal approximation to the probability that $\bar{p} \leq \frac{1}{12}$.
11. Use Table 7-3 to approximate the probability that $\bar{p}$ is between 0.4 and 0.6 when $n = 10$ and $p = \frac{1}{2}$.

7-4 RANDOM SAMPLING

Up to now, we have discussed the probabilities of samples having various $\bar{p}$'s among all possible samples. But what does that distribution over all possible samples have to do with any specific sample survey? After all, exactly one $\bar{p}$ appears when the survey is carried out. What if it is carried out so that the $\bar{p}$ actually falls into one of the 0+ cells, far from the true but unknown p?

To get an answer to this question, we must all make our peace with uncertainty. We have to get used to the uncomfortable existence where we cannot be sure how things will work out. Most of us have made our peace with this notion in everyday life, but we may still believe that in intellectual matters certainty is achievable. In mathematics this often seems possible, and in some translations into real life we can come very close. For example, in counting our change, we are nearly, but not quite, certain. In other translations from mathematics to real life, like the one we discuss next, we must settle for being fairly sure, because certainty may be either impossible or impractically expensive.

When we computed the frequencies of the samples producing the different $\bar{p}$'s for the Williams family, we assigned each sample the same weight, 1, summed all the $\overline{p}$'s, and divided by the total number of possible samples. To assure that the sort of theory that we discussed in earlier sections of this chapter applies here as well, we need to arrange equal weights for the samples. We can accomplish this through the mechanism of random sampling.

Suppose that all the items in the population we want to sample are listed from 1 through N. We can imagine writing the numbers on cards, stirring or shuffling these cards, and then drawing our sample of n by selecting n cards from the well-shuffled stack. This is an attractive idea, and one often used, but it is very impractical. Cards are difficult to shuffle, and after one gets beyond hundreds of them, as in the great money lotteries held in many states and countries, elaborate equipment is required to carry out the shuffling. Even when we shuffle a standard deck of 52 cards, they tend to stick together in little clumps.

What we use instead of shuffled cards are random numbers. By special electronic processes, by mathematical methods, or by their combination, we produce random numbers that nearly enough have certain properties that we require. Examples of such properties are approximately equal numbers of 0's, 1's, 2's, . . . , 9's and absence of dependence of one digit on other digits, as we now explain. We want each digit to have a $\frac{1}{10}$ chance of coming up next, no matter what has happened before. If the previous digit is a 3, we should be just as likely to get any digit next, including 3, as if the previous digit were a 5 or any other digit.

We have many ways to use random numbers to select samples. Let us mention two:

Method 1: Direct Selection. Suppose the population consists of 75 items numbered 01, 02, . . . , 10, 11, . . . , 74, 75, and we wish to draw 5 items without replacement. We use a table like Table 7-4 or Table A-7 in

TABLE 7-4

One hundred random digits

Line number	*Column number* 1–10		11–20	
1	15544	80712	97742	21500
2	01011	21285	04729	39986
3	47435	53308	40718	29050
4	91312	75137	86274	59834
5	12775	08768	80791	16298

Appendix III in the back of the book, taking two-digit numbers in succession. Suppose the first few are these:

14
26
62
90
49
88
26
27

Then the first item chosen is number 14.

The next item chosen is number 26.

The third item chosen is number 62.

The 90 cannot be used, for we have no number higher than 75.

The fourth item is number 49.

We reject 88 as too high.

Item 26 has already been used and cannot be used again because we are sampling without replacement.

Item 27 is the fifth and final item in the sample.

And that completes the sample.

Starting. Some people take elaborate precautions in deciding on the first number in a sequence of random numbers chosen from a table. What is important is to write down exactly what the rules for starting and continuing are and to follow them strictly. After using a set of random numbers, mark them off in the table, and then you are ready to start the next problem at the place where you left off the last problem. People who worry about starting are worrying about the great bias that might be caused in choosing, say, one particular number because of preferences of the chooser, as when the number is chosen to settle a bet. In statistical work, we ordinarily use quite a few random numbers in a study—hundreds or thousands. Even so, good practice means having an explicit rule about the start and the continuation.

Method 2: Ordering. Sometimes it is more convenient to assign the random numbers to the items and then let their sizes do the choosing, as we next explain. If this is done, we might well use a several-digit number.

To draw a sample of 5 from 75 items using method 2, we associate with each of the 75 items a random number, say a five-digit number. We might then choose as our sample of 5 the 5 items associated with the 5 smallest five-digit random numbers. In Table A-7 in Appendix III in the back of the book we have 1000 one-digit numbers, or, if we prefer, 200 five-digit numbers. Let us number them vertically so that the first column gives us the first 20 random numbers. The first four columns will give us 80, but we don't need the last 5 numbers in column 4 because we have only 75 items. Now we locate the 5 smallest five-digit numbers in the first 3¾ columns. They are as follows:

Item Number	Random Number
2	01011
16	07851
42	04729
50	00116
60	01634

Thus our sample consists of the 5 items numbered 2, 16, 42, 50, and 60.

Ties. Sometimes we get ties when we use this method, but they can be broken when necessary by adding further random digits to the number. If we had been using three-digit random numbers, we might have had to choose between the 474 in line 3 of the first five-digit column and the 474 in line 16 of the second five-digit column. We just add digits from the table to the right until the tie is broken. Here the next digit gives 4743 for the first 474 and 4745 for the second. If we run out of digits going to the right, we use the number at the beginning of the line.

We now have two methods for drawing a random sample from a population, and these methods should approximate closely the equal weights used in the frequency-of-samples approach. Now we can speak of the probability that our sample gets close to the true p. The probabilities are approximated by our normal approximation for the two-category problems.

Thus, in practice, we make an active effort to produce a random sample by using random number tables.

PRACTICAL ADVICE

Small Problems. If the problem is a modest one that will yield readily to simple random sampling, using random numbers and lists may be quite adequate, and the methods described here can be helpful both for drawing the sample and for assessing its uncertainty. To tie our work back to Chapters 5 and 6, let us consider how to carry out a small sample survey.

Sample survey: What fraction of people in *Who's Who in America* (or some comparable survey) attended or worked at Harvard University at some time during their careers? Reading every item would be a terrible chore, and our hurry to do it probably would lead to poor-quality work.

Design: For our purpose, a modest sample size like $n = 100$ might do well. The proportion of Harvard-related people probably is somewhere between zero and 10 percent; suppose we guess 5 percent. Then the standard deviation of $\bar{p}$ is

$$\sigma = \sqrt{\frac{p(1-p)}{n}} = \sqrt{\frac{.05(.95)}{100}} \approx .02.$$

In many problems, the sample mean falls within ± 2 standard deviations of the true mean about 95 percent of the time. Thus, with a sample of size 100, we hope 95 percent of the time to come within 2σ or 4 percent of the true p.

We might pick 100 pages at random and begin with the first biography starting after the middle of the first column. If we want to be a little fussier, we can choose a random column from the page chosen. We cannot begin with an item that crosses the middle of the first column, because longer items are more likely to cross it than shorter ones. People with longer biographies may have been more places and thus may be more likely to have been at Harvard.

Before actually carrying out the survey, we may wish to add a few more properties we would like to record for the 100 chosen (for example, sex, age, profession). One danger in any study is that one can get bogged down in collecting a great deal of information and then never get finished. Do a few things in the first round, but don't necessarily try to do everything.

Large Surveys. The ideas given here are intended to introduce the reader to what can be done. If you need to take a sample of the nation's citizens for some reason, then you will do well not to try to do it yourself. First, a great deal of expertise has gone into developing such technology. It requires many people suitably placed geographically, as well as help in setting up the questionnaires, having the questions quickly tested in the field, sending out the forms, doing the interviewing, and gathering the data and processing it. Consequently, because we cannot afford to have a good national sample survey system set up just for one survey, it will be wise to purchase the services of a reliable organization that has already established a national system. In taking this step, one is purchasing not only the help of the field staff but also the expertise of the organization.

EXAMPLE 6 *Draft lotteries.* Lotteries to determine the order for calling up men for military service have been a common way to provide fair play in the United States. At the same time, until recently, we have had a lot of trouble with the method. For example, in the 1940 draft lottery there were 9000 numbers put into opaque capsules and stirred around in a fish bowl. Some capsules broke open, and the slips fell to the bottom. Six numbers wound up missing, and serial numbers in the group from 1 to 2400 were rarely drawn in the first 2000 draws. In the draft lottery for the year 1970, birthdays were used, and so only 366 numbers (including one for February 29) were randomized. A peculiar plan for putting the numbers into the box for the drawing again resulted in what seemed to be a bias. The average number for a given month would be $(1 + 366)/2 = 183.5$. But there seemed to be a severe trend by month, as shown in Fig. 7-5. The line slopes in the direction one might expect, because the months were put into the box in order, with December numbers going in last and therefore being more likely to be on top and to be drawn first.

Every such drawing has produced numerous complaints of bias until the 1971 drawing, when the National Bureau of Standards, with a panel of former presidents of the American Statistical Association, designed the

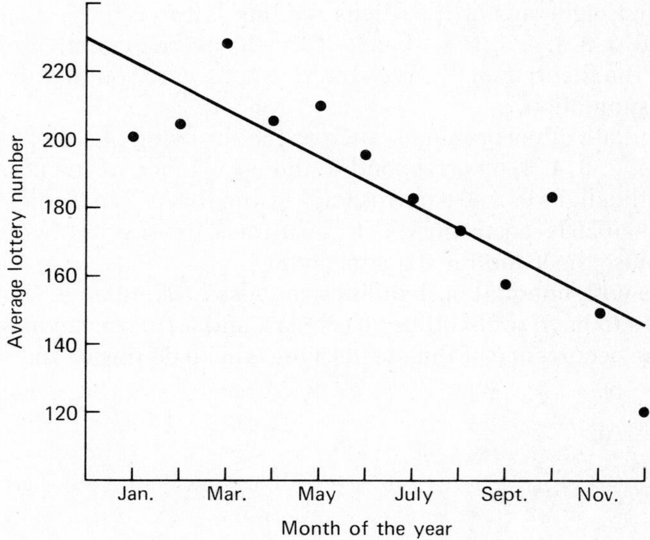

Figure 7-5 The 1970 draft lottery.

Source: S. E. Fienberg (1971). Randomization and social affairs: The 1970 Draft Lottery. *Science*, Vol. 171, pp. 255–261, 22 January 1971. Copyright 1971 by the American Association for the Advancement of Science.

drawing. Balls numbered 1 to 365 (no leap year) for the birthdays were placed in a drum in a random order. Then another set of balls for the ordering of the birthdays was randomized and put into a second drum. The balls in each drum were mixed thoroughly. Finally, a ball was drawn from each drum, and these two balls were matched, thus determining the position of that birthday in the ordering. This combination of randomization through the use of random numbers and physical mixing seems to have solved both the fairness problem and the problem of lack of randomness, because there have been no complaints about trends in the resulting numbers (Fienberg, 1971).

Simulations. Random numbers can be used to simulate probability experiments. For example, to study the variation in the number of heads produced in tossing a coin five times, we could proceed as follows. Let the event "head" correspond to the digits 0, 1, 2, 3, 4, and the event "tail" to 5, 6, 7, 8, 9. Then if we used sets of five digits to correspond to five tosses, we might use the random numbers as arranged in bursts of five in Table 7-4. The first set, 15544, translates into head, tail, tail, head, head, or 3 heads in all. The first eight sets of five digits reading left to right and from the top down yield: 3, 3, 2, 4, 5, 3, 3, 1 heads. Thus the average number of heads is 3 instead of the theoretical 2½. We would usually carry out hundreds of such trials in a simulation.

To simulate other problems, such as the throwing of a die, we might let the digits 1, 2, 3, 4, 5, 6 correspond to the occurrence of the corresponding faces and the digits 0, 7, 8, 9 correspond to "no throw." In Table 7-4, the last five digits—16298—correspond to three throws, the results of which are 1, 6, and 2 because the 9 and the 8 do not count.

Events with unequal probabilities can also be simulated. If events A, B, and C were to have probabilities 0.03, 0.22, and 0.75, we can assign pairs of digits to the occurrence of the events. One way to do this would be

A: 00, 01, 02
B: 03, 04, . . . , 24
C: 26, 27, . . . , 99

From our table, the first four digits would produce first a 15 meaning B and then a 54 meaning C.

PROBLEMS FOR SECTION 7-4

In the following problems, use Table A-7 in Appendix III in the back of this book for drawing samples.

1. Describe two different ways to use the random numbers to draw a sample of size 25 from an infinite binomial population with $p = \frac{1}{2}$.

2. Use the methods described in Problem 1 to actually draw the samples both ways, and compute $\bar{p}$ for each.

3. Using columns 21 through 30, make a tally of the number of times each of the digits 0, 1, 2, . . . , 9 appears. Discuss your results.

4. Using some of the digits in columns 31 through 34, generate 25 binomial samples of size $n = 5$, with $p = 0.2$. Explain your method. Make a table to show the relative frequencies of the resulting values of $\bar{p}$.

5. Take a random sample of 20 pages in this book. Explain your method. Record the number of each page of the sample that has at least one figure or table on it. Estimate the true proportion of pages in the book that have figures or tables.

6. Carry out the survey of 100 people drawn from *Who's Who in America*, keeping track of sex, age, profession, and affiliations with universities. Report the proportion of women, the average year of birth for men and for women in the sample, the distribution of professions, and the proportion reported as having been affiliated with Harvard University at some time in their careers.

Use the following information for Problems 7 and 8. A group of 10 patients is to be randomly assigned to two treatments, half to treatment *A* and half to treatment *B*.

7. How would you use method 1 to make the assignments?

8. How would you use method 2 to make the assignments?

9. How could you use the random numbers to simulate the outcome for six tosses of an ordinary six-sided die?

10. Suppose someone claims that when a die is tossed 4 times, four different numbers appear in at least 60 percent of the sets of 4 tosses. Use the random number table to simulate 100 sets of 4 tosses and see if this estimate is plausible.

7-5 SUMMARY OF CHAPTER 7

1. For binomial distributions, the probability that $\bar{p}$ falls in a given range can be obtained from binomial tables, or the probability, for large samples, can be approximated using Tables 7-2 and 7-3 and Table A-1 in

Appendix III in the back of the book for the standard normal distribution and

$$\sigma = \sqrt{\frac{p(1-p)}{n}}$$

as the unit of distance.

2. The normal approximation can be improved by using as the location of the endpoint of an interval a position midway between the extreme $\bar{p}$ inside the interval and the next more extreme $\bar{p}$ outside the interval. The endpoints of the interval are found by computing $\bar{p} \pm 1/(2n)$.
3. Random samples can be drawn from a finite population by the following two methods (among others): (1) numbering the items from 1 to N and using random numbers with as many digits as N has; (2) assigning each item a random number and choosing the n items with the smallest numbers. It is best to use several digits with this method.
4. Random numbers can be used to simulate probability experiments. By doing many simulations, we can often get answers to problems whose mathematics are especially difficult.

SUMMARY PROBLEMS FOR CHAPTER 7

1. Why do we want to use the normal approximation to the binomial distribution? Give two reasons.
2. For $n = 25$ and $p = 0.2$, use the normal approximation to calculate the probability that $\bar{p}$ is within 2 standard deviations of p.
3. Continuation. If you have not already done so, use the midpoint approach to improve the approximation in Problem 2.
4. For $n = 36$ and $p = 0.2$, use the normal approximation (with midpoint rule) to calculate how often $\bar{p}$ is within 1 standard deviation of p. What about within 2 standard deviations?
5. Continuation. For $n = 36$ and $p = 0.8$, what proportion of $\bar{p}$'s are within 1 standard deviation of p (approximately)?
6. For $n = 96$ and $p = 0.6$, find the proportion of $\bar{p}$'s between 0.53 and 0.72 (approximately).
7. In county A, on the average, 45 babies are born each day, and in county B about 15 babies are born each day. As you know, about 50 percent of all babies are boys, but the exact percentage of baby boys varies from day to day. Sometimes it may be higher than 50 percent, sometimes

lower. For a period of 1 year, each county recorded the days on which more than 60 percent of the babies born were boys. Which county do you think recorded more such days?

a) the larger county,
b) the smaller county,
c) about the same.

Briefly explain your answer.

8. Suppose that we are going to select juries of size 6 at random from a panel of 35 men and 15 women. Using the normal approximation, find the probability that the number of women jurors will be 0 or 1.

9. A target shooter ordinarily makes 30 percent bull's-eyes. Use the normal approximation to estimate the probability that he will make more than 50 percent bull's-eyes in his next 20 shots.

10. If a coin is tossed 1000 times, approximate the probability that it will come up heads exactly 500 times.

11. Explain one way to use random numbers to draw a random sample of size 10 from a population of size 102.

REFERENCES

S. E. Fienberg (1971). Randomization and social affairs: the 1970 draft lottery. *Science* 171:255–261.

F. Mosteller and R. E. K. Rourke (1973). *Sturdy Statistics*. Addison-Wesley, Reading, Mass.

F. Mosteller, R. E. K. Rourke, and G. B. Thomas, Jr. (1970). *Probability with Statistical Applications*, second edition. Addison-Wesley, Reading, Mass.

8

Probability

Learning Objectives

1. Getting acquainted with the ideas of probability experiments and sample spaces
2. Gaining skill in using the basic rules of probability
3. Calculating probabilities for compound events
4. Adjusting probabilities in the light of additional evidence (Bayes' formula)
5. Using the notion of a random variable and examining the properties of the distribution of a random variable, such as its mean, variance, and standard deviation

★ This chapter is optional because the treatment of probability in other chapters is adequate for handling the statistical material.

8-1 INTRODUCTION

In everyday language we use the word **probability** to describe situations or events that do not occur with certainty. In this sense we can speak of the probability that it will rain tomorrow, or the probability that an electric appliance is defective, or even the probability of war. In Chapter 6 we indirectly introduced the notion of probability in a more technical sense as population frequencies. This frequency approach to probability usually is associated with events that can be repeated again and again, and under essentially the same conditions.

By studying probability at this point, we achieve several goals. First, we learn how to compute probabilities of complex events formed from two or more elementary events. Second, we solidify the notion of a population and the probability distribution of numbers generated by samples drawn from it. Third, we introduce properties of distributions that we use later to approximate probabilities when the problems become too complex for exact solution.

We now give a more formal presentation of some of the basic ideas of probability that we have already encountered in Chapters 6 and 7.

8-2 PROBABILITY EXPERIMENTS AND SAMPLE SPACES

In recreational games we have many examples of events whose outcomes are uncertain and of events that present equally likely cases. An ordinary 52-card deck of playing cards consists of four suits of 13 cards each: spades ♠, hearts ♡, diamonds ♢, and clubs ♣. Each suit of 13 contains an ace, 2, 3, . . . , 10, a jack (J), a queen (Q), and a king (K). When a card is drawn from a well-shuffled face-down deck of ordinary playing cards, 52 outcomes are possible:

A♠, 2♠, . . . , K♠, A♡, 2♡, . . . , K♡, A♢, 2♢, . . . , K♢, A♣, 2♣, . . . , K♣.

We believe in equal likelihood for drawing each card because the cards have been manufactured to be physically identical in shape and have identical backs, and because of the shuffling.

In dealing with uncertain outcomes such as drawing cards or flipping coins, we shall speak of **probability experiments.**

We call the list of possible outcomes the **sample space** of a probability experiment.

For drawing a card from a deck, the sample space might be the list of 52 cards. More than one form of list is possible, depending somewhat on our knowledge and the events whose probabilities we want to compute. For example, were we interested only in the events red (R) and black (B) then R,B would be an adequate sample space. If the outcome were 7♣, the event would be B.

EXAMPLE 1 *College lectures.* A college course consists of two lectures per week on separate days, among Monday, Tuesday, Wednesday, Thursday, and Friday. Describe the sample space of possible times a student might choose. In your school, are the outcomes equally likely?

SOLUTION. The sample space consists of a list of all possible pairs of days:

Monday, Tuesday
Monday, Wednesday
Monday, Thursday
Monday, Friday
Tuesday, Wednesday
Tuesday, Thursday
Tuesday, Friday
Wednesday, Thursday
Wednesday, Friday
Thursday, Friday

In most courses, lectures are not given on consecutive days. Thus the outcomes

Monday, Tuesday
Tuesday, Wednesday
Wednesday, Thursday
Thursday, Friday

will each be less likely than each of the remaining six. If we look at the course catalogue, we will find relatively fewer of these.

Events consist of collections of outcomes.

We often want to calculate the probability that an event occurs. For example, when we draw a card, what is the probability that the card drawn is an ace? Four possible outcomes are aces, and the rest are not, and so we say that the probability of an ace is $4/52$. If the event is "ace," then we write

$$P(\text{ace}) = \frac{4}{52}.$$

The probability of not getting an ace is

$$P(\text{not ace}) = \frac{48}{52}.$$

Since either "ace" or "not ace" must happen, the total of the two probabilities is 1.

For any probability experiment whose outcomes are equally likely, we can assign to an event A the probability

$$P(A) = \frac{\text{number of outcomes favorable to event A}}{\text{number of possible outcomes of the experiment}} \tag{1}$$

If the outcomes or **elementary events** are equally likely, then we can find the probability by simply counting the number of elementary events. Hence it is important to construct the sample space carefully.

EXAMPLE 2 *Hearts.* A card is drawn from a well-shuffled deck of playing cards. What is the probability that it is a heart?

SOLUTION. Among the 52 cards are 13 hearts.

$$P(\text{heart}) = \frac{\text{number of hearts}}{\text{number of cards}} = \frac{13}{52} = \frac{1}{4}.$$

Because we are counting outcomes, probabilities run from 0 to 1, where 0 means that the event is certain not to occur, and 1 means that it is certain to occur.

EXAMPLE 3 *Joker.* What is the chance that the joker is drawn from the 52-card pack of Example 2? What is the chance that we do not draw the joker?

SOLUTION. The pack has no joker, and so the number of elementary events favorable to the joker is 0, and

$$P(\text{joker}) = \frac{0}{52} = 0.$$

Similarly, the probability that we do not draw the joker is

$$P(\text{no joker}) = \frac{52}{52} = 1.$$

EXAMPLE 4 *Two coins.* You toss a dime and a nickel separately. Find the probability that both come up heads. Also find the probability that exactly 1 comes up heads.

SOLUTION. Let us write out the sample space of the outcomes:

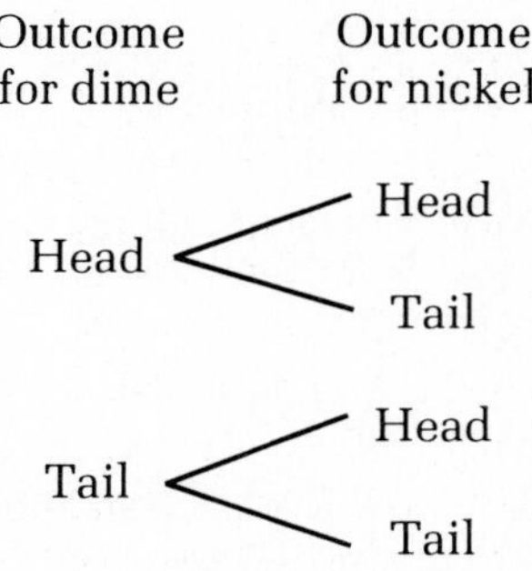

The symmetry of the coins gives each of them probability ½ of coming up heads. Whatever side shows for the dime, the outcome for the nickel is equally likely to be heads or tails. Thus the sample space consists of 4 outcomes. We find it reasonable to consider these 4 outcomes as equally likely when a dime and a nickel are tossed:

Dime	Nickel
Head	Head
Head	Tail
Tail	Head
Tail	Tail

Thus

$$P(2 \text{ heads}) = 1/4,$$
$$P(1 \text{ head and } 1 \text{ tail}) = 2/4.$$

The sample space helps us organize our thinking in probability problems, whether the outcomes are equally likely or not.

PROBLEMS FOR SECTION 8-2

1. Write out sample spaces for the outcomes of the following probability experiments:

a) rolling a die

b) rolling two dice (one red and one green)

c) tossing three coins: a penny, a nickel, and a dime

2. One card is drawn from a well-shuffled 52-card deck. List suitable sample spaces for the experiment if we are interested in (a) the suit drawn, (b) whether or not a face card (Jack, Queen, or King) is drawn.

3. Three slips of paper labeled *a*, *b*, and *c* are placed in a hat. Two slips are drawn. List a sample space of outcomes.

4. Six identical tickets are labeled 1, 2, 3, 4, 5, 6 and placed in a box. After the tickets are thoroughly mixed, two tickets are drawn from the box. List a sample space for this probability experiment.

5. Four letters are chosen at random from the word *start*. What is the probability that they can be used to spell the word *tart*? The word *star*? (Choosing four letters is equivalent to omitting one.)

6. In a contest at the corner grocery store, a prize is to be given to the person whose name is on a card drawn from a box with well-mixed cards containing the names of all 110 of today's customers. Ten different members of your class make purchases at the grocery store today. What is the probability that some member of the class will win the prize?

7. *Accidents.* In a 4-year period, some drivers had automobile accidents in 2 different years and no accidents in the other 2 years. List the sample space for the experiment of driving for 4 years. (For example, if N means no accident and A means an accident occurred during the calendar year, then $NNAA$ means no accident in the first 2 years and an accident in each of the third and fourth years.) Among drivers who had accidents in exactly 2 years, find the probability that at least one accident occurred in the first 2 years.

8. *Choices.* Three individuals, A, B, and C, each secretly chose at random one of the others to work with at a later task. Find the probability that all three are chosen. In writing out the eight outcomes in the sample space, you may find the following tree diagram helpful:

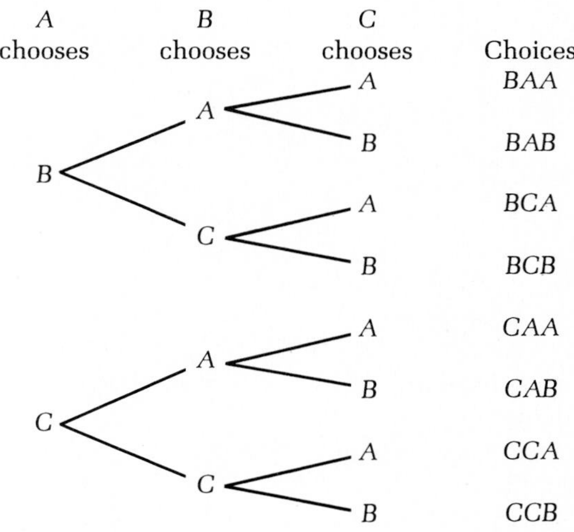

9. Continuation. Find the probability for Problem 8 that exactly two are chosen.

10. Continuation. Find the probability for Problem 8 that exactly one is chosen.

11. Continuation. By summing the probabilities for Problems 8, 9, and 10, check that their total is 1.

Sampling without and with replacement. In some sampling problems, when we draw an item from the population we do not replace it before the next draw. Thus when we draw one ace from a pack of ordinary playing cards and do not replace it, the next draw has 3 chances in 51 or a probability of $3/51$ of also being an ace, because one ace is now missing from the pack. Had we replaced the first ace and reshuffled, the probability of the ace on the second draw would be $4/52$, just as it was before the first draw. When we sample with replacement, the probabilities are identical on each individual draw.

In some sampling problems, such as tossing a coin to get head or tail, the result of the first toss (or drawing) does not remove an item from the population of potential heads and tails, and so we are essentially sampling with replacement.

12. *Sampling without replacement.* A box of 12 transistor radios has 2 radios that are defective. If two radios are randomly drawn from the box, what is the probability that neither radio is defective? You may find the following sample space helpful, where N stands for nondefective and D for defective.

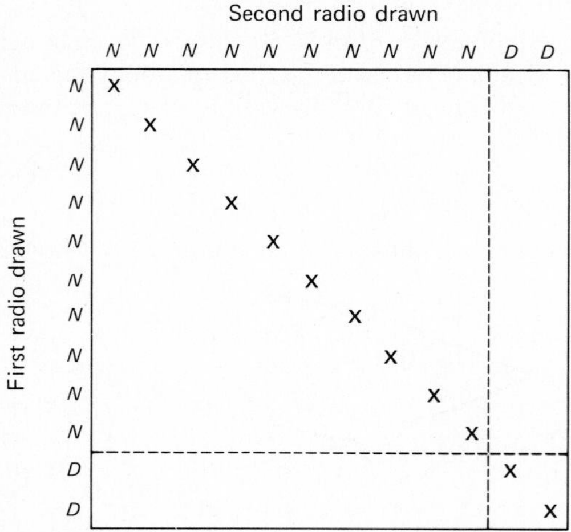

13. Continuation. Find the probability that exactly one of two radios in the sample in Problem 12 is defective.
14. Continuation. Find the probability that both the radios drawn in the sample in Problem 12 are defective.
15. Continuation. Check that the total probability is 1 by adding up the three probabilities found in Problems 12, 13, and 14.
16. *Sampling with replacement.* In Problem 12, suppose that after drawing one radio at random and inspecting it, you replace it and draw at random a second radio. Construct the sample space for this experiment, and compare it with that for Problem 12.

8-3 INDEPENDENT EVENTS

In everyday language, if we say that two events are independent, we mean that the chance of occurrence of one event is not affected by the occurrence of the other. We used this idea to solve the problem of tossing a nickel and a dime in Example 4 in Section 8-2.

If H_D means "head on the dime" and H_N means "head on the nickel," we observe that

$$P(H_D) = \frac{1}{2}, \quad P(H_N) = \frac{1}{2},$$

and

$$P(2 \text{ heads}) = P(H_D \text{ and } H_N) = \frac{1}{4}.$$

Because

$$\frac{1}{2} \times \frac{1}{2} = \frac{1}{4},$$

we note that for this problem

$$P(H_D \text{ and } H_N) = P(H_D)P(H_N).$$

The key to the multiplication of the probabilities is the fact that the outcome from tossing the nickel has nothing to do with the outcome from tossing the dime; that is, the two events are independent.

To take a more general example, suppose A is the event that a newly manufactured car has a faulty brake, with $P(A) = 1/1000$, say, and B is the event that the color of the car is red, with $P(B) = 1/10$, say. The manufacture of the brake and the choice of color have nothing to do with each other. And so, out of a stream of 10,000 cars, 10 will have a faulty brake, on the average, and of these 10 cars, 1 will be red, again, on the average. Multiplying the probabilities gives

$$P(A)P(B) = \left(\frac{1}{1000}\right)\left(\frac{1}{10}\right) = \frac{1}{10{,}000}$$

for the stream of cars.

Definition: The events A and B are independent if and only if

$$P(A \text{ and } B) = P(A)P(B) \qquad (2)$$

EXAMPLE 5 Marksman 1 gets a bull's-eye 20 percent of the time. Marksman 2 gets one 40 percent of the time. If both shoot at the target once, what is the probability that both get bull's-eyes?

SOLUTION. We assume that they do not influence one another's accuracy. Then if A is the event that marksman 1 gets a bull's-eye and B that marksman 2 does, we use the product rule to get

$$P(\text{both get bull's-eyes}) = P(A)P(B) = .2 \times .4 = .08.$$

And so, on the average in 8 of every 100 times that both shoot, both get bull's-eyes.

EXAMPLE 6 One day in four, Nancy forgets her early morning chore of putting plants out in the sun. But one day in five the sky is overcast, and her oversight will not matter. What is the chance that she forgets and that it does not matter?

SOLUTION. Assuming that her oversight is not linked to the weather, the probability that both events happen is $\frac{1}{4} \times \frac{1}{5} = \frac{1}{20}$.

EXAMPLE 7 *Chair cushions.* A department store holds a clearance sale, and you want to purchase a pair of matching cushions for an old rocking chair. There are two boxes of leftover cushions, one box with 8 seat cushions (3 are white, 4 are blue, and 1 is red) and the other box with 7 back cushions (2 are red and 5 are green). You draw a cushion at random from each box. What is the probability that you select a white seat cushion and a red back cushion? What is the probability that the cushions are the same color?

SOLUTION 1. Counting in the sample space. Let us write out the possible pairings conveniently in an 8×7 array using the first letter of the color as an abbreviation. The sample space in Table 8-1 shows 6 outcomes with white seats and red backs among the 56 possible pairs. These outcomes are circled in the table. Because the seat and back cushions are randomly selected, we regard these 56 outcomes as equally likely. Thus

$$P(WR) = \frac{6}{56} = \frac{3}{28}.$$

Similarly, the only matching cushions are both red, and

$$P(RR) = \frac{2}{56} = \frac{1}{28}.$$

SOLUTION 2. Independence approach. The probability that the seat cushion is white is

$$P(W) = \frac{3}{8},$$

TABLE 8-1
Sample space of possible combinations of 7 back cushions (2 red, 5 green) and 8 seat cushions (3 white, 4 blue, 1 red)

Color of seat cushion	**Color of back cushion** R_1	R_2	G_1	G_2	G_3	G_4	G_5
W_1	*WR*†	*WR*	WG	WG	WG	WG	WG
W_2	*WR*	*WR*	*WG*	*WG*	*WG*	*WG*	*WG*
W_3	*WR*	*WR*	*WG*	*WG*	*WG*	*WG*	*WG*
B_1	*BR*	*BR*	*BG*	*BG*	*BG*	*BG*	*BG*
B_2	*BR*	*BR*	*BG*	*BG*	*BG*	*BG*	*BG*
B_3	*BR*	*BR*	*BG*	*BG*	*BG*	*BG*	*BG*
B_4	*BR*	*BR*	*BG*	*BG*	*BG*	*BG*	*BG*
R	*RR*	*RR*	*RG*	*RG*	*RG*	*RG*	*RG*

† Color of seat cushion appears first and color of back cushion second.

and the probability that the back cushion is red is

$$P(R) = \frac{2}{7}.$$

Because the seat and back cushions are chosen independently (randomly),

$$P(WR) = P(W \text{ and } R) = P(W)P(R) = \frac{3}{8} \times \frac{2}{7} = \frac{6}{56} = \frac{3}{28}.$$

Similarly,

$$\begin{aligned} P\,(\text{matching cushions}) &= P\,(\text{red seat and red back}) \\ &= P\,(\text{red seat})\,P\,(\text{red back}) \\ &= \frac{1}{8} \times \frac{2}{7} = \frac{2}{56} = \frac{1}{28}. \end{aligned}$$

Thus the two methods give the same result. Furthermore, even when we do not have equally likely cases, if we have independence we can still multiply to find the probability that two events both occur.

EXAMPLE 8 Suppose that $\frac{1}{10}$ of cars are "lemons" (cars with several defects) and $\frac{1}{20}$ of wildcat oil drillings are successful. If we buy a car and invest in a wildcat drilling venture, what is the probability that both outcomes are favorable?

SOLUTION. Let L be the event buying a "lemon" and $\overline{L}$ the event buying a defect-free car, and let G be the event that the drilling is successful and $\overline{G}$ the event that it is not. The car-buying and oil-drilling ventures are so separate that independence is a reasonable assumption. We want $P(\overline{L}$ and $G)$. Using independence, we have

$$P(\overline{L} \text{ and } G) = P(\overline{L})P(G).$$

Because $P(L) = 0.1$ and the total probability for L and $\overline{L}$ must be 1, we have $P(\overline{L}) = 1 - P(L) = 0.9$. Also, $P(G) = 0.05$. Therefore, we get

$$P(\overline{L})P(G) = (.9)(.05) = .045.$$

MORE THAN TWO INDEPENDENT EVENTS

When we deal with more than two events, if all are mutually independent and no pattern of occurrence of any events changes the probability of the others, the multiplication rule still holds. If events A, B, and C are independent, then

$$P(A \text{ and } B \text{ and } C) = P(A)P(B)P(C).$$

EXAMPLE 9 *Three-girl family.* What is the probability that in a family with three children, all three are girls? We assume that a boy (B) and a girl (G) are equally likely at each birth.

SOLUTION. Let G_1, G_2, and G_3 represent girls on the first, second, and third births, respectively. We assume independence of outcomes at each birth. We also assume that the probability of a girl is the same at each birth, and we find

$$P(G_1 \text{ and } G_2 \text{ and } G_3) = P(G_1)P(G_2)P(G_3) = \frac{1}{2} \times \frac{1}{2} \times \frac{1}{2} = \frac{1}{8}.$$

EXAMPLE 10 The Pennsylvania Daily Lottery. Like many states, Pennsylvania holds a daily lottery. People buy lottery tickets and try to guess which three-digit number (000 through 999) will be picked each evening at 7 P.M. during a televised drawing. The winning lottery number is produced from three separate machines (one for each digit) that contain 10 Ping-Pong balls, labeled 0 through 9. The balls are blown about in a container by a jet of air and mixed. Then one is sucked through an opening at the top of the machine. What is the probability that the number drawn on any given day will be 123? What about 666?

SOLUTION. The three machines with Ping-Pong balls are identical and physically separate. Thus we treat them as being independent. For each machine, the mixing of the balls assures that the 10 digits 0 through 9 are equally likely. Thus we have

$$\frac{1}{10} = P(0) = P(1) = P(2) = P(3) = P(4) = P(5) = P(6) = P(7) = P(8) = P(9).$$

Because the machines are identical, we can use the same probabilities over again for each digit in the three-digit number. Using the product rule, we have

$$\begin{aligned} P(123) &= P(1)P(2)P(3) \\ &= \frac{1}{10} \times \frac{1}{10} \times \frac{1}{10} \\ &= \frac{1}{1000}. \end{aligned}$$

Similarly,

$$P(666) = \frac{1}{1000}.$$

EXAMPLE 11 *Was the Lottery fixed?* On one day in April 1980 the daily lottery number picked was 666. The next day the Pittsburgh newspapers reported that an unusually large number of tickets for 666 had been purchased and that there were rumors that the lottery had been fixed. Part of people's concerns was that all three digits in the number were the same. What is the probability that this will happen? Should we be surprised when we see a repeat-digit number?

TABLE 8-2
Counts of the number of times the 1000 numbers 000 through 999 were selected in the first 1108 drawings in the Pennsylvania Daily Lottery

Number of times number was drawn	*Count of numbers*
0	316
1	380
2	212
3	70
4	16
5	6†

Check on number of drawings:
$1(380) + 2(212) + 3(70) + 4(16) + 5(6) = 1108$

† Example: 6 numbers were drawn 5 times.

SOLUTION. There are 10 numbers with all three digits the same: 000, 111, . . . , 999. Each has probability $1/1000$ of being picked. The total probability of getting one of these 10 numbers is

$$10 \times \frac{1}{1000} = \frac{1}{100}.$$

Thus in 1000 drawings we expect to see roughly $1000 \times (1/100) = 10$ such repeated-digit numbers. The actual observed number of same-digit numbers that occurred was 14, for an observed proportion of 0.0126, close to the expected value of $1/100 = 0.01$ we computed here.

Table 8-2 shows that, out of the first 1108 drawings in the Pennsylvania Lottery, 316 three-digit numbers were never drawn and that no number was drawn more than 5 times.

PROBLEMS FOR SECTION 8-3

1. There are two bowls, *A* and *B*. Bowl *A* contains 3 white, 5 red, and 4 black marbles. Bowl *B* contains 2 white, 3 red, and 6 black marbles. All the marbles are

the same size. One marble is picked at random from each bowl. What is the probability that both marbles will be white?

2. Continuation. What is the probability that both marbles will be of the same color?
3. Continuation. What is the probability that one marble will be white and the other black? (Hint: The white marble could come from either bowl A or bowl B.)
4. Two dice, one red and one green, are rolled. What is the probability that they will both turn up 6?
5. Two cards are drawn from a 52-card deck with replacement (that is, the first card drawn is replaced and the deck is reshuffled before the second card is drawn). Find the probability that the first card is a spade ♠ and the second is a heart ♡. Find the probability that both are queens.
6. Suppose that the probability that a new tire will have blemishes is 0.15 and further that the probability that a new car wheel is bent is 0.02. For a tire on a new wheel, what is the probability that the tire is blemished and the wheel is bent?
7. Continuation. Find the probability that neither the tire is blemished nor the wheel is bent.
8. If the probability is 0.15 that each tire on a new car has blemishes, what is the probability that all 4 tires have blemishes?
9. Five coins (penny, nickel, dime, quarter, half-dollar) are tossed. What is the probability that all five come up tails? What is the probability of five tails if all five coins are pennies?
10. *Pennsylvania Daily Lottery*. If the probability of a number being drawn is 1/1000 and 1108 have been drawn, why has every one of the possible numbers not been chosen?
11. Use Table 8-2 to estimate the probability that the same number will be drawn at least three times in 1108 draws in the Pennsylvania Daily Lottery.
12. The 10 digits (0 through 9) are equally likely to be drawn on any draw. When two digits are drawn, with replacement, what is the probability that they are different digits?
13. A good bowler has a 50 percent chance of getting a "strike" (all 10 pins knocked down by the first ball). He needs 12 consecutive strikes to bowl a perfect game, scoring 300. What is his chance of a perfect game?

8-4 DEPENDENT EVENTS

In the examples in the preceding sections we have sometimes had independence between events. With some extensions, the sample-space approach will work even when events are not independent. If two events are not independent, we say that they are **dependent.**

EXAMPLE 12 *Fuse box.* An electrician plans to put 2 new fuses into the fuse sockets in a new house. He has a box of 4 fuses. He thinks that all 4 fuses are good, but actually one has already blown. They all look alike. When he puts two fuses from the box into the sockets, what is the probability that both will be good ones?

SOLUTION. Let us number the fuses from 1 to 4 and for convenience call the blown fuse number 4. The sample space consists of the possible pairs of fuses he might put into the sockets:

Pair	**Condition**	
1 2	good	good
1 3	good	good
1 4	good	blown
2 3	good	good
2 4	good	blown
3 4	good	blown

Because the fuses are indistinguishable, it seems reasonable that all pairs are equally likely. We get

$$P(\text{both fuses good}) = \frac{\text{number favorable cases}}{\text{total cases}} = \frac{3}{6} = \frac{1}{2}.$$

Note: If we take the view that

$$A = \text{good fuse in first socket,}$$
$$B = \text{good fuse in second socket,}$$

we might try the multiplication rule and get

$$P(A \text{ and } B) = P(A)P(B) = \frac{3}{4} \times \frac{3}{4} = \frac{9}{16} \neq \frac{1}{2}$$

($\neq$ is the symbol for "not equal to"). What is the matter? Lack of independence. If we draw a good fuse first, that leaves 2 good fuses and 1 blown fuse for the second socket. So $P(A) = 3/4$, but $P(B \text{ if } A \text{ occurred}) = 2/3$. Notice that

$$P(A)P(B \text{ if } A \text{ occurred}) = \frac{3}{4} \times \frac{2}{3} = \frac{2}{4} = \frac{1}{2},$$

which agrees with the result of the sample-space calculation.

Later in this chapter we shall find that this rule works in general:

$$P(A \text{ and } B) = P(A)P(B \text{ if } A \text{ occurs}) \qquad (3)$$

In Chapter 6 we discussed the idea of drawing several items more or less simultaneously from a population of finite size. We called this **sampling without replacement.** Whenever we sample objects without replacement from a specific collection of objects, the results are dependent. That is what was happening in the fuse example.

PROBLEMS FOR SECTION 8-4

1. When two dice are rolled, what is the probability that they show different numbers?
2. Two cards are drawn without replacement from an ordinary 52-card deck. Find the probability that both are kings. Find the probability that the first card is a spade and the second card is black. Compare the two probabilities.
3. A box of 12 transistor radios has 2 defective radios. If two radios are randomly drawn from the box without replacement, what is the probability that neither radio is defective? Use formula (3) to make the calculation.
4. Continuation. In Problem 3, use formula (3) to find the probability that both radios are defective.
5. Continuation. In Problem 3, use formula (3) twice to find the probability that exactly one of the radios is defective.
6. *Random numbers.* Digits 0 through 9 are equally likely to appear on each spin of a spinner. After each spin, the digit is recorded. If a two-digit number is so constructed, use formula (3) to compute the probability that both digits are the same.
7. *Lunch seats.* Two people sit down to lunch at a square lunch table with room for just one on a side. Use formula (3) to compute the probability that they will sit together at a corner instead of across from one another if they sit down at random.

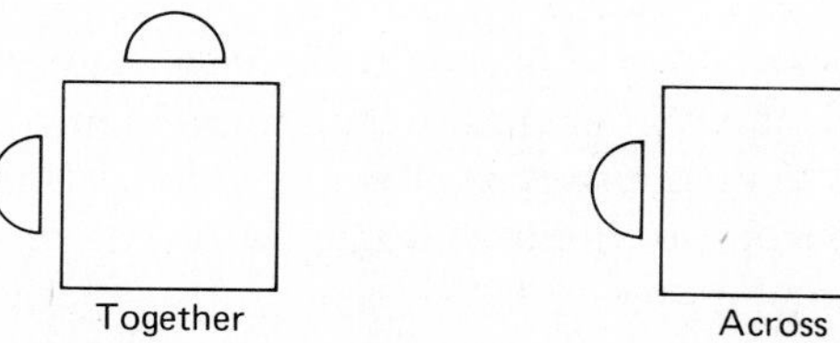

8. *Matching events and dates.* In a multiple-choice question, a student is asked to match three dates with three events. Use formula (3) to compute the probability that sheer guessing will produce three correct answers. (Note: Once two are correctly matched, the third is automatically correct.)
9. *Blood types.* Among 10 people, 6 have blood type *U*, and 4 have blood type *V*. Use formula (3) twice to find the probability that two people drawn randomly without replacement from this group include both blood types.
10. *Advertisements and products.* The probability that a magazine subscriber will read an advertisement is 0.03, and the probability that anyone who reads an advertisement will try the product is 0.006. Find the fraction of subscribers who will buy the product.

8-5 PROBABILITIES FOR COMBINED EVENTS

We think of the outcomes listed in our sample space as elementary events from which other events are constructed. In selecting one card from the deck of cards, we had 52 elementary events or outcomes. The event "ace" is composed by putting together the four elementary events A♠, A♡, A♢, and A♣. The probability of the combined event is the sum of the probabilities of the elementary events, because they do not overlap:

$$P\,(\text{ace}) = P(\text{A}\spadesuit) + P(\text{A}\heartsuit) + P(\text{A}\diamondsuit) + P(\text{A}\clubsuit)$$

$$= \frac{1}{52} + \frac{1}{52} + \frac{1}{52} + \frac{1}{52} = \frac{4}{52},$$

as we found earlier.

> The probability of an event consists of the sum of the probabilities of its elementary events.

When the events are equally likely, we merely apply the formula for the ratio of the counts.

The Combined Event "A and B". If we have two events *A* and *B*, they may or may not have elementary events in common. Figure 8-1(a) shows that the events *A* and *B* have elementary events in common. If both event *A* and event *B* occur, then one of the elementary events they have in common must occur. We call this combined event "*A* and *B*". Let us represent an elementary event by a point in the sample space. Figure 8-1(a) shows 10

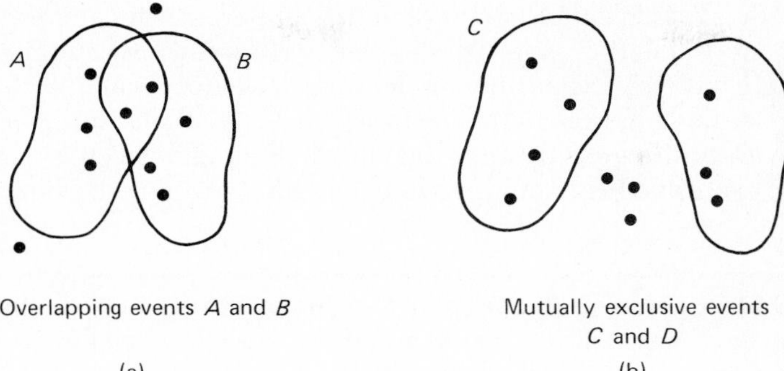

Figure 8-1 Collection of elementary events and associated overlapping events A and B and mutually exclusive events C and D.

points in the sample space; 5 points belong to A and 5 points to B. Two points are common to both A and B. Two belong to neither A nor B. If all 10 points are equally likely, then

$$P(A) = P(B) = \frac{5}{10} = \frac{1}{2}$$

and

$$P(A \text{ and } B) = \frac{2}{10} = \frac{1}{5}.$$

When events have points in common, we say that they overlap.

EXAMPLE 13 *Colored cushions.* In the chair-cushion problem of Example 7, let M be the event that the pair has a white seat cushion and N the event that the back cushion is red. Find $P(M \text{ and } N)$.

SOLUTION. The event "M and N" consists of the 6 WR cases in the upper left corner of Table 8-1, and so $P(M \text{ and } N) = 6/56 = 3/28$.

Suppose two events do not overlap. Then, if one occurs, the other cannot. We say that such events are **mutually exclusive,** and their overlap has probability 0. In Fig. 8-1(b), the events C and D are mutually exclusive, and thus $P(C \text{ and } D) = 0$. Elementary events are mutually exclusive.

Warning. We must not confuse "mutually exclusive" with "independent." When two events have probabilities different from 0, they cannot be both mutually exclusive and independent. If 20 percent of cars are brown and 5 percent are station wagons, and these properties are independent, then when we multiply the positive probabilities together, $\frac{1}{5}$ and $\frac{1}{20}$, we must get a strictly positive number, which happens to be $\frac{1}{100}$. Thus, independence of events with positive probabilities makes mutual exclusiveness impossible. Mutual exclusiveness would require that the probability be zero.

EXAMPLE **14** *Cushions.* In our chair-cushion example, let M be the event that the seat and back cushions have the same color and N be the event that the seat cushion is white. Find $P(M \text{ and } N)$.

SOLUTION. In the only two outcomes having cushions with the same color, the color is red. Thus the event "M and N" contains no elementary events, and $P(M \text{ and } N) = 0$.

The Event "A or B". Sometimes we want to talk about all of the elementary events that are in A or B or both. We label this combination of elementary events "A or B". We speak of the inclusive "or." In Figs. 8-1(a) and 8-1(b),

$$P(A \text{ or } B) = \frac{8}{10},$$

$$P(C \text{ or } D) = \frac{7}{10}.$$

EXAMPLE 15 *Three children.* Find the probability that a family with three children has a girl first or a boy third or both.

SOLUTION. Staying in the framework of the three-child family, we can lay out the cases in our sample space in a two-by-two array:

	Girl Third	**Boy Third**
Girl First	GGG GBG	GGB GBB
Boy First	BGG BBG	BGB BBB

We see from the table that a girl is first everywhere in the first row and a boy is third everywhere in the second column. Thus there are 6 favorable entries out of 8, and so

$$P(\text{girl first or boy third}) = \frac{6}{8} = \frac{3}{4}.$$

If two events A and B are mutually exclusive, then all the elementary events in "A or B" must fall either in A or in B, and thus

$$P(A \text{ or } B) = P(A) + P(B).$$

EXAMPLE 16 Continuing with Example 14 on cushions, we have

$$M = \text{same-color cushions,}$$
$$N = \text{seat cushion is white.}$$

Find $P(M \text{ or } N)$.

SOLUTION. We have that $P(M) = 2/56$ and $P(N) = 21/56$. Because M and N are mutually exclusive,

$$P(M \text{ or } N) = P(M) + P(N) = \frac{23}{56}.$$

When events are not mutually exclusive, we cannot find the probability that one event or the other occurs by simply adding the probabilities. Looking at Fig. 8-1(a), we see that adding together $P(A)$ and $P(B)$ double counts the elementary events in the overlap. Thus we need to subtract off the probability of the overlap. This produces a general rule for adding probabilities:

GENERAL ADDITION RULE

$$P(A \text{ or } B) = P(A) + P(B) - P(A \text{ and } B) \qquad (4)$$

EXAMPLE 17 *Blood test.* An automatic testing machine scores the concentration of a chemical in a blood sample as 1, 2, 3, 4, 5, or 6. For the population of healthy people, these scores are equally likely. For people with a certain disease, the probability of a 6 is very high. To check the machine, the technician scores two healthy people. If either scores a 6, he rechecks the settings on the machine. What is the probability that he does this needlessly? (That is, when the settings are correct, he rechecks.)

SOLUTION. Let the events be:

A: the first person checked scores a 6.

B: the second person checked scores a 6.

A and B: both score a 6.

Clearly, event A and event B overlap:

$$P(A) = \frac{1}{6} = \frac{6}{36}, \quad P(B) = \frac{1}{6} = \frac{6}{36},$$

and

$$P(A \text{ and } B) = \frac{1}{6}\left(\frac{1}{6}\right) = \frac{1}{36}.$$

Using formula (4), we have

$$P(A \text{ or } B) = \frac{6}{36} + \frac{6}{36} - \frac{1}{36} = \frac{11}{36}.$$

When the machine is properly set, he will recheck the settings more than 30 percent of the time.

Complements. $A, \overline{A}$. If A is an event in the sample space, it has a complement $\overline{A}$ consisting of all the other elementary events in the sample space. The sample space S consists of all the elementary events in A or in $\overline{A}$, and so

$$S = (A \text{ or } \overline{A}).$$

Therefore, S is an event with $P(S) = 1$. Because A and $\overline{A}$ have no elementary events in common, they are mutually exclusive, and

$$P(A \text{ and } \overline{A}) = 0.$$

We also note that

$$P(A \text{ or } \overline{A}) = P(A) + P(\overline{A}).$$

But because $P(S) = 1$ and $S = (A \text{ or } \overline{A})$,

$$P(A) + P(\overline{A}) = 1,$$

or

$$P(\overline{A}) = 1 - P(A).$$

We have already used this idea in Example 8, where we needed to get $P(\overline{L})$, the probability of buying a defect-free car.

EXAMPLE 18 *Not ace.* Find the probability that a card drawn from a 52-card deck is not an ace.

SOLUTION. We have $P(\text{ace}) = 1/13$. And so, $P(\text{not ace}) = 1 - P(\text{ace}) = 12/13$.

PROBLEMS FOR SECTION 8-5

For Problems 1 through 7, refer to the sample space for the two blood-test experiments given in Table 8-3. Find the probability that:

1. The sum of the two scores is exactly 7. Is not exactly 7.
2. Each test scores either 3 or 4.
3. Neither score is 3 or 4.
4. At least one score is 3 or 4.
5. At least one score is more than 4.

TABLE 8-3

Sample space for probability experiment involving blood tests for two healthy people

Score of first person	Score of second person 1	2	3	4	5	6
1	(1, 1)	(1, 2)	(1, 3)	(1, 4)	(1, 5)	(1, 6)
2	(2, 1)	(2, 2)	(2, 3)	(2, 4)	(2, 5)	(2, 6)
3	(3, 1)	(3, 2)	(3, 3)	(3, 4)	(3, 5)	(3, 6)
4	(4, 1)	(4, 2)	(4, 3)	(4, 4)	(4, 5)	(4, 6)
5	(5, 1)	(5, 2)	(5, 3)	(5, 4)	(5, 5)	(5, 6)
6	(6, 1)	(6, 2)	(6, 3)	(6, 4)	(6, 5)	(6, 6)

6. Both scores are more than 4.
7. Only one score is more than 4.
8. Suppose that two events A and B are mutually exclusive: $P(A) = 0.2$ and $P(B) = 0.7$. Find
   a) $P(A \text{ and } B)$
   b) $P(A \text{ or } B)$
   c) $P(\bar{A})$
   d) $P(\bar{A} \text{ and } B)$
9. In a factory, 0.4 of the workers are women, 0.7 of the workers have production jobs (the rest have clerical jobs), and 0.3 of the workers are women working in clerical positions. A worker is chosen at random. What is the probability that the worker is either a woman or works in a clerical job or both?
10. A pair of dice is rolled. By relabeling the rows and columns of a table similar to Table 8-3, set up the sample space for this experiment. Find the probability that: (a) the sum of the values on the two upturned faces is 7, (b) the sum is at least 11, and (c) the sum is at least 8.
11. Continuation. Find the probability that either the sum is greater than 8 or the sum is an odd number.

8-6 CONDITIONAL PROBABILITY

Sometimes we have extra information about a probability experiment that changes the probabilities. This is one of the helpful bases of inference in science. This section shows how to use this extra information to adjust the probabilities in the light of additional evidence. It is helpful to begin by considering the case of equally likely outcomes.

EQUALLY LIKELY OUTCOMES

EXAMPLE 19 *Two children.* A family has two children. If they are not both girls, what is the probability that exactly one is a boy?

SOLUTION. The original sample space consists of four elementary events:

First Child	Second Child
B	B
B	G
G	B
G	G

We take these to be equally likely. We are told that they are not both girls. This eliminates the element GG and leaves 3 elements: BB, BG, GB. Assuming that these are still equally likely, we get

$$P(BG \text{ or } GB, \text{ if not } GG) = \frac{2}{3}.$$

The assumption that the elementary events are still equally likely is the key. We make this assumption if we can believe that the assumptions leading to the original sample space have not been changed by the new information. As an example of such a change, if the reason we are sure that the family does not have two girls is that a population has been formed of only all-boy families, the assumption would be mistaken.

The probability dealt with in Example 19 is called a conditional probability, because the added condition (no all-girl family) changed the sample space. *All probabilities are conditional on the sample space used,* although we usually do not emphasize this.

> The notation
>
> $$P(A \mid B)$$
>
> means the probability of the event A given that the event B occurs. Read the vertical bar | as "given."

When the event B in $P(A \mid B)$ consists of the original sample space, S, then

$$P(A \mid S) = P(A).$$

Sometimes the new information may eliminate some elementary events favorable to the event of interest.

EXAMPLE 20 *Two children.* In a family of two children, the second child is a boy. What is the probability that the family has exactly one boy?

SOLUTION. The easy answer is that the first child has a probability of ½ of being a girl, and so that is the conditional probability.

We will learn more from a formal approach. The original sample space has four elements,

$$S: \quad BB, BG, GB, GG.$$

The reduced sample space, S', that goes with the information that the second child is a boy is

$$S': \quad BB, GB.$$

The original event, A, is the event "exactly one boy":

$$A: \quad BG, GB.$$

By focusing on the reduced sample space, we change A to event

$$A': \quad GB.$$

For equally likely events, we compute

$$P(A' \mid S') = \frac{1}{2}.$$

We note that

$$A' = (A \text{ and } S'),$$

the elementary events common to A and S'.

Essentially, all we are doing is redistributing the total probability of 1 among the points remaining in the sample space. We do this whether or not the elementary events are equally likely. Thus, we define conditional probability as follows:

> If A and B are events and $P(B) \neq 0$, then the conditional probability of A, given that B occurs, is
>
> $$P(A \mid B) = \frac{P(A \text{ and } B)}{P(B)} \tag{5}$$

Often, formula (5) and more elaborate versions of it are called **Bayes' formula.**

EXAMPLE 21 *Marital status.* In a population of adults the proportions of people in various marital states are

Single:	30 percent
Married:	50 percent
Divorced:	20 percent

A person is selected at random. If that person is not divorced, find the probability he or she is single.

SOLUTION. Let A = single, B = not divorced. Then (A and B) = A, and so using Bayes' formula from Eq. (5), we find

$$P(A \mid B) = \frac{P(A \text{ and } B)}{P(B)} = \frac{P(A)}{P(B)} = \frac{.3}{.8} = \frac{3}{8}.$$

Note that in the reduced sample space (single, married) we have increased the probabilities of being single and of being married proportionately so that their total is 1. Because their probabilities only added to 0.8, we must multiply each by 1.0/0.8 or $^{10}/_{8}$ and get $(^{10}/_{8}).3 = ^{3}/_{8}$ and $(^{10}/_{8}).5 = ^{5}/_{8}$ and, of course, the new total probability is 1. It must be, because the original probability was 0.8, and we multiply each probability by its reciprocal $^{10}/_{8}$ to get a total of 1.

EXAMPLE 22 *Fixing the Pennsylvania Daily Lottery.* In Examples 10 and 11 we discussed the occurrence of the number 666 and whether the Lottery was fixed. Some evidence was produced to suggest that someone had used a hypodermic needle to inject white latex paint into all of the Ping-Pong balls except those numbered 4 and 6 in each machine. Then, when the balls were blown into the air, only those with 4's and 6's on them were blown high enough to be sucked through the tops of the machines. If this did happen, find the probability of picking 666; that is, find the conditional probability of 666 given that the Lottery was fixed to produce only 6's and 4's.

SOLUTION. Let A = 666 and B = only 4's and 6's can occur. The event B consists of 8 of the original 1000 elementary outcomes:

444, 446, 464, 644, 466, 646, 664, 666

Thus $P(B) = ^{8}/_{1000}$. In this example, the event A is an elementary outcome in B, and so $P(A \text{ and } B) = P(A) = ^{1}/_{1000}$. And

$$P(A \mid B) = \frac{P(A \text{ and } B)}{P(B)} = \frac{P(A)}{P(B)} = \frac{1/1000}{8/1000} = \frac{1}{8}.$$

Thus, if the Lottery had been fixed, the probability of picking 666 would have been $^{1}/_{8}$ instead of $^{1}/_{1000}$.

THE GENERAL CASE

We turn now to more general events A and B, but we continue with the same definition of conditional probability.

Figure 8-2 The winning number is drawn, under the watchful eye of a lottery official.

EXAMPLE 23 *Colorblindness.* In a large population composed of 40 percent men and 60 percent women, 6 percent of the men are colorblind, as are 1 percent of the women. Find the probability that a colorblind person is a man.

SOLUTION. Let A be the event of being a man, B the event of being colorblind. We want the value of $P(A|B)$. We lay out a 2×2 (read: two-by-two) table showing all the possible outcomes and their probabilities when a sample of one person is drawn, as in Table 8-4.

Table 8-4 gives the probabilities for each of the four elementary events "A and B", "A and $\overline{B}$", "$\overline{A}$ and B", "$\overline{A}$ and $\overline{B}$", as well as the marginal probabilities $P(A)$, $P(\overline{A})$, $P(B)$, $P(\overline{B})$. They are called **marginal probabilities** because they appear on the margin of the table. The really new information provided by the table is that in this population the probability of being colorblind is $P(B) = 0.03$. The probability of being male and colorblind is

$$.4 \times .06 = .024 = P(A)P(B) = P(A \text{ and } B).$$

TABLE 8-4
Distribution of colorblindness in an adult population

	B *Colorblind*	$\overline{B}$ *Not colorblind*	Total
A: *Men*	$P(A \text{ and } B) = .024$	$P(A \text{ and } \overline{B}) = .376$	$P(A) = .400$
$\overline{A}$: *Women*	$P(\overline{A} \text{ and } B) = .006$	$P(\overline{A} \text{ and } \overline{B}) = .594$	$P(\overline{A}) = .600$
Total	$P(B) = .030$	$P(\overline{B}) = .970$	$P(S) = 1$

The conditional probability of being male, given colorblindness, is

$$P(A \mid B) = \frac{P(A \text{ and } B)}{P(B)} = \frac{.024}{.03} = .8.$$

Thus, in this population, four of five colorblind people are male.

What is the probability that a noncolorblind person is male?

$$P(A \mid \overline{B}) = \frac{P(A \text{ and } \overline{B})}{P(\overline{B})} = \frac{.376}{.970} = .388.$$

Thus, noncolorblind persons are a little less likely to be male (0.388) than members of the population as a whole, 0.400. Because there are relatively more males in the colorblind group, 0.800 as compared to 0.400, we expect this reversal in the noncolorblind group.

We can also use the idea of displaying joint probabilities in tables when there are more than two possibilities.

EXAMPLE 24 *Party affiliation.* In Table 8-5 we have the proportions of married couples in a population with various party affiliations for both the husband and the wife. Find the probability for a randomly selected couple from this population that the wife is a Democrat, given that the husband is a Democrat.

SOLUTION. Let A = wife is a Democrat; B = husband is a Democrat. Then

$$P(A) = .70,$$
$$P(B) = .51,$$

TABLE 8-5
Party affiliations for married couples

	Wife			
Husband	*Democrat*	*Republican*	*Other*	*Total*
Democrat	.40	.10	.01	.51
Republican	.25	.15	.00	.40
Other	.05	.01	.03	.09
Total	.70	.26	.04	1.00

and

$$P(A \mid B) = \frac{P(A \text{ and } B)}{P(B)} = \frac{.40}{.51} = .784.$$

Knowing that the party affiliation of the husband is Democrat increases the probability that the wife is Democrat from 0.70 to 0.784.

PROBLEMS FOR SECTION 8-6

1. In Example 22, what is the probability that exactly two of the three digits in the number picked are 6's, if the Lottery is fixed to produce only 4's and 6's?
2. One card is drawn from a well-shuffled 52-card deck. Given that the card is a spade, what is the probability that it is a face card (a jack, queen, or king)?
3. Two cards are drawn from a well-shuffled 52-card deck without replacement. What is the probability of the second card being an ace, given that the first card is an ace?
4. A committee consists of 10 members: 4 men and 6 women. A person is selected at random to run the next committee meeting, and a different person is selected at random to keep the minutes. Find the probability that the person who keeps the minutes is a woman, given that a man is selected to run the meeting. What if a woman is selected to run the meeting?
5. Two dice are rolled. Find the probability that the sum is 8, given that the numbers on the two dice are different.
6. Continuation. Find the probability that the sum is less than 5, given that at least one die shows 1 or 2.
7. Using the probabilities for party affiliation in Table 8-5, for a randomly selected couple find the probability that the husband is a Republican, given that the wife is a Republican.

8. Continuation. What is the probability that the wife is a Republican, given that the husband is?

9. *Pennsylvania Daily Lottery*. Suppose an attempt is made to fix the Lottery by weighting all of the balls except those numbered 7, 8, and 9 in each machine. Given such a fix, what is the probability of a triple being selected, that is, 777, 888, or 999?

10. Continuation. To ensure that each drawing is random, Lottery officials require that each official drawing be followed by an unofficial drawing with the same set of balls. Given that all but the 7, 8, and 9 balls are weighted so as to prevent 0 through 6 from being drawn, find the probability that the same three-digit number will be selected on both the official and unofficial drawings.

11. Continuation. Devise a more accurate test to help ensure that the official drawing is random.

12. *Class project*. Investigate the procedures used in an official lottery in your state or in a nearby state. Is there a check on the mechanism used for the drawing to ensure that the numbers are drawn at random?

8-7 ASSIGNING PROBABILITIES USING THE CHAIN RULE

Now that we know how to assign probabilities when conditions have been given, we can more easily handle some problems we treated earlier. We have the formula

$$P(A|B) = \frac{P(A \text{ and } B)}{P(B)}.$$

Because $P(B) \neq 0$, we can multiply through by $P(B)$ and write

$$P(A \text{ and } B) = P(B)P(A|B) \qquad (6)$$

In this formula, we conditioned on B. We could have conditioned on A and written

$$P(A \text{ and } B) = P(A)P(B|A).$$

These two formulas are very helpful in computing compound probabilities. Let us return to the two-fuse example.

EXAMPLE 25 *Two fuses.* The electrician plugs in 2 fuses chosen randomly from a box of 4 fuses, with 1 defective. Find the probability that both fuses are good.

SOLUTION. Let A be the event "the first fuse is good" and B the event "the second fuse is good." Then the event "A and B" is "both fuses are good." We want $P(A \text{ and } B)$. We can use the formula

$$P(A \text{ and } B) = P(A)P(B \mid A).$$

The probability that the first fuse is good is

$$P(A) = \frac{3}{4}.$$

Once one good fuse has been drawn from the box, we are left with a new sample space with 2 good fuses and 1 bad fuse. The new probability that a fuse is good is

$$P(B \mid A) = \frac{2}{3}.$$

To compute the probability that both fuses are good, we merely multiply:

$$\begin{aligned} P(A \text{ and } B) &= P(A)P(B \mid A) \\ &= \frac{3}{4} \times \frac{2}{3} \\ &= \frac{2}{4} = \frac{1}{2}. \end{aligned}$$

This is what we found earlier in Example 12 in Section 8-4 and verified by direct count of equally likely elements.

The conditional probability formula (6) can be extended to more than two events. For example, with three events, A, B, C, we have the following:

$$P(A \text{ and } B \text{ and } C) = P(A)P(B \mid A)P(C \mid A \text{ and } B) \tag{7}$$

We call this the **chain rule.**

EXAMPLE 26 *Omelet.* A box contains 12 eggs, 4 of which are bad. Find the chance that an omelet made from three eggs drawn from this box has no bad eggs.

SOLUTION. We need all three eggs to be good. Let us draw them out one at a time. Let A be the event that the first egg is good, B the event that the second egg is good, and C the event that the third egg is good. We want

$$P(A \text{ and } B \text{ and } C) = P(A)P(B \mid A)P(C \mid A \text{ and } B).$$

We have $P(A) = 8/12$, $P(B \mid A) = 7/11$, $P(C \mid A \text{ and } B) = 6/10$. Finally,

$$P(A \text{ and } B \text{ and } C) = \frac{8}{12} \times \frac{7}{11} \times \frac{6}{10} = .255.$$

Thus the chance of getting a good omelet is a little over 1 in 4.

Another way to compute this probability is to divide the number of ways to succeed by the total number of possible outcomes. We have $8 \times 7 \times 6$ ways of succeeding and $12 \times 11 \times 10$ ways of drawing three eggs. The ratio is the required probability.

Earlier we discussed the notion of independence from a practical point of view. We need now to relate it to our conditional probability work. Recall that when we regarded A and B as independent, we got the probability of "A and B" from the product $P(A)P(B)$. Our new conditional probability formula gives

$$P(A \text{ and } B) = P(A)P(B \mid A).$$

To make what we did earlier agree with this formula, we need the following:

Independence of A and B: $P(B \mid A) = P(B)$	(8)

This requirement is very natural. If event B is to be independent of event A, then we want

$$P(B \mid A) = P(B \mid \overline{A}) = P(B).$$

This seems to be a natural assumption when we are tossing coins or dice or drawing cards with replacement from a deck.

When items are produced in mass production, we usually suppose that the properties such as defectiveness and nondefectiveness of successive items are independent.

EXAMPLE 27 *Defective radios.* As they come off a production line, 10 percent of radios are defective in some respect. Find the probability that among the next three radios exactly one is defective.

SOLUTION. Our elementary events might be the outcome defective, D, or nondefective, N, for the successive three radios. Among the 8 possible elementary events, 3 have a defective: DNN, NDN, and NND. Let us compute the probability for DNN. Let

A be the event that the first radio is D,

B be the event that the second radio is N, and

C be the event that the third radio is N.

Then we want the probability of "A and B and C". The chain rule says that

$$\begin{aligned} P(A \text{ and } B \text{ and } C) &= P(A)P(B \mid A)P(C \mid A \text{ and } B) \\ &= P(A)P(B)P(C) \\ &= .1(.9)(.9) = .081. \end{aligned}$$

The other two orders give the same probability. Thus the total probability is

$$P(\text{exactly 1 defective}) = 3(.081) = .243,$$

or about one-quarter.

PROBLEMS FOR SECTION 8-7

1. A bowl contains 5 black and 10 red marbles. Two marbles are drawn at random, one after another without replacement. Compute
   a) $P(\text{2nd marble red} \mid \text{1st marble black})$
   b) $P(\text{2nd marble red})$
   c) $P(\text{1st marble red} \mid \text{2nd marble red})$
2. Continuation. Use your answer from Problem 1(a) to compute $P(\text{1st marble black and 2nd marble red})$.
3. Continuation. Use your answers from Problems 1(b) and 1(c) to compute $P(\text{both marbles red})$.

4. Three cards are drawn with replacement from a well-shuffled 52-card deck. Find the probability that (a) all three are kings, (b) all three are spades, and (c) the first two are spades and the third is a heart.
5. In dealing cards from a 52-card deck, what is the probability that the first heart is the fourth card dealt? The fifth card?
6. *Automobile accidents.* Each year 0.025 members of a class of automobile drivers have accidents, independently from year to year. Find the probability that a member of this class has no accidents for two years and an accident in the third year.
7. *Defective radios.* A box of 12 transistor radios has 2 defective radios. A sample of 3 different radios is drawn randomly from the box. Find the probability that none is defective, using formula (7).
8. Continuation. In Problem 7, find the probability that both defectives are in the sample.
9. *Matching events and dates.* In a multiple-choice question, a student is asked to match 5 dates with 5 events. Find the probability that he gets all 5 correct if he is only guessing.
10. *Blood types.* Suppose that among 10 hospital employees, the blood types U, V, and W occur in 5, 3, and 2 employees, respectively. A random group of three of these people are on duty at the hospital one night. What is the probability that all three blood types are represented? (Note that there are 6 possible orders: UVW, $UWV, \ldots, WVU$.)

8-8 RANDOM VARIABLES

In previous chapters we have discussed many frequency distributions, and in Section 6-6 we discussed the probability distribution of $\bar{p}$. Let us review an example and establish some language widely used in discussions of probability.

When probabilistic experiments produce numbers, these numbers are called **random variables.** They have associated probability distributions whose properties we want to know. For example, we want the average value the number will take, we want to know how much it varies from sample to sample, and we want to compute probabilities when several samples are drawn. We turn now to a more careful discussion of this language.

EXAMPLE 28 $N = 10$ and $p = 0.4$. From a population of size $N = 10$, with 4 items having a special quality ($p = 0.4$), 5 items are randomly drawn without replacement. We call the observed proportion in the sample with the special quality $\overline{p}$. Because 0, 1, 2, 3, or 4 (but not 5) items with the special quality can be drawn, the possible values of $\overline{p}$ are 0.0, 0.2, 0.4, 0.6, and 0.8. The value 1.0 cannot occur. The probabilities of occurrence of these quantities are given by the following probability distribution.

Values of $\overline{p}$:	.0	.2	.4	.6	.8	1.0
Probability of $\overline{p}$:	.024	.238	.476	.238	.024	.000

For example, the probability that $\overline{p} = 0$ is

$$\frac{6 \times 5 \times 4 \times 3 \times 2}{10 \times 9 \times 8 \times 7 \times 6} = .024.$$

Before the drawing, we could not know the value $\overline{p}$ would take, though we knew it would be a number between 0 and 1.0, and even what values were possible. Before the probability experiment occurs, $\overline{p}$ is called a random variable. After the experiment has been performed, $\overline{p}$ takes one of its possible values. Instead of thinking about drawing the sample of size 5 from the 10 items, if we are primarily concerned with the proportion having the special quality, we can focus on the probability distribution of $\overline{p}$. Then we are drawing one sample of size 1 from the distribution of $\overline{p}$'s, and we need not think about the 5 items. We attend to the $\overline{p}$, which does concern us.

More generally, when we have a drawing from a probability distribution that produces a numerical value, we call the quantity to be drawn a random variable. Often we use capital letters, such as X, to designate random variables. These capital letters remind us that they are placeholders for the value that will occur when the experiment is performed, and that value will be designated by x. Thus, with this convention, x's are the values X's will take. Each x has an associated probability.

Random variables are like other variables you are acquainted with, and they have the additional property that we may know something about the probabilities that they take on particular values, or fall into a given interval.

We want to compute certain properties of the distribution of random variables, such as their means, variances, and standard deviations, because we need this information to approximate probabilities in later chapters.

The first property we discuss is the **mean** or **expected value** of the random variable. Often the Greek letter μ (pronounced "mu") is used to designate the mean. Sometimes the name of the random variable is given in a subscript, as μ_X, to emphasize what variable is being discussed. We computed these for a variety of distributions of $\overline{p}$ in Chapter 6.

THE MEAN OF X OR THE EXPECTATION OF X: *E*(X)

The letter *E* stands for "expectation." We compute the mean of X as follows:

Step 1. Multiply each possible value of *X* by its probability.

Step 2. Add these products to get the mean.

As a formula, the two steps combined are:

$$\mu_X = E(X) = \text{sum of } [(\text{value}) \times (\text{probability of value})] \tag{9}$$

EXAMPLE 29 Mean for $N = 10$ and $p = 0.4$. Using the data given in the preceding example, compute the mean value of $\overline{p}$. (That is, $X = \overline{p}$.)

SOLUTION. We compute as follows:

Value × Probability
.0 × .024 = .0000
.2 × .238 = .0476
.4 × .476 = .1904
.6 × .238 = .1428
.8 × .024 = .0192
1.0 × .000 = .0000
E (X) = .4000

Thus we see in this example that $\mu = E(X) = p$. We have mentioned earlier that this means that $\overline{p}$ is an unbiased estimate of p.

THE VARIANCE, σ^2, OF A RANDOM VARIABLE

The mean, μ, of the probability distribution of a random variable sums up in one number an idea of the sizes of the numbers in the distribution. It does not give an idea about the variability of the numbers.

As measures of variability, we discussed in Section 6-2 the variance and the standard deviation of $\overline{p}$. The same idea applies to the measured variable X.

If the variable X has mean μ, we define the variance, σ^2, as the mean squared deviation of the values from the mean μ. To compute it from the definition, we use the following steps:

Step 1. For each x, compute the deviation from the mean, $x - \mu$.

Step 2. Square the deviation for each $x - \mu$.

Step 3. Multiply each squared deviation by its probability.

Step 4. Sum these products to get the variance.

Step 5. Take the square root of the variance to get the standard deviation, σ, of X.

As a formula, the first four steps combined are

$$\sigma^2 = \text{sum of } [(\text{value minus mean})^2 \times (\text{probability of value})] \tag{10}$$

EXAMPLE 30 $N = 10$ and $p = 0.4$. Using the data in the previous two examples, find the variance and standard deviation of $\overline{p}$.

SOLUTION. From Example 29, we have $\mu = 0.4$. Let us follow the steps one by one:

Step 1: Deviations	Step 2: Squares	Step 3: Probability × Square
$.0 - .4 = -.4$	.16	$.024 \times .16 = .00384$
$.2 - .4 = -.2$	.04	$.238 \times .04 = .00952$
$.4 - .4 = .0$	.00	$.476 \times 0 = .00000$
$.6 - .4 = .2$	.04	$.238 \times .04 = .00952$
$.8 - .4 = .4$	.16	$.024 \times .16 = .00384$
$1.0 - .4 = .6$	.36	$.000 \times .36 = .00000$
		Step 4: sum $= .02672$
		$=$ variance $= \sigma^2$
		Step 5: square root $= \sqrt{.02672} = .163$
		$= \sigma$

Thus our measure of variability for the distribution of $\bar{p}$ is $\sigma = 0.163$, which agrees to three decimal places with the formula from Chapter 6:

$$\sigma = \sqrt{\frac{p\,(1 - p)}{n}\left(\frac{N - n}{N - 1}\right)} = \sqrt{\frac{.4 \times .6}{5} \times \frac{5}{9}} = .163.$$

We expect about 95 percent of the distribution to be within 2 standard deviations of μ. In this problem,

$$\mu \pm 2\sigma = .400 \pm .326 = .074, .726.$$

This interval includes $\bar{p} = 0.2$, 0.4, and 0.6. Their total probability is 0.238 + 0.476 + 0.238 = 0.952, about as close to 95 percent as we can get with such a clumpy distribution.

PROBLEMS FOR SECTION 8-8

1. What is a random variable?

2. An ordinary die is tossed, and the number, X, on its top face is recorded.

a) Write out the sample space for the random variable X.
b) What is $P(X = 1)$?
c) What is $P(X = x)$ for $x = 1, 2, 3, 4, 5, 6$?
d) Write out the probability distribution for the random variable.
e) Compute the mean value of X, $E(X)$.

3. Two kinds of elements, 2 of one kind and 3 of another, are arranged in a row at random. Consecutive elements of the same kind form a run; for example, *ABABB* has four runs, and *AAABB* has two runs. The probability distribution of the number, X, of runs of elements is as follows:

Value of X:	2	3	4	5
$P(X = x)$:	.2	.3	.4	.1

Find the mean and standard deviation of X, the number of runs.

4. *Turning points.* If among three successive numerical measurements the middle one is the least or greatest of the three, it is called a turning point. If we have four different numbers and all orders are equally likely, the distribution of the number of turning points is as follows:

Number of turning points, X:	0	1	2
$P(X = x)$:	$2/24$	$12/24$	$10/24$

Find the mean number of turning points and the standard deviation of the distribution.

5. Continuation. Write out the 6 possible orders of the numbers 1, 2, and 3, and obtain the distribution of the number of turning points if the orders are equally likely. Find the mean number and the standard deviation.

6. The number of accidents that occur at a particular intersection between 4:30 and 6:30 P.M. on Fridays is 0, 1, 2, or 3, with corresponding probabilities 0.94, 0.03, 0.02, and 0.01. Find the expected number of accidents; that is, if X is the number of accidents, find $E(X)$ (a) during one such Friday period, (b) during 100 such periods. Find the standard deviation for case (a).

7. A farmer estimates that during the coming year his hens will produce 10,000 dozen eggs. He further estimates that, after taking into account his various costs and the seasonal price fluctuations, he may gain as much as 6 cents per dozen, or lose as much as 2 cents per dozen, and that the probabilities associated with these possibilities are as follows:

Gain (in cents per dozen):	6	4	2	0	−2
Probability:	.20	.50	.20	.06	.04

What should he estimate to be his expected gain (a) in cents per dozen, (b) on the 10,000 dozen?

8. *Dates and events.* In a history examination, a student must pair 3 dates with 3 events. If one knows nothing about the period, the six possible arrangements are equally likely. Find the distribution of X, the number of correct answers when he chooses an arrangement at random, and compute its mean and variance.

9. The numbers 1, 2, and 3 are equally likely on each trial. You draw a number, record it, and replace it. Continue until you draw one of the three numbers twice. Verify that the distribution of the number of draws until a repeat is as follows:

Number of draws, X:	2	3	4
Probability, $P(X = x)$:	$3/9$	$4/9$	$2/9$

Find the mean μ_X and the standard deviation σ_X.

10. If the numbers from 1 to n are equally likely to be drawn, show that the mean number is $(n + 1)/2$.

11. Four identical light bulbs are temporarily removed from their sockets and placed in a box. The bulbs are drawn at random from the box and put back into the sockets. What is the expected number of bulbs that will be replaced in their original sockets?

12. The possible values of a random variable X are the integers from n through $n + m$. If these possibilities are equally likely, find $E(X)$.

8-9 SUMMARY OF CHAPTER 8

The equation numbers given here are the same numbers as those used for these equations in the text sections.

1. We usually associate the frequency approach to probability with events that can be repeated again and again, and under essentially the same conditions. When dealing with uncertain outcomes, we speak of probability experiments. We call a list of possible outcomes a sample space of the probability experiment.
2. For any probability experiment having outcomes that are equally likely, we can assign to an event A the probability

$$P(A) = \frac{\text{number of outcomes favorable to event } A}{\text{number of possible outcomes of the experiment}}. \qquad (1)$$

3. We can summarize several of the results about probability in the following set of rules:

RULE 1 Positiveness. The probability assigned to each event is positive or zero:

$$P(A) \geq 0$$

RULE 2 Certainty. The probability of the entire sample space S is 1:

$$P(S) = 1$$

RULE 3 Additivity. If A and B are mutually exclusive events, then the probability of either A or B is the sum of their probabilities. That is,

$$P(A \text{ or } B) = P(A) + P(B)$$

RULE 4 General additivity rule. For any two events A and B,

$$P(A \text{ or } B) = P(A) + P(B) - P(A \text{ and } B) \qquad (4)$$

RULE 5 Complements. If A is an event, it has a complement $\overline{A}$ consisting of all the other elementary events in the sample space, and

$$P(A) + P(\overline{A}) = 1$$

RULE 6 Independence. The events A and B are independent if and only if

$$P(A \text{ and } B) = P(A)P(B) \qquad (2)$$

RULE 7 Conditional probability. If A and B are events and $P(B) \neq 0$, then the conditional probability of A, given that B occurs, is

$$P(A|B) = \frac{P(A \text{ and } B)}{P(B)} \qquad (5)$$

If $P(A) \neq 0$, then the conditional probability of B, given A, is

$$P(B|A) = \frac{P(A \text{ and } B)}{P(A)}$$

RULE 8 Chain rule. For three events A, B, and C, we use the conditional probability formula to get

$$P(A \text{ and } B \text{ and } C) = P(A)P(B \mid A)P(C \mid A \text{ and } B) \quad (7)$$

Formula (7) extends to more than three events, and also the order can be changed. For example,

$$P(A \text{ and } B \text{ and } C) = P(B)P(C|B)P(A|B \text{ and } C)$$

4. When we sample from a probability distribution that produces a numerical value, we call the quantity to be drawn a random variable.
5. If X is a random variable, the mean μ_X of X or the expectation of X, $E(X)$, is

$$\mu_X = E(X) = \text{sum of [(value)} \times \text{(probability of value)]}. \quad (9)$$

6. The variance, σ^2, of a random variable X is the mean squared deviation of the values from the mean μ_X:

$$\sigma^2 = \text{sum of [(value minus } \mu_X)^2 \times \text{(probability of value)]}. \quad (10)$$

7. The standard deviation of X is σ.

SUMMARY PROBLEMS FOR CHAPTER 8

1. When an ordinary six-sided die is tossed, why do we expect the sides to be about equally likely to appear as the top face?
2. A long-range weather forecaster assumes that the outcomes for Cleveland, Ohio, "rain on May 1" and "rain on May 15" are independent events. (a) Explain why these events may be independent. (b) State practical objections you have to this assumption. (c) If it is true, and the probabilities are 0.1 and 0.05, respectively, find the probability that it rains on both dates.
3. The letter *a* is written on one card, the letter *r* on another, and the letter *t* on a third. The cards are drawn randomly one at a time without replacement. (a) What is the sample space of the experiment? (b) What is the probability that the resulting sequence of three letters spells an English word?

4. Piece parts consisting of three types of pieces, A, B, and C, are randomly assembled to form a subassembly. If any of the three parts are defective, the subassembly must be rejected. The probabilities of defectives for the three parts are $P_A = 0.02$, $P_B = 0.04$, and $P_C = 0.03$. What fraction of subassemblies must be rejected?

5. Some potentially lucrative, but very uncertain, investments can be made independently. Each has the probability of 0.1 of being a success. As an investment program, a firm invests in 10 of these. Find the probability that the firm gets at least one success.

6. Continuation. In Problem 5, find the probability that the firm gets exactly one success.

7. If one bank check in 10,000 is forged, if 5 percent of all checks are postdated, and if 80 percent of forged checks are postdated, find the probability that a postdated check is forged.

8. In an election for chairman from among five candidates, two men and three women, all choices are equally likely. Once the election is over, a vice-chairman is chosen from the remaining candidates by putting the four names into a hat and drawing one at random. Find the probability that both offices will be filled by women (a) by listing the sample spaces of outcomes, (b) by using the conditional probability argument.

9. In a box of 12 transistor radios, 3 are defective. At each sale the dealer picks a radio at random from the box. Find the probability that the first defective radio sold is the fourth radio sold.

In Problems 10 through 13, you are given uniform distributions on successive integers. That is, each of a set of integers has the same probability of occurring. If L is the number of successive equally likely integers, each with probability $1/L$, the variance of the distribution is $\sigma^2 = (L^2 - 1)/12$. Verify that this formula works in each problem, and find what fraction of the observations fall within 1 standard deviation of the mean.

10.

$P(x)$:	$\frac{1}{3}$	$\frac{1}{3}$	$\frac{1}{3}$
x:	1	2	3

11.

$P(x)$:	$\frac{1}{3}$	$\frac{1}{3}$	$\frac{1}{3}$
x:	2	3	4

Explain why you need not recalculate in this problem for σ if you have already done it in Problem 10.

12.

$P(x)$:	$\frac{1}{5}$	$\frac{1}{5}$	$\frac{1}{5}$	$\frac{1}{5}$	$\frac{1}{5}$
x:	-2	-1	0	1	2

13. Apply the proposed formula when all the probability is at one point, say $x = 11$.

14. Find the mean, variance, and standard deviation of the following distribution:

$P(x)$:	.4	.3	.2	.1
x:	0	1	2	3

Find how much probability is (a) within 1 standard deviation of the mean, (b) within 2 standard deviations.

For Problems 15 through 18, use the following information. In a factory the distribution of the number of accidents, X, an employee has in a year is closely approximated by the following:

$P(x)$:	.75	.22	.03
x:	0	1	2

The outcomes for successive years are independent.

15. Find the mean, variance, and standard deviation of the number of accidents per year.

16. Find the probability that an employee has 2 accidents in a year if it is known that the employee has had at least one accident.

17. Write out the sample space and associated probabilities for the outcomes for a 2-year period.

18. Continuation. Use the table you created in Problem 17 to compute the probability that an employee has a total of exactly 2 accidents in the 2-year period.

REFERENCE

F. Mosteller, R. E. K. Rourke, and G. B. Thomas, Jr. (1970). *Probability with Statistical Applications*, second edition. Addison-Wesley, Reading, Mass.

9

Comparisons for Proportions and Counts

Learning Objectives

1. Standard forms for critical ratio and for confidence limits
2. Handling tests of binomial problems that have ties
3. Comparing binomial proportions
4. Chi-square for contingency tables 2 × 2 and larger
5. Approximate normality of chi-square distribution with many degrees of freedom

9-1 THE NEED FOR COMPARISONS

Although we often hear that data speak for themselves, their voices can be soft and sly. Usually, analysis can help us tell what we can reliably believe, provided that we have not run too far afoul of the difficulties described in Chapter 5. We now deal with comparisons and with the sizes of differences that can be believed in investigations whose outcomes are based on counts.

Whether we deal with data drawn from experiments, observational studies, or sample surveys, we will wish to make comparisons either (a) of observed outcomes with standard values (for example, has the seasonally adjusted unemployment rate risen above last year's level?) or (b) between groups (for example, is the divorce rate for married men 20 to 29 years of age higher than that for men 30 to 39?).

We speak of comparison with a standard when some natural or desirable level invites comparison. Sometimes the standard is a census figure, such as the percentage of juveniles unemployed or last year's production of pig iron. Or it may be a goal, such as a reduction of 5 percent in the cost of producing electricity. Or it may be the improvement required for a new process to break even financially. Such comparisons are extremely common in practice. They suffer from the difficulty that we often have no idea how big a difference to expect.

EXAMPLE *Homicide rate.* If we hear a report that the homicide rate is up 200 percent, we know that (a) the direction of the change is unsatisfactory, but we do not know (b) whether the change is large. A rise from 1 to 3 homicides per year could well come from natural variation, but a rise from 100 to 300 requires explanation. For numbers in between, we need more delicate appraisals.

EXAMPLE *Absolute versus relative size.* Most of us think of a million dollars as a large sum of money, but people in the Office of Management and Budget call 100 million dollars "a tenth," because they deal in billions. So, for them, a million dollars is a small part of the roundoff error.

To give evidence of the reality of the change in a process, we compare the change with a measure of its variability and refer the result to a suitable

table to help us with the interpretation. Without going into details, we use two main methods:

1. Tests of significance
2. Confidence intervals

TEST OF SIGNIFICANCE

In a test of significance, the statistic we use measures the change in terms of its underlying variability. And we usually then convert to a probability level or P value by referring to an appropriate table to find the probability of getting a difference as extreme as or more extreme than the one observed if the process is identical in mean with that of the comparison value. If, for example, we know the distribution to be binomial, we go directly to a table of the binomial. In many problems we use normal approximations. For these normal approximations we need the concept of critical ratio.

The Big Idea. The sort of calculation we often use has the form

$$\text{critical ratio} = \frac{\text{statistic} - \text{standard}}{\text{standard deviation of statistic}},$$

and for many situations such a quantity runs between $+2$ and -2 in value about 95 percent of the time when the statistic has a mean that equals the standard. If the mean of the statistic differs from the standard, then the ratio is more likely to fall outside the limits ± 2, thus producing evidence favoring a difference. This chapter illustrates many variations on this theme.

CONFIDENCE LIMITS

For any choice of confidence—for example, 95 percent—we use tables to construct confidence limits, such as the following:

$$\text{statistic} \pm \text{tabled value} \times (\text{standard deviation of statistic}).$$

In good-sized samples, the "tabled value" is often near 2. The confidence statement says that if we make such calculations over and over in different samples, then 95 percent of the time the interval between the lower limit and upper limit will contain the true mean value of the statistic. Different samples usually produce different intervals.

The advantage of the confidence statement over the test of significance is that it gives some idea of the variability of the statistic on the original scale of measurement. In many ways, the two approaches, when both can be carried out, contain equivalent information, provided we have available both the numerator and the denominator of the critical ratio. The investigator will have both parts, but may publish only the ratio or the P value. Then some information is lost for the reader of the work.

To carry out either method of comparison, we need to know more about the standard deviation of the statistic, and then we can apply the foregoing ideas concretely in a variety of circumstances. Sometimes the data are binomial, as in this chapter, sometimes measurements, as in Chapter 10. Sometimes we have one sample, sometimes two. Sometimes we have matched data, sometimes independent samples. These variations lead to many formulas, but their basic ideas are already given above in our descriptions of tests of significance and confidence limits.

CAUTIONARY REMARKS

1. A 95 percent confidence interval does *not* mean that the statistic from a new sample from the same population has a 95 percent chance of falling into the old interval.
2. If we reject the equality of the mean of the statistic with the standard at the 95 percent level, we do not say that the probability is 95 percent that the mean of the statistic does not equal the standard. All we say is that if the mean of the statistic and standard were equal, a difference smaller than that observed would occur at least 95 percent of the time. As a practical matter, we do not suppose that the mean and standard are exactly equal to many decimal places, but they might be near enough so that compared with sampling variation their difference seems small. If we accept the hypothesis of equality of mean and standard, this does not mean that we are 95 percent sure that they are equal. We are rather sure that they are not exactly equal.

PROBLEMS FOR SECTION 9-1

Answer the items as true (T) or false (F), or fill in the blanks as appropriate.

1. We make comparisons only for data from experiments. __________
2. The two main sources of comparison are __________ and __________ .
3. An example of a standard of comparison is __________ .
4. A doubling in the number of suicides suggests a change in the suicide rate when __________ .
5. To assess a difference between a group mean and a standard, we use as a unit of measurement __________ .
6. The two methods for assessing changes discussed here are __________ and __________ .

7. A critical ratio is ________ .
8. Critical ratios usually lie between 2 and −2 when ________ .
9. The form of a confidence limit is ________ .
10. We get the same 95 percent confidence intervals from every sample ________ .
11. When the 95 percent confidence interval includes zero, the test of significance at the 95 percent level accepts the equality of the mean and the standard ________ .

9-2 MATCHED PAIRS—ONE WINNER

Great advantage accrues when different treatments can be compared on similar items, because then the differences among items are controlled. When two treatments are pitted directly against each other so that one or the other will have performed better at the end of the trial, we say there is one winner for that pairing. To improve precision we use several pairings. Some examples are as follows:

1. In metallurgy, we compare the effects of two different temperature treatments on the hardness of two pieces of metal cut from the same bar.
2. In agriculture, two kinds of fertilizer are compared on side-by-side plots.
3. In medicine, two different treatments are used for twins suffering from the same disease.
4. In nutrition experiments, different diets are fed to littermates, often to more than two in the litter.
5. In chemistry, different treatments are applied to different portions of the same liquid that has been thoroughly mixed.

EXAMPLE *Rash.* Persons suffering from poison ivy rash on their hands and forearms are to be treated by one of two lotions, A and B. To improve the experimental control, one arm is treated with A, the other with B. Using two treatments for one individual offers great control, with no variations in genetics, body, diet, environment, and behavior. Twenty individuals enter the experiment. Half are *randomly* assigned to A on the left side and B on the right side, and vice versa for the other half. After a fixed number of days, a physician who does not know which treatment has been assigned to which side scores the sides in terms of which looks better. Then the assignment of the treatments is related to the outcomes, and it is found that treatment A came out better in 15 of the 20 individuals tested. What can we make of this finding? What can we believe?

SOLUTION A. Binomial. We suppose that treatment A is superior to treatment B in some proportion of the cases p, where p is unknown. If $p > \frac{1}{2}$, then treatment A is preferred to treatment B; if $p < \frac{1}{2}$, treatment B is preferred. We doubt that $p = \frac{1}{2}$, exactly, unless both treatments are totally ineffective or totally effective. If they are both effective, it is unlikely that their effectiveness will be exactly equal.

If the count of successes for treatment A is sufficiently high, we reject the idea that $p \le \frac{1}{2}$ in favor of $p > \frac{1}{2}$. The largest p that does not favor treatment A is $p = \frac{1}{2}$. We ask what the chance of getting 15 or more successes is when $p = \frac{1}{2}$, noticing that when $p > \frac{1}{2}$ the chances get larger.

We can find this probability from Table A-5 in Appendix III in the back of the book, the cumulative binomial for $p = \frac{1}{2}$. We can use the relation, $P(a \ge 15) = P(a \le n - 15)$ and find $P = 0.021$. This value of 0.021 is the P value. Values of p larger than $\frac{1}{2}$ give larger values of P. We ask whether we would rather believe (a) that an event of rarity about 0.021 occurred or (b) that treatment A does better than treatment B, that is, $p > \frac{1}{2}$. In this situation, many people prefer to believe that treatment A is better.

The value $p = \frac{1}{2}$ is called the **null hypothesis** in this problem because it implies no difference in performance between the treatments. The value 0.021 is the **descriptive significance level** for the outcome.

SOLUTION B. Normal approximation. As an alternative to using the binomial table, we can use the normal approximation to the binomial from Chapter 7. We compute the value of the critical ratio

$$z = \frac{x - \frac{1}{2} - np}{\sqrt{np(1-p)}},$$

that is,

$$z = \frac{15 - \frac{1}{2} - 10}{\sqrt{20 \times \frac{1}{2} \times \frac{1}{2}}} = \frac{4.5}{2.24} = 2.01.$$

The normal table, either Table 7-2 or Table A-1 in Appendix III in the back of the book, shows that the probability that $Z > 2.01$ is 0.022. This is a very close approximation to the correct binomial answer, 0.021.

Notation. We use Z as the value that will result from the investigation, a random quantity, and z as a specific number. Thus we speak of the probability that Z exceeds z. Often the value z is that actually achieved in an investigation.

PROBLEMS FOR SECTION 9-2

1. Use the normal approximation for a binomial with $n = 40$ and $p = ½$ to find approximately the probability that $X \geq 30$. Approximate the probability that $X \leq 10$.

2. On a true-false test, a student flips a coin to select an answer to each question. What is the probability of getting 13 or more correct out of 20 questions? Use Table A-5 in Appendix III in the back of the book.

3. Continuation. Use the normal approximation to get the probability, and compare your answer with that in Problem 2.

★ **4.** Suppose the better team in a World Series has a 0.6 chance of winning single games and that outcomes of games are independent. Although the maximum length of the Series has usually been 7 games, other odd numbers could be chosen if we want to be sure to get a winner. In a 1-game series, the better team here would have probability 0.6 of winning. What is the smallest size of Series that would ensure that the better team had at least probability 0.9 of winning the Series? (Hint: The probability of winning a series of length $2m - 1$ can be regarded as the probability of getting m or more successes in $2m - 1$ trials, even though the full $2m - 1$ trials may not need to be played in order to find the winner.)

5. A new method of teaching is compared with a standard method in 20 pairs of schools. The new method performs better in 14 of the pairs. Can this result have readily occurred by chance if the methods are equally good? Give the P value.

6. What is a null hypothesis.?

7. In Problem 5, what is the null hypothesis?

8. What is meant by the descriptive level of significance?

9. What is the value of the descriptive level of significance in Problem 5?

10. In this section, what do the symbols Z and z mean?

9-3 MATCHED PAIRS: 0,1 OUTCOMES

A standard design that takes advantage of the matching idea used in Section 9-2 differs slightly because the outcome is not necessarily one winner.

EXAMPLE *Hunting caps.* As hunting-cap colors, flaming red and goldenrod yellow are to be tested. In a standard outdoor hunting setting, a sample of viewers get a quick look at a landscape where caps of one or the other color are exposed. Each viewer gets a quick look at the landscape with one cap, then a quick look at the landscape with the other cap. "Success" is spotting the cap; "failure" is not spotting it. That is why we speak of 0,1 outcomes. We assign 1 to success, 0 to failure. Each viewer has one of four outcomes:

(1,1) spotted both caps
(1,0) spotted the red, but not the yellow
(0,1) spotted the yellow, but not the red
(0,0) spotted neither

We sum up the outcomes of the experiment in Table 9-1. What does the table reveal?

TABLE 9-1

Outcomes for hunting-cap experiment (trials of 100 viewers)

	Yellow		
Red	***Spotted***	***Not spotted***	***Total***
Spotted	70	5	75
Not spotted	15	10	25
Total	85	15	100

Discussion. First, the table shows that most viewers saw both caps, and even more saw at least one. A few viewers saw neither cap.

What can we say about relative visibility for these two colors? It looks as if yellow is better, 85 to 75. But what can we say about the rarity of such an event as 85 versus 75?

Consider the times when the viewers spotted both caps. This information tells us nothing about comparative visibility for these colors, although it does say something about absolute visibility: Both were usually seen.

Similarly, the results for viewers who saw neither cap tell us nothing about comparative visibility for these colors; they give only information about absolute invisibility.

The 1,0 cases mean that red "won" and yellow "lost," and the 0,1 cases mean that yellow "won" and red "lost." These counts tell us about the comparative successes of the colors when there is a difference. Essentially, we ask about the probability p that yellow wins, *when there is a winner*. We have $n = 20$ cases with winners, and the probability that yellow wins at least 15 when $p = \frac{1}{2}$ is 0.021, as in Section 9-2. We have reduced this problem to that of Section 9-2, and the binomial approach or the normal approximation to it now applies.

Because $P = 0.021$ indicates a moderately rare event, we conclude that $p > \frac{1}{2}$ and that yellow seems to be more readily seen than red.

PROBLEMS FOR SECTION 9-3

1. Young boys tend to discuss school problems more with one parent than with the other; the same is true of friendship problems. A sample of 200 boys who discuss their problems with one parent or the other breaks down as shown in Table 9-2. Are the data consistent with the idea that approximately as many of the boys who talk with their mothers about school talk with their mothers about friendships, or is there a statistically significant difference? Show your calculations.

TABLE 9-2

Boys' discussions with parents

	School		
Friendship	*Mother*	*Father*	*Total*
Mother	110	20	130
Father	30	40	70
Total	140	60	200

2. A study of individuals who changed party affiliations between two elections showed the following table of counts:

First Election	Second Election: Democrat	Second Election: Republican
Democrat	42	8
Republican	12	21

Was there a significant shift toward the Democrats between the first election and second election?

3. Players A and B engage in a chess match consisting of 12 games. If A wins 4 and B wins 2, and 6 are tied, test the significance of the difference in their performances.
4. In each of 20 well-separated plots, two kinds of cabbages were grown, varieties A and B. The numbers of cabbages produced were equal in 6 plots. In 9 plots, type A had the most, and in 5 plots, type B had the most. Test for a difference in performance of the two types.

9-4 INDEPENDENT SAMPLES: 0,1 OUTCOMES

Sometimes matching is not convenient, but we still have 0,1 outcomes. Experimenters then can use a design that has two independent samples, one of which gets treatment A, the other treatment B. Ideally, the assignment of individuals to the treatment is randomized. The independence occurs because what happens in one of the samples is entirely separate from what happens in the other. Similar situations arise in sample surveys and observational studies when groups are to be compared.

EXAMPLE *Vitamin C for prevention of colds.* In a Toronto study, two groups of volunteers were tested to appraise Linus Pauling's theory that large doses of vitamin C tend to prevent the common cold and to reduce its impact when it does occur. One group received placebo (pills with no active ingredient, used as a control), and the other group received vitamin C at 1 gram per day (a large dose). Each group had about 400 volunteers (and for the illustration we treat this as exactly 400). Over a period of about 2 months, 26 percent of the vitamin C group and 18 percent of the placebo group had no colds. The vitamin C group seems to have done somewhat better. How can we appraise this more formally? We can determine how much better the vitamin C group did than the control group and how often this difference could occur by chance if vitamin C had no effect.

Notation. Let p_1 be the true proportion of volunteers who would get no colds if we have infinitely many volunteers using vitamin C, and let p_2 be the corresponding true value for infinitely many taking placebos. We want to know whether $p_1 - p_2 > 0$ or $p_1 - p_2 \le 0$.

Because we do not have an infinite sample, we do not know p_1 or p_2, but only their estimates, the observed proportions of people with no colds, $\bar{p}_1 = 0.26$ and $\bar{p}_2 = 0.18$ in this experiment.

Let us suppose that the true p_1 and p_2 are both equal to, say, p. Then, from our study of the binomial distribution in Chapters 6 and 7, we know that the variances of $\bar{p}_1$ and of $\bar{p}_2$ are

$$\text{Var } \bar{p}_1 = \frac{p(1-p)}{n_1} \quad \text{and} \quad \text{Var } \bar{p}_2 = \frac{p(1-p)}{n_2},$$

where n_1 and n_2 are the sample sizes for the two groups. We need the following facts:

The variance of the difference between two independent estimated proportions is the sum of their variances:

$$\text{Var}(\bar{p}_1 - \bar{p}_2) = \text{Var } \bar{p}_1 + \text{Var } \bar{p}_2 = \frac{p_1(1-p_1)}{n_1} + \frac{p_2(1-p_2)}{n_2}$$

If $p_1 = p_2 = p$,

$$\mathit{Var}(\bar{p}_1 - \bar{p}_2) = p(1-p)\left(\frac{1}{n_1} + \frac{1}{n_2}\right)$$

Although we do not know p, we can estimate it as

$$\bar{p} = \frac{\text{total successes}}{n_1 + n_2}.$$

In our example, the successes are the people who got no colds. And so,

$$\text{estimated Var}(\bar{p}_1 - \bar{p}_2) = \bar{p}(1-\bar{p})\left(\frac{1}{n_1} + \frac{1}{n_2}\right).$$

We plan to compute the critical ratio,

$$z = \frac{\text{statistic} - \text{standard}}{\text{estimated standard deviation of statistic}},$$

where

$$\text{statistic} = \overline{p}_1 - \overline{p}_2,$$

and because we set $p_1 = p_2 = p$,

$$\text{standard} = p_1 - p_2 = p - p = 0.$$

And so, finally, we compute

$$z = \frac{\overline{p}_1 - \overline{p}_2 - 0}{\sqrt{\text{estimated Var}(\overline{p}_1 - \overline{p}_2)}}$$

and refer the value of z to a normal distribution to appraise the result. If $p_1 = p_2 = p$, then Z will, for good-sized n's, fluctuate according to a standard normal distribution, approximately. If p is near ½ and the n's are larger than 10 each, then the normal approximation works well. If p is very small (that is, near 0) or very large (near 1), we need substantially larger n's for the normal approximation to work well.

SOLUTION. Let us apply this approach to our example:

$$\overline{p} = \frac{1}{2}(.26 + .18) = .22,$$

$$\overline{p}_1 - \overline{p}_2 = .26 - .18 = .08,$$

$$\text{estimated Var}(\overline{p}_1 - \overline{p}_2) = .22(.78)\left(\frac{1}{400} + \frac{1}{400}\right) = .000858,$$

$$\text{estimated standard deviation}(\overline{p}_1 - \overline{p}_2) = \sqrt{.000858} = .0293,$$

$$z = \frac{.08}{.0293} = 2.73.$$

We refer z to the standard normal distribution, Table A-1 in Appendix III in the back of the book. The probability of getting a result this extreme or more extreme if the p's are actually equal is about 0.0032. If $p_1 > p_2$, then the event will be less rare, and we prefer to believe that $p_1 > p_2$, rather than that $p_1 = p_2$, and that a rare event occurred. Thus, although the gain (8 percent) was not large, it looks real in this experiment. But we must consider additional matters, because the results may not be due solely to the vitamins.

A surprising result was found in one of the several experiments done by the Toronto group. Some participants tried to figure out whether they were getting vitamin C by opening their capsules and tasting. Among those who were given placebos but believed that they were taking vitamin C, there were fewer colds than among those who were given vitamin C but believed that they were taking placebos. Thus, perhaps the results were influenced by how the participants believed things should work out.

When those who tasted and correctly guessed what they were getting were removed from the analysis, there was no significant difference in outcome between the vitamin group and the placebo group. The investigators did the test using the revised proportions, and they found a descriptive level of significance larger than 0.05, a standard cutoff value people often use for tests of significance.

Thus, there is much more to handling an experiment than running it, recording results, and computing the probability level as we did earlier. We must consider the treatment. Is it what the subjects get, or what the subjects think they get? And is the treatment what the investigators think the subjects are getting? Thus, when deciding what to believe, we need to know more than the result of the significance test.

We prefer to report the actual descriptive level of significance, because it carries more information than the words *significant difference* or *no significant difference at the 5 percent level.*

To be more precise about this language, what we have done is to break the possible outcomes of an investigation into two sets—a region that contains, say, 95 percent of the outcomes when the fluctuations are purely owing to chance, and a region where more extreme outcomes occur 5 percent of the time. When an observed outcome falls into the 5 percent region, we report that the result is significant at the 5 percent level. When it falls into the 95 percent region, we say that the result is not significant at the 5 percent level. The use of 5 percent is arbitrary, but it is widely used in the scientific literature.

Association and Independence in 2 × 2 Tables. When the two binomials leading to a 2 × 2 table have equal values of p, that is, $p_1 = p_2$, we say that there is no association between the two attributes being studied. And we say that the attributes are independent. If the p's are unequal, we say that the attributes are associated. We found in the vitamin C example reason to suppose that there was an association between the treatment given and success in avoiding colds.

In showing exact lack of association, a 2 × 2 table has rows with counts that are proportional and similarly has columns with counts that are proportional. For example, consider this 2 × 2 table:

	A	**not-*A***
B	100	300
not-*B*	400	1200

The counts in the second row are 4 times those in the first, and the counts in the second column are 3 times those in the first. The column percentages are equal:

$$\bar{p}(B|A) = \frac{100}{500} = .2,$$

$$\bar{p}(B|\text{not-}A) = \frac{300}{1500} = .2.$$

Similarly, the row percentages are equal:

$$\bar{p}(A|B) = \frac{100}{400} = .25,$$

$$\bar{p}(A|\text{not-B}) = \frac{400}{1600} = .25.$$

When we test for equality of the true p's in such tables, we are testing for lack of association between attributes A and B. Equivalently, we are testing for independence between A and B.

The 2×2 table could come from a binomial sample of 500 A's, so that $\bar{p}_1 = \bar{p}(B|A)$, and a binomial sample of 1500 not-A's, so that $\bar{p}_2 = \bar{p}(B|\text{not-}A)$. Or the binomial samples might run across the rows.

PROBLEMS FOR SECTION 9-4

1. *Unemployment ratio.* Two independent labor-force surveys with samples of 25,000 each, taken one year apart, show a change from 5 percent to 4 percent unemployment. Do we have strong grounds for believing that the unemployment rate has gone down?
2. *Opinion changes.* A polling agency found that in two independent samples of 2000 each, taken 3 months apart, the percentage responding favorably to Great Britain in an attitude question increased from 50 to 60. Set confidence limits on the true difference in proportions. Use the standard deviation of the difference from this section and the confidence-limit idea from Section 9-1.
3. Use the data in the forecasts gathered by G. H. Moore by a certain forecasting method for the period 1947 to 1965 to test whether the stock market more often rises when a rise is predicted than when a fall is predicted. Forecasts for two to four quarters (6–12 months) beyond the present quarter:

	Forecast	
Actual	**Rise**	**Fall**
Rise	23	10
Fall	7	5

4. *Correlation of cardiac symptoms.* For 322 patients who are essentially cardiac invalids, the following table shows the distribution of the relationship between degree of mitral insufficiency and degree of valvular calcification. Test for association between the two symptoms.

Mitral Insufficiency	Valvular Calcification Low	High	Total
Low	142	75	217
High	45	60	105
Total	187	135	322

5. *Sterile bandages.* After surgeons began using clean bandages instead of dirty bandages for wounds, Joseph Lister went further and tried bandaging while using antiseptic methods, which we shall call sterile bandaging. After surgery, results like those in the following table were found:

Condition of Wound	Bandaging Sterile	Ordinary
Well	14	9
Blood poisoning	1	6

How strong is the evidence in favor of sterile bandaging in these data?

6. *Survival by season.* The following table gives, by season, numbers of 100-day survivors and deaths following x-ray therapy for bronchial carcinoma. Is the seasonal effect statistically significant at the 5 percent level?

Season	Survivors	Deaths
Winter	30	30
Nonwinter	109	42

9-5 INDEPENDENCE OF ATTRIBUTES

When people or objects or events possess two attributes, we have the situation described in Section 9-4 for the vitamin C experiment and illustrated earlier by the example of the hunting caps. In the latter example,

TABLE 9-3
Outcomes for hunting-cap experiment

Red	*Yellow* Spotted	*Yellow* Not spotted	*Total*
Spotted	70	5	75
Not spotted	15	10	25
Total	85	15	100

each situation could lead to spotting or not spotting the red cap and the yellow cap. We represent the outcome again in a 2×2 table, Table 9-3. We notice that the proportion of times that red is spotted by viewers who also spot yellow is $70/85 = 0.824$, and the proportion of times red is spotted by viewers who do not spot yellow is $5/15 = 0.333$. These two proportions are substantially different. This fact gives evidence that the outcome for the red cap was associated with or depended on the outcome for the yellow. If the two situations have the same proportions of spotting red, we say that the rate of spotting red is not associated with or is independent of the outcome for yellow, as discussed in Section 9-4.

TABLE 9-4
Hunting-cap experiment where sightings were independent of color

Red	*Yellow* Spotted	*Yellow* Not spotted	*Total*
Spotted	68	12	80
Not spotted	17	3	20
Total	85	15	100

In Table 9-4 we illustrate a set of outcomes having the same proportions of spottings. The proportion of spottings of red when yellow is not spotted is $12/15 = 0.8$, and that for red no matter which is sighted is $80/100 = 0.8$.

We should note, and you will be asked to verify in Problem 4, that we get equal sighting proportions for yellow whether or not red is spotted. This is a feature of independence in a 2×2 table of counts:

INDEPENDENCE

If the row proportions are independent of the columns, then the column proportions are independent of the rows.

If we have independence and know the row and column totals, what numbers should be expected in the cells of the 2×2 table?

	A	not-A	**Totals**
B			r
not-B			$n - r$
Totals	c	$n - c$	n

For independence, we want the same proportion of B's when A is present as when not-A is present. To get the upper left-hand cell, we take the proportion r/n of c and get as the expected count

$$\frac{r \times c}{n}.$$

Similarly, in the upper right-hand cell we get

$$\frac{r \times (n - c)}{n}.$$

In the lower left and lower right we get

$$\frac{(n - r) \times c}{n} \quad \text{and} \quad \frac{(n - r) \times (n - c)}{n}.$$

These formulas can all be summed up by noting that when we have independence, the expected count in a contingency table is estimated by the formula:

$$\frac{\text{row total} \times \text{column total}}{\text{grand total}} \qquad (1)$$

One reason we call such a table a contingency table is that the probabilities are contingent on which row (or which column) is chosen. As we saw in Chapter 8, $P(A|B)$, $P(A|\overline{B})$, and $P(A)$ may all differ.

Even when independence holds, we do not expect to get such results in experiments, because sampling fluctuations alone will throw the proportions off a bit. Nevertheless, these expected counts of formula (1) associated with independence offer a baseline against which to measure departures.

EXAMPLE For the vitamin C example, we have the counts in Table 9-5(a). The expected count for the upper left-hand corner cell is

$$\frac{\text{row total} \times \text{column total}}{\text{grand total}} = \frac{400 \times 176}{800} = 88.$$

We can get the rest of the expected counts by subtraction, because the row and column totals must add up to the marginal totals for the expected counts. The table of expected counts is shown as Table 9-5(b).

The table of residuals, where

$$\text{residual} = \text{observed} - \text{expected},$$

is shown as Table 9-5(c). We note that except for sign, all residuals have the same value. This feature applies to all 2×2 tables, but not to tables with more rows or columns. We use these residuals in the next section to assess independence, first in a 2×2 table and then in a larger table.

TABLE 9-5
Results of vitamin C experiment in Toronto, Canada

TABLE 9-5(a)
Observed counts

	Vitamin C	*Placebo*	*Totals*
No cold	104	72	176
Cold	296	328	624
Totals	400	400	800

TABLE 9-5(b)
Expected counts

	Vitamin C	*Placebo*	*Totals*
No cold	88	88	176
Cold	312	312	624
Totals	400	400	800

TABLE 9-5(c)
Residuals

	Vitamin C	*Placebo*
No cold	16	−16
Cold	−16	16

PROBLEMS FOR SECTION 9-5

1. What do we mean by independence in a 2×2 table?
2. In the following 2×2 table, find the expected counts of teenagers in each cell, when 200 teenagers are classified by ages of parents.

	Mother's age		
Father's age	**Younger than 45**	**Over 45**	**Totals**
Younger than 45			80
Over 45			120
Totals	90	110	200

3. If we do not expect independence to hold in observed 2×2 tables, why is the concept useful?
4. In Table 9-4, show that the sighting of a yellow cap is independent of the sighting of a red cap.

Find the expected counts under the independence assumption and the residuals for the data given in the following problems:

5. Problem 3 of Section 9-4.
6. Problem 4 of Section 9-4.
7. Problem 5 of Section 9-4.
8. Problem 6 of Section 9-4.

9-6 CHI-SQUARE TEST FOR INDEPENDENCE

An index we use to appraise departure from independence is the chi-square statistic. For each cell we compute

$$\frac{(\text{observed count} - \text{expected count})^2}{\text{expected count}}$$

and add these up over all the cells in the table. If we denote the observed count by O and the expected count by E, then the formula is

$$\chi^2 = \Sigma \frac{(O - E)^2}{E},$$

where Σ means to sum over all cells and χ is the Greek letter chi. The larger the $O - E$ residuals, the larger the numerator. The E in the denominator says that larger expected counts should be allowed to have larger residuals for the same amount of departure. Large values of chi-square in a table of given size throw doubt on independence. In a 2×2 table, the average value of χ^2 (read chi-square) when the rows and columns are independent is 1.

EXAMPLE Compute chi-square for the vitamin C example of Section 9-4 (see Table 9-5).

SOLUTION. Using the table of expected counts and the table of residuals, we get

$$\begin{aligned} \chi^2 &= \frac{(16)^2}{88} + \frac{(-16)^2}{88} + \frac{(16)^2}{312} + \frac{(-16)^2}{312} \\ &= 256\left(\frac{1}{44} + \frac{1}{156}\right) \\ &= 7.46. \end{aligned}$$

First, let us note that $\sqrt{7.46} = 2.73$, which is the z we got in testing the difference in proportions. For 2×2 tables it is a general proposition that

$$\chi^2 = z^2.$$

This gives us an easy way to relate χ^2 to the normal table. Table A-3 in Appendix III in the back of the book is a table appropriate for chi-square itself. The 2×2 table is said to have one degree of freedom (1 d.f.), because given the marginal totals, once we write in the value for one cell we are forced to fill in all the remaining three cells by subtraction automatically.

Our next example illustrates a 2×2 problem in which we do not have two samples.

EXAMPLE *Amoebas and intestinal disease.* When an epidemic of severe intestinal disease occurred among workers in a plant in South Bend, Indiana, doctors said that the illness resulted from infection with the amoeba *Entamoeba histolytica.* There are actually two races of these amoebas, large and small, and the large ones were believed to be causing the disease. Doctors suspected that the presence of the small ones might help people resist infection by the large ones. To check on this, public health officials chose a random sample of 138 apparently healthy workers and determined if they were infected with either the large or small amoebas. Table 9-6 gives the resulting data. Is the presence of the large race independent of the presence of the small one?

TABLE 9-6

Presence of amoebas in apparently healthy workers in a plant in South Bend, Indiana

	Large race		
Small race	*Present*	*Absent*	*Total*
Present	12	23	35
Absent	35	68	103
Total	47	91	138

Source: J. E. Cohen (1973). Independence of Amoebas. In *Statistics by Example: Weighing Chances,* edited by F. Mosteller, R. S. Pieters, W. H. Kruskal, G. R. Rising, and R. F. Link, with the assistance of R. Carlson and M. Zelinka, p. 72. Addison-Wesley: Reading, Mass. Reproduced by the permission of the author and publisher.

SOLUTION. We test the conjecture of independence using χ^2. We do the computation using the general formula and the expected values, assuming independence:

	Large Race		
Small Race	Present	Absent	Total
Present	11.92	23.08	35
Absent	35.08	67.92	103
Total	47	91	138

$$\chi^2 = \frac{(12 - 11.92)^2}{11.92} + \frac{(23 - 23.08)^2}{23.08} + \frac{(35 - 35.08)^2}{35.08} + \frac{(68 - 67.92)^2}{67.92}$$

$$= (.08)^2\left(\frac{1}{11.92} + \frac{1}{23.08} + \frac{1}{35.08} + \frac{1}{67.92}\right)$$

$$= .0064 \times .1704 = .0011.$$

From the chi-square table, Table A-3 in Appendix III in the back of the book, we see that for 1 d.f., $P(\chi^2 \geq 0.0011)$ is so large that we cannot read it accurately from this table. A table having more decimal places gives an approximate probability of 0.97. Thus the table agrees closely with the hypothesis of independence, indeed, extremely closely.

PROBLEMS FOR SECTION 9-6

1. The members of a random sample of 400 boys 16 to 19 years of age are classified as employed or unemployed, and those in a second random sample of 600 men 20 to 24 years of age are similarly classified. The data are as follows:

Age	Unemployed	Employed	Totals
16–19	80	320	400
20–24	36	564	600

Compute χ^2 to test whether the true proportions of unemployed are the same for the two age groups.

2. *Hail suppression.* In a hail-suppression experiment in Switzerland, cloud seeding was intended to reduce the amount and frequency of hail. Seeding was randomly carried out on stormy days. The following data were obtained on days with maximum wind velocity between 40 and 80 kilometers per hour.

Condition	Days without Seeding	Days with Seeding
Hail	5	15
No hail	55	57
Total	60	72

Use χ^2 to test whether seeding seems to suppress hail.

3. *Lizards of Bimini.* Ecologists studying lizards often are interested in relationships among the variables that can be used to describe a lizard's habitat. Two such variables are the height and the diameter of the perch used by the lizard. The following data are from observations made on 164 adult male lizards in Bimini by an ecologist, T. Schoener:

	Perch Diameter		
Perch Height	Narrow	Wide	Total
High	32	11	43
Low	86	35	121
Total	118	46	164

Use χ^2 to determine if perch height and diameter are related.

4. Apply the chi-square test to the data of Problem 5 following Section 9-4.

Compute χ^2 and comment on the independence of the variables in Problems 5 and 6.

5. Use the data in Table 9-4.
6. Use the data in Problem 4 following Section 9-4.
7. We reject independence when we get large values of χ^2. For what values of z do we reject independence?
8. Check that the one-sided 5 percent level for χ^2 is the two-sided 5 percent level for z from a normal table, using $\chi^2 = z^2$ for one degree of freedom.
9. When $z = 1$, $\chi^2 = 1$, for one degree of freedom. Check that the tail areas from the normal table and the chi-square tables agree for this value, approximately. Show your work.

10. Use the normal table to check that $P \approx 0.97$ for χ^2 in the amoeba example.
11. Note that the chi-square table, Table A-3 in Appendix III in the back of the book, gives 0.95 for $\chi^2 = 0.00$. Clearly, if $\chi^2 = 0$, then $P = 1$. Explain.

9-7 THE $r \times c$ TABLE

In tables of counts larger than the 2×2 table, we may also wish to test for independence or compare the relative proportions for rows (or columns). The χ^2 approach generalizes at once. We use the same formula as before, but we must sum over more cells.

EXAMPLE *Air accidents.* For a period of observation, Table 9-7 shows the numbers of U.S. Air Force pilots involved in accidents, classified by type of service. Use the chi-square approach to test for independence between type of service and occurrence of accidents.

SOLUTION. The argument we used to find the expected counts in 2×2 tables for the independence situation continues to apply. The formula is still

$$\frac{\text{row total} \times \text{column total}}{\text{grand total}}.$$

Table 9-7 shows the expected counts as well as raw counts. We now

TABLE 9-7
Numbers of U.S. Air Force pilots having accidents, by type of service

	Bomber	*Transport*	*Liaison*	*Totals*
Observed				
No accidents	26,307	36,244	8,183	70,734
Accidents	291	389	148	828
Totals	26,598	36,633	8,331	71,562
Expected				
No accidents	26,290	36,209	8,235	70,734
Accidents	308	424	96	828
Totals	26,598	36,633	8,331	71,562

compute χ^2:

$$\chi^2 = \frac{(17)^2}{26{,}290} + \frac{(35)^2}{36{,}209} + \frac{(-52)^2}{8{,}235} + \frac{(-17)^2}{308} + \frac{(-35)^2}{424} + \frac{(52)^2}{96} = 32.4.$$

Once we get χ^2, the next question arises: How many degrees of freedom are there? In earlier discussions of degrees of freedom in this book we noted that they correspond to available pieces of information. For the 2 × 2 table, once we computed one expected value the remaining expected values could be computed by subtraction from the marginal totals—the one expected value to be computed corresponds to the one degree of freedom for the 2 × 2 table. For a 2 × 3 table, the following skeleton table represents the situation schematically.

✓	✓	×	fixed
×	×	×	fixed
fixed	fixed	fixed	fixed

The cells with check marks can be freely filled in. Once these two cells are filled, the entries for the rest of the cells (marked ×) are fixed, because the marginal totals for the expected values must add up to the observed totals. Thus there are two degrees of freedom. (We could have checked any two cells not in the same column to be freely filled in.)

The value of $\chi^2 = 32.4$, computed earlier, is referred to Table A-3 in Appendix III in the back of the book for two degrees of freedom, and it corresponds to a point beyond 0.001 for the probability in the upper tail. A value of χ^2 as large as the observed value can occur less than once in 1000 times if accident and type of service are independent. Note that most of the 32.4 (indeed, 28.2 of it) came from the lower right-hand corner. This suggests that the liaison service is having relatively more accidents than the others. It also suggests that when a calculation has one very large contribution to chi-square, it may be unnecessary to compute the rest of the terms to test independence.

Larger Tables. In general, when we are testing for independence in an $r \times c$ table, formulas for expected values and χ^2 are the same as before, and there are $(r - 1)(c - 1)$ degrees of freedom for χ^2. The reasoning leading to this number of degrees of freedom is similar to that in our 2 × 3 table example. The expected values have the same marginal totals as the observed counts. Thus we need only fill in $c - 1$ entries in the first row, and then the other one is found by subtraction. This can be done for $r - 1$ of the rows, and then the entries for the remaining row are found by subtraction from the column totals.

The mean and variance of χ^2 with k degrees of freedom are k and $2k$, respectively. When k is large, χ^2 is approximately normal, and

$$z = \frac{\chi^2 - k}{\sqrt{2k}}$$

can be used as an approximate standard normal deviate for testing.

To illustrate, suppose the observed value of χ^2 is 12, with eight degrees of freedom. The approximation gives $z = (12 - 8)/\sqrt{16} = 1$, and using a normal table, we estimate the probability of a larger value as 0.16. Linear interpolation in the chi-square table gives 0.16. Because of this approximate normality, Table A-3 in Appendix III in the back of the book contains entries primarily for fewer than 30 degrees of freedom.

Our use of the chi-square distributions with the statistic χ^2 is an approximation, just as the normal distribution is used to approximate the binomial.

EXAMPLE *Occupation and aptitude.* In a famous long-term follow-up study conducted by the National Bureau of Economic Research involving 4353 volunteers for the pilot, navigator, and bombardier training programs of the U.S. Army Air Corps during World War II, investigators collected the data in Table 9-8 on aptitude (measured during training) and occupation in 1969. The four occupations listed here were selected from a somewhat longer list of occupations. Carry out a test to determine whether or not occupation is independent of aptitude.

SOLUTION. The details of the computation of χ^2 are left as an exercise. The value is $\chi^2 = 35.8$. Because the table has 5 rows and 4 columns, there are $(5 - 1)(4 - 1) = 12$ degrees of freedom. Using Table A-3 in Appendix III in the back of the book, we find

$$P(\chi^2_{12} > 32.91) = 0.001.$$

Thus if we have independence, a value as large as or larger than the observed value of χ^2 can occur only rarely, and we conclude that aptitude and occupation are related.

TABLE 9-8

Cross-classification of 4353 World War II volunteers by occupation and aptitude

Aptitude		*Occupation*				Total
		O1†	O2	O3	O4	
Low	A1	122	30	20	472	644
	A2	226	51	66	704	1047
	A3	306	115	96	1072	1589
	A4	130	59	38	501	728
High	A5	50	31	15	249	345
Total		834	286	235	2998	4353

† O1 = self-employed, business; O2 = self-employed, professional; O3 = teacher; O4 = salaried, employed.
Source: A. E. Beaton (1975). The influence of education and ability on salary and attitudes. From *Education, Income, and Human Behavior*, edited by F. T. Juster, p. 377. McGraw-Hill, New York. Copyright © 1975 by the Carnegie Foundation for the Advancement of Teaching and the National Bureau of Economic Research. Used with the permission of McGraw-Hill Book Company.

PROBLEMS FOR SECTION 9-7

1. In a study of social status in jury deliberations, jurors were asked which of four occupational levels they would prefer to compose a jury if a member of their family were on trial. These choices were rated on a four-point scale and compared to the subject's own occupational level. About two-thirds of the subjects "switched" levels, choosing one, two, or three levels higher (or lower) than their own. The data for those who switched, slightly modified and recast, were as follows:

	Levels Higher or Lower			Total
	1	2	3	
Chose higher	84	48	18	150
Chose lower	36	12	2	50
Total	120	60	20	200

Carry out a test for independence using these data. Explain what you take the result to mean.

2. Continuation. What feature of the data suggests a trend? Assuming that there is a trend, how would you interpret it?

3. Among three groups of people, chosen from different geographic regions in one country, the distribution of hair colors is as follows:

	Dark Hair	Light Hair	Red Hair
Group *A*	9	9	2
Group *B*	7	20	3
Group *C*	15	21	15

Do these data indicate that hair color is dependent on geographic region in this country?

4. Dr. Benjamin Spock, author of a well-known book on baby care, and other people were initially convicted of conspiracy in connection with draft resistance during the Vietnam war. They appealed the conviction, one of their grounds being the sex composition of the jury panel. The jury itself had no women, but chance and challenges could account for that. Although defense attorneys might have contended that the jury list (from which jurors are chosen) should contain 55 percent women, as in the general population, they did not. Instead, they complained that whereas six judges in that court averaged 29 percent women in their jury lists, the seventh judge (before whom Spock was tried) had fewer, not just on this occasion but systematically. The following data show the numbers of women on jury lists for the seven judges for a 30-month period (Judge G tried the Spock case).

Judge	Number on Jury List	Number of Women
A	354	119
B	730	197
C	405	118
D	226	77
E	111	30
F	552	149
G	597	86

Carry out an appropriate test to check on the claim of the defense attorneys.

5. *Social mobility in Great Britain.* In an investigation of social mobility in Great Britain from one generation to another, occupations were divided into five

classes (1 = highest; 5 = lowest). A sample of 3497 men was selected for the study and assessed according to their own occupations and their fathers' occupations. The results are summarized in the following table:

	Son's Status					
Father's Status	**1**	**2**	**3**	**4**	**5**	**Totals**
1	50	45	8	18	8	129
2	28	174	84	154	55	495
3	11	78	110	223	96	518
4	14	150	185	714	447	1510
5	0	42	72	320	411	845
Totals	103	489	459	1429	1017	3497

a) Suppose we assume that son's occupation is independent of father's occupation. Write down the expected values of the four cells in the upper left corner of the table with these marginal totals.
b) The investigator started to compute χ^2 for this table, but after looking at the expected and observed values for the upper left-hand corner cell, he abandoned the calculation. Why was that reasonable?
c) Continuation. Reviewing the pattern of observed data and all the expected values, interpret in a couple of sentences what the comparison shows, and state how the data differ from independence.

6. Carry out the calculation of chi-square for Table 9-8.

7. If chi-square has 25 degrees of freedom and the value $\chi^2 = 37.65$, find approximately the significance level of the result

a) using Table A-3 in Appendix III in the back of the book, and
b) using the normal approximation.

8. Do Problem 7 with 50 degrees of freedom, observing $\chi^2 = 49.33$.

9. Examine Table A-3 in Appendix III in the back of the book for $P = 0.50$ as the number of degrees of freedom grows large. What does it suggest to you?

9-8 SUMMARY OF CHAPTER 9

1. For a test of significance, we usually calculate

$$\text{critical ratio} = \frac{\text{statistic} - \text{standard}}{\text{standard deviation of statistic}}.$$

2. Confidence limits for a quantity often take the form

statistic ± tabled value × (standard deviation of statistic).

The tabled value often is near 2 for 95 percent confidence.

3. For matched pairs, one winner, we test whether or not $p = \frac{1}{2}$ in a binomial problem. Either use the binomial tables or enter the normal table with the critical ratio

$$z = \frac{x - \frac{1}{2} - n(\frac{1}{2})}{\sqrt{n(\frac{1}{2})(\frac{1}{2})}}.$$

4. For comparing two binomial proportions, p_1 and p_2, we enter the normal table with the critical ratio

$$z = \frac{\bar{p}_1 - \bar{p}_2}{\sqrt{\text{estimated Var}(\bar{p}_1 - \bar{p}_2)}},$$

where

$$\text{estimated Var}(\bar{p}_1 - \bar{p}_2) = \bar{p}(1 - \bar{p})\left(\frac{1}{n_1} + \frac{1}{n_2}\right)$$

and

$$\bar{p} = \frac{\text{total successes}}{n_1 + n_2}.$$

5. Independence in a 2 × 2 table implies equality of proportions for the rows and columns.

6. In two-way tables of counts we often use the model of independence as a baseline from which to measure departures.

7. Expected values for counts in a two-way table under the model of independence follow this multiplication rule:

$$\text{expected count} = \frac{\text{row total} \times \text{column total}}{\text{grand total}}.$$

8. To assess how close the expected counts follow the observed, we compute the chi-square statistic:

$$\chi^2 = \sum_{\text{all cells}} \frac{(\text{observed} - \text{expected})^2}{\text{expected}}.$$

9. The number of degrees of freedom associated with an $r \times c$ table of counts equals $(r - 1)(c - 1)$.

10. We refer the statistic χ^2 computed for an $r \times c$ table to the tables of the chi-square distribution with $(r - 1)(c - 1)$ degrees of freedom.

11. The chi-square distribution is an approximation to the actual distribution of χ^2 for tables of counts. A chi-square distribution has mean k and standard deviation $\sqrt{2k}$, where k is the number of degrees of freedom.

12. For large k,

$$Z = \frac{\chi^2 - k}{\sqrt{2k}}$$

is approximately normally distributed.

SUMMARY PROBLEMS FOR CHAPTER 9

1. What form does a critical ratio take?

2. Continuation. What is its value if the statistic has value 10, the standard has value 5, and the standard deviation of the statistic has value 10?

3. Continuation. Find the probability of a larger value of the critical ratio, assuming approximate normality and assuming that the standard is the true mean of the statistic.

4. Assuming approximate normality for the statistic, display the form of the 95 percent confidence limits for the statistic.

5. Continuation. Apply the numerical values given in Problem 2 to obtain numerical values for the confidence limits in Problem 4.

6. In skeet shooting, with shots at two clay pigeons released simultaneously, a man hit both pigeons 25 times, hit neither 20 times, hit with only his first shot 25 times, and hit with only his second shot 30 times. Test whether he is better with his second shot than with his first.

7. Use the data in Problem 6 to test the difference in proportions of first-shot hits when the second shot was a hit and when it was a miss.

8. Use the data of Problem 6 to get χ^2 for independence between the outcome for the first shot and the outcome for the second shot.

9. Check numerically whether or not χ^2 from Problem 8 equals z^2 from Problem 7.

10. Show, using checks (✓) for freely filled-in cells and crosses ($\times$) for cells whose values are forced by the totals, how to count the degrees of freedom in a 3×3 contingency table, and check the result against the formula.

11. In a large contingency table, the calculation gives χ^2 almost exactly equal to k, the degrees of freedom. Approximately what is the significance level for χ^2? How do you know?

REFERENCES

F. Mosteller and R. E. K. Rourke (1973). *Sturdy Statistics.* Addison-Wesley, Reading, Mass.

F. Mosteller, R. E. K. Rourke, and G. B. Thomas, Jr. (1970). *Probability with Statistical Applications,* second edition. Addison-Wesley, Reading, Mass.

Comparisons for Measurements

Learning Objectives

1. Understanding distributions of sample means for measurements
2. Central limit theorem
3. Tests of hypotheses for measurements:
   a) One sample
   b) Two matched samples
   c) Two independent samples
4. Confidence limits for measurements
5. Effect of not knowing the variances

10-1 DEALING WITH MEASUREMENTS

The problems of estimation, variability, and testing that we treated in Chapter 9, in which outcomes have only two categories, arise again when our data come as measurements instead of counts. We now extend our methods to handle these situations. In a way, we dealt with measurements when we treated $\overline{p}$ for the binomial. We computed its mean and its variance across samples and computed its distribution for sampling with and without replacement. And so we have made a dent in the theory of measurement in the special case of discrete distributions. Indeed, from that experience, we might guess how to compute the variance of a sample mean of measurements drawn from a finite population either with or without replacement.

EXAMPLE *Four measurements.* Find the mean and variance of the average $\overline{X}$ of samples of size 2 when they are drawn with replacement and when they are drawn without replacement from a population consisting of two 1's, one 2, and one 4.

START OF SOLUTION. Looking back at our formulas for the binomial case and for sampling without replacement, we note that the formula can be written

$$\text{Var } \overline{p} = \frac{\text{variance in the population}}{\text{sample size}} \text{(improvement factor)}.$$

The variance in the population was computed from the frequencies of "measurements" 1 and 0 in the population and turned out to have the neat formula $p(1 - p)$. We cannot expect a neat formula for the variance for measurements, but we might guess that the variance of $\overline{X}$ will turn out to obey the same general formula as does the variance of $\overline{p}$. Let us see what happens. To get the variance for the population, we first compute the population mean, which for measurements is usually denoted by the Greek letter μ (mu). It corresponds to the "m" of the word mean.

DEFINITION Population mean for measurements:

$$\mu = \frac{\text{sum of measurements}}{N}$$

In our example with the four measurements 1, 1, 2, and 4,

$$\mu = \frac{1 + 1 + 2 + 4}{4} = 2.$$

Subscripts on μ and σ. In some problems, when we speak of means and variances, more than one mean or variance may be part of the discussion. Then, to avoid confusion, we use the variable whose mean or variance is being computed as a subscript. For example, the mean and variance of Z may be written μ_Z and σ_Z^2. In the current discussion it is important to distinguish between the variance of X and the variance of $\bar{X}$. We recall from Chapter 8 how to compute the variance of a random variable X when all values have equal probabilities. To get the population variance of X, σ_X^2, we square the deviations from the mean, sum the squares, and divide by N:

DEFINITION Population variance for measurements:

$$\text{Var } X = \sigma_X^2 = \frac{\text{sum of } (x - \mu)^2}{N}$$

For our example,

$$\text{Var } X = \sigma_X^2 = \frac{(1-2)^2 + (1-2)^2 + (2-2)^2 + (4-2)^2}{4}$$

$$= \frac{1 + 1 + 0 + 4}{4} = \frac{6}{4} = 1.5.$$

Here we generalize from binomial variances to those for measurements when n is the sample size, σ_X^2 is the variance of the population of measurements, and N is the size of the population:

FORMULAS FOR VARIANCE OF $\bar{X}$

With replacement:	Without replacement:
$\sigma_{\bar{X}}^2 = \text{Var } \bar{X} = \dfrac{\sigma_X^2}{n}$	$\sigma_{\bar{X}}^2 = \text{Var } \bar{X} = \dfrac{\sigma_X^2}{n}\left(\dfrac{N - n}{N - 1}\right)$

And, of course, in both cases we expect

$$\text{average of } \overline{X} = \mu$$

EXAMPLE CONTINUED. Next we enumerate the samples of size 2 for our four-item population, and check the mean and variance formulas just given.

SOLUTION. The $\overline{x}$'s for the samples of size 2 can be laid out as shown in Table 10-1. The circles indicate the samples that occur only with replacement. With replacement, we have as our variance of 16 sample means

$$\text{Var } \overline{X} = \frac{\text{sum of } (\overline{x} - \mu)^2}{16}$$

$$= \frac{4(1-2)^2 + 4(1.5-2)^2 + 4(2.5-2)^2 + (2-2)^2 + 2(3-2)^2 + (4-2)^2}{16}$$

$$= \frac{12}{16} = \frac{3}{4}.$$

The generalization given in the box on page 267 for the variance when sampling with replacement was

$$\text{Var } \overline{X} = \frac{\sigma_X^2}{n} = \frac{1.5}{2} = \frac{3}{4}.$$

This checks our generalization for this example.

TABLE 10-1

Sample means for samples of size 2

First observation	Second observation: 1	1	2	4
1	(1)†	1	1.5	2.5
1	1	(1)	1.5	2.5
2	1.5	1.5	(2)	3
4	2.5	2.5	3	(4)

†The circled numbers arise in sampling with replacement.

For sampling without replacement, we use the numbers above the diagonal in Table 10-1 and get

$$\text{Var } \overline{X} = \frac{\text{sum of } (\overline{x} - \mu)^2}{6}$$

$$= \frac{(1 - 2)^2 + 2\,(1.5 - 2)^2 + 2\,(2.5 - 2)^2 + (3 - 2)^2}{6}$$

$$= \frac{3}{6} = \frac{1}{2}.$$

From our generalization we get

$$\text{Var } \overline{X} = \frac{\sigma_X^2}{n}\left(\frac{N - n}{N - 1}\right) = \frac{1.5}{2}\left(\frac{4 - 2}{4 - 1}\right) = \frac{3}{6} = \frac{1}{2}.$$

Again, this checks the generalization for this example. (We could have used all the uncircled numbers, and then we would have had to use 12 in the denominator instead of 6.)

We shall not prove the generalizations, but they are true, and we shall use them.

The Sample Variance. The results just given tell us how precise our sampling results are when we sample measurements. Of course, we may not know σ^2 for the population of measurements, and so then we have to estimate it, just as we use $\overline{p}(1 - \overline{p})$ instead of $p(1 - p)$ when we do not know p. We use the notation s^2 for

$$\text{sample variance} = s^2 = \frac{\text{sum of } (x - \overline{x})^2}{n - 1}.$$

The capital Greek letter sigma, Σ, is used to indicate "sum of" symbolically. So we have

SAMPLE VARIANCE

$$s^2 = \frac{\Sigma(x - \overline{x})^2}{n - 1}$$

In large samples the distinction between n and $n - 1$ in the denominator makes little difference. Using $n - 1$ gives s^2 the rather neat property that the average of s^2 over all samples is σ^2, and so the statistic s^2 is an unbiased estimate of σ^2. If we divide by n instead of $n - 1$ we get, on the average,

$$\left(\frac{n-1}{n}\right)\sigma^2 = \left(1 - \frac{1}{n}\right)\sigma^2,$$

which is a little smaller than σ^2. Unbiasedness, although a pleasant property, is not the primary reason for using $n - 1$ rather than n; the reason is that certain tables we often need to use are built on the assumption that the formula for sample variance has $n - 1$ in its denominator.

Degrees of Freedom. When we hear people speak of **degrees of freedom** in connection with the number of observations, they are referring to an idea related to the calculation of s^2. We have used the sample $\bar{x}$ as the location from which to measure deviations in computing the sample variance, and so we have arbitrarily arranged to make the sum of the deviations $(x - \bar{x})$ add up to zero, which it ordinarily would not do if we used the location value μ. This constraint on the deviations is called a loss of a degree of freedom. Its most severe effect occurs in the sample of size 1. Then x is the only observation, $\bar{x} = x$, and the deviation is $x - \bar{x} = 0$. But when we square this and try to divide by $n - 1$, we find ourselves illegally trying to divide by zero. Thus, with only one observation, we have no way of using only sample information to compute s^2 as an estimate of σ^2. When we have 1 observation and want to estimate σ^2, we lose the only degree of freedom we have by estimating μ by x, and we say that we have no degrees of freedom left for estimating σ^2.

When we want to estimate σ^2, we have to "pay" a degree of freedom for having to estimate μ, and this leaves us $n - 1$ degrees of freedom, or essentially $n - 1$ observations' worth of information for estimating σ^2. In Chapter 9, we lost 3 degrees of freedom in the 2×2 table because of constraints imposed by marginal totals.

The idea of degrees of freedom keeps popping up in more complicated problems, and so it is important to become comfortable with it. Later we lose additional degrees of freedom by estimating more things.

NOTATION

When we speak of the variable X or the mean $\bar{X}$, we refer to the special variable whose value is yet to be measured—we call it a random variable. The random variable fluctuates from sample to sample. Thus we speak of $\bar{X}$ as having an average value of μ and a variance σ_X^2/n. When we observe an actual value, however, we may use $\bar{x}$. Thus we usually use a capital letter to refer to the random variable and a lower-case letter to refer to an observed value, as we did in Section 8-8 on random variables.

To speak of the variance of x (an observed value) is to speak of the variance of some number like 2.5, and its variance is zero because 2.5 does not vary. Occasionally we break with this notation. We have used $\bar{p}$ as an observed proportion and spoken of the variance of $\bar{p}$ as $p(1 - p)/n$. Strictly speaking, we should have introduced the random variable $\bar{P}$ with variance $p(1 - p)/n$, and then $\bar{p}$ would have been a specific observed number. We did not do this because it would have created an extra notation problem that at the time might have seemed a nuisance.

COMPUTATION OF $\Sigma(x - \bar{x})^2$ AND $\Sigma(x - \mu)^2$

By an algebraic development we can prove that for n measurements

$$\Sigma(x - \bar{x})^2 = \Sigma x^2 - n\bar{x}^2$$

This right-hand form is sometimes convenient for calculation.

For example, if the measurements are 1, 1, 2, and 4, then $\bar{x} = 2$, and

$$\Sigma(x - \bar{x})^2 = (1 - 2)^2 + (1 - 2)^2 + (2 - 2)^2 + (4 - 2)^2 = 6.$$

But

$$\Sigma x^2 - n\bar{x}^2 = 1^2 + 1^2 + 2^2 + 4^2 - 4(2)^2 = 22 - 16 = 6,$$

thus checking the equality for this example.

PROBLEMS FOR SECTION 10-1

1. Recall that $\sigma_{\bar{X}} = \sqrt{\text{Var } \bar{X}}$. In sampling with replacement, how does doubling the sample size affect $\sigma_{\bar{X}}$?
2. Use Table 10-1 to check for the example that

$$\text{average of } \bar{X} = \mu$$

for samples of size 2 when (a) sampling without replacement, (b) sampling with replacement.

3. For a population consisting of measurements 0, 2, 4, and 6, make a table like Table 10-1 and use it to compute the variance of $\bar{X}$ for samples of size 2, sampling with replacement.
4. Continuation. Compute the variance σ_X^2 for the population of four measurements given in Problem 3, and check that $\sigma_{\bar{X}}^2$, using the formula, agrees with that obtained arithmetically in Problem 3.

5. Given two measurements x_1 and x_2 with average $\overline{x}$, show that

$$\Sigma(x - \overline{x})^2 = \frac{(x_1 - x_2)^2}{2}.$$

6. We want to check for an example the unbiasedness of

$$s^2 = \frac{\Sigma(x - \overline{x})^2}{n - 1}$$

for a sample of size 2. Using the 16 samples represented in Table 10-1 and the formula stated in Problem 5, verify that the average sample variance for sampling with replacement equals σ_X^2.

7. In your own words, what is a degree of freedom?

8. Illustrating your statements with Table 10-1, distinguish between σ_X^2, $\sigma_{\overline{X}}^2$, and s^2.

9. How would you estimate $\sigma_{\overline{X}}^2$ from a sample?

10. Without making detailed calculations, explain why the variance of the numbers 9, 10, 11, 12, and 13 will be the same as that for 15, 16, 17, 18, and 19.

11. Regarding the numbers 1, 2, 3, 4, 5, 6, 7, 8, 9, 10, and 11 as a population, find their variance.

★ 12. Continuation. The variance of N successive integers is

$$\sigma^2 = \frac{N^2 - 1}{12}.$$

Verify that the formula works for the numbers of Problem 11.

★ 13. Continuation. A discrete uniform distribution consisting of the N successive integers 1, 2, . . . , N has range $N - 1$. How large is this range in standard deviation units? About how large is it when N is large?

10-2 APPLYING THE NORMAL APPROXIMATION TO MEASUREMENTS

The approximation we used in Chapter 7 for estimating the probability that $\overline{p}$ falls near p can also be used to estimate the probability that a sample mean $\overline{X}$ falls near the population mean μ. The general idea is that when measurements are independently taken from the same distribution, the sample mean is approximately distributed in a manner described well by the normal, Table A-1 in Appendix III in the back of the book.

EXAMPLE *Sampling inspection.* A manufacturing machine produces cylinder-shaped objects whose lengths are supposed to measure 2.0000 centimeters. The standard deviation of these lengths for the type of machine being used is $\sigma = 0.0008$ centimeter. In a sample of 4 measurements (a sample size often used for quality-control work), what is the chance that the sample mean $\overline{X}$ is within 0.0010 centimeter of 2.0000 when the machine is properly set?

SOLUTION. The population mean $\mu = 2.0000$. The standard deviation of sample means $\overline{X}$ is

$$\sigma_{\overline{X}} = \frac{\sigma}{\sqrt{n}} = \frac{.0008}{\sqrt{4}} = .0004.$$

Thus 0.0010 is $2.5\sigma_{\overline{X}}$ from μ. Table 7-2 or Table A-1 in Appendix III in the back of the book will show that the probability is 0.988 that $\overline{X}$ is at least this close to μ.

How this sort of information is used in practice is illustrated by the following example.

EXAMPLE *A discordant sample mean.* In the setup described in the preceding example, the operator measures the lengths of four parts, and they exceed 2.0000 centimeters by .0013, .0017, .0012, and .0028 centimeters. What does the supervisor conclude?

SOLUTION. He observes that $\overline{x} = 2.0018$ centimeters, and he recognizes that very rarely would a mean so far out occur if the process had a $\mu = 2.0000$ and $\sigma = .0008$, because $\overline{x}$ is over $4\sigma_{\overline{X}}$ from μ. He decides that either an extremely rare event has occurred or that the process is no longer behaving according to the μ and σ it is supposed to follow. The range among the four measured values, .0015, is about 2σ, not an unusual amount, and so the supervisor may well decide that the setup of the machine no longer has $\mu = 2.0000$. Because the machine does drift away from the μ originally set, the decision probably will be to reset the machine, rather than to suppose that a very unusual sample has been drawn.

If the process is generating cylinders, we can think of the machine making a never-ending sequence, and so we can think of an infinite population of cylinders. If the situation is one in which a manufactured lot of N items is being assessed by a random sample of size n, then one might want to consider the "without replacement" improvement factor.

This use of the normal approximation for proportions and for measurements requires some justification. The reason the approximation works is a result from probability called the central limit theorem:

THE CENTRAL LIMIT THEOREM

If a population of measurements has mean μ and variance σ^2, then in a random sample of n measurements drawn with replacement, as n grows large, the probability that the sample mean $\overline{X}$ is within a distance $z\sigma_{\overline{X}}$ of μ is given by the standard normal distribution as shown in Table 7-2 and Table A-1 in Appendix III in the back of the book.

Comment. The central limit theorem approximation works for sampling without replacement as well. The essential conditions are that (a) the sample size n be large and (b) the size of the remainder of the population also be large. The need for the first condition is the same for both kinds of sampling, to overwhelm the shape of the original distribution through the averaging process. The second condition is to make sure that n does not get so close to N that the number of observations missing will rule the shape of the actual distribution. For example, when the sample size n is $N - 1$, the distribution of $\overline{X}$ is entirely controlled by the single missing observation, because then

$$\overline{x} = \frac{(\Sigma x) - x}{N - 1},$$

where Σx is the sum of all the measurements in the population and the isolated x is the one value not in the sample. Because Σx is a constant, the variation in $\overline{X}$ depends entirely on how observations vary in the original population. For example, if the x's were 101 equally spaced integers, running from 0 to 100, and if we drew a sample of size $n = 100$, then the values of $\overline{x}$ would be equally spaced, running from 49.50 to 50.50 by intervals of 0.01. The values of $\overline{x}$ then are distributed according to an equally spaced and equally likely discrete distribution that is called the uniform distribution and is not approximately normal. The approximate normality of means (or sums) of variables helps us approximately solve many problems in which the amount of arithmetic for an exact solution would be prohibitive.

PROBLEMS FOR SECTION 10-2

1. Why do Table 7-2 and Table A-1 in Appendix III in the back of the book play such an important role in statistical work?

Data for Problems 2 and 3: Paratroopers complete with equipment average 200 pounds per man, with a standard deviation of 10 pounds, and their weights are approximately normally distributed.

2. What fraction of equipped paratroopers weigh between 190 and 210 pounds?

3. A safely loaded glider can carry up to 2100 pounds. If 10 paratroopers are randomly assigned to a glider, what is the chance it will be safely loaded?

4. The process leading to the single random digits in Table 7-4 makes their occurrences equally likely. Graph the theoretical frequency distribution of these 10 integers (0, 1, 2, 3, 4, 5, 6, 7, 8, 9). (Note that the graph is a collection of points falling on a horizontal line.)

5. Make a table like Table 10-1 for sums of two numbers from the random digits 0 through 9 (drawn with replacement), and then sketch the frequency distribution for their sums running from 0 to 18. (The total frequency is 100; the result will look rather triangular: Δ.)

6. In Problems 4 and 5 we worked with the actual probabilities for the equally likely single-digit integers. When we add up several random digits, this approach using probabilities becomes cumbersome. An alternative is to approximate the actual distribution with a large number of samples. From the random numbers in Table A-7 in Appendix III in the back of the book add up the five digits in each sequence in columns 1–5 if your surname begins with one of the letters A through D (to get 20 sums), 6–10 if your name begins with E through I, 11–15 for J through N, 16–20 for O through S, 21–25 for T through Z. Then make a frequency distribution for your 20 sums.

7. Continuation. Class problem. Get the frequency distribution for the 100 sums from Problem 6 and note how it is beginning to be shaped more like a normal distribution and less like a uniform distribution.

8. Continuation. The population variance for sums of five independent random digits is

$$\sigma^2_{\text{sum of 5}} = 5\sigma^2_X.$$

Find the numerical value of the theoretical standard deviation.

9. Continuation. Find how many of the 100 sums in Problem 7 are within 1 standard deviation of the mean, and compare that result with the theoretical result suggested by the normal (Table 7-2 or Table A-1 in Appendix III in the back of the book).

10. A sugar distributor filling 5-pound bags of sugar says that his machine delivers 5.04 pounds, on the average, with a standard deviation of 0.02 pounds. A weights-and-measures inspector takes at random four 5-pound bags, totals their weight, and gets 19.98 pounds. The distributor says: "Oh well, only one bag in forty will be underweight." The inspector says: "I understand that, but this result still doesn't agree with your weighing description." Explain both the distributor's remark and the inspector's remark.

10-3 MATCHED PAIRS, MEASUREMENT OUTCOMES, AND SINGLE SAMPLES

Matching adds control not only in the winner or 0,1 situations of Chapter 9 but also in measurement problems.

EXAMPLE *Marijuana and changes in intellectual performance.* To investigate the effects of marijuana on the intellectual performance of subjects who had had no previous experience with marijuana, investigators tested for changes in intellectual performance following the smoking of an ordinary cigarette and following the smoking of a marijuana cigarette. The results are shown in Table 10-2. Our general method of assessing results is to look at the average difference and compare it with its standard deviation to see if the result is far from the 0 standard. We use

$$z = \frac{\overline{D}}{\text{estimated standard deviation of } \overline{D}}.$$

In our example, $\overline{D} = 6$. We also need to estimate the standard deviation of $\overline{D}$. That is given by the following formula:

$$s_{\overline{D}} = \text{estimated standard deviation of } \overline{D} = \sqrt{\frac{\Sigma(D - \overline{D})^2}{n(n-1)}}$$

where n is the number of cases and $\overline{D}$ is the average difference. A computing formula is

$$s_{\overline{D}} = \sqrt{\frac{\Sigma D^2 - n\overline{D}^2}{n(n-1)}}.$$

For our example, we find from Table 10-2 that

$$s_{\overline{D}} = \sqrt{\frac{1074 - 9(6^2)}{9(8)}} = 3.23.$$

TABLE 10-2

Changes in performance on intellectual tests after smoking ordinary and marijuana cigarettes

Subject	*Ordinary* x†	*Marijuana* y	*Differences* D = x − y	D^2
1	−3	5	−8	64
2	10	−17	27	729
3	−3	−7	4	16
4	3	−3	6	36
5	4	−7	11	121
6	−3	−9	6	36
7	2	−6	8	64
8	−1	1	−2	4
9	−1	−3	2	4
		Totals	54	1074
			$\overline{D} = 6$	

$$\frac{\Sigma D^2 - n\overline{D}^2}{n(n-1)} = \frac{1074 - 9(6^2)}{9(8)} = \frac{750}{72} = 10.4$$

Estimated standard deviation of $\overline{D} = \sqrt{10.4} = 3.22$

$$t = \frac{\overline{D}}{s_{\overline{D}}} = \frac{6}{3.22} = 1.86$$

95 percent confidence limits:

$$\begin{aligned}\overline{D} \pm t_8\, s_{\overline{D}} &= 6 \pm 2.31(3.22) \\ &= 6 \pm 7.4 = (-1.4, 13.4)\end{aligned}$$

†Positive x's and y's represent improvements in performance.

Source: Adapted from A. T. Weil, N. E. Zinberg, and J. M. Nelsen (1968). Clinical and psychological effects of marijuana in man. *Science* 162: 1234–1242.

Therefore, we compute

$$z = \frac{\overline{D}}{s_{\overline{D}}} = 1.86.$$

From the normal table, we conclude that the probability of a difference this large or larger when the true mean change for an infinite number of subjects is zero is about 0.03 (one-tailed). Thus it looks as if smoking a marijuana cigarette may produce a bigger loss in intellectual performance than smoking an ordinary cigarette.

A More Accurate Approach. The normal distribution approach that has served us so well with large samples of known variability is not entirely satisfactory when we do not know the variability and even less satisfactory when the samples are small. We can adjust for this smallness by using a special table, called the t table, that allows for sample size.

The reason the normal table is not entirely adequate is that it depends on knowing the population variability of the measurements exactly. If we have infinitely many measurements, then we can estimate the variance exactly, but in practice we have fewer measurements, and the quotient Z no longer fluctuates exactly according to the normal shape even when the raw measurements come from a normal distribution. Instead, each sample size has associated with it its own distribution, called Student's t distribution. ("Student" was the pseudonym of the inventor, W. S. Gosset, a statistician who worked for a brewery.) In one-sample problems, we index the t distributions by $n - 1$, the degrees of freedom. In our example, with $n = 9$, we have 8 degrees of freedom, and so we wish to refer z to the t distribution with eight degrees of freedom whose percentage points appear in Table 10-3. The ratio we computed is, assuming that the true mean of D is zero,

$$t_{n-1} = \frac{\overline{D}}{s_{\overline{D}}}.$$

The distribution of t is more variable than that of Z because we have to estimate the denominator $s_{\overline{D}}$.

For 8 degrees of freedom, we can compare some values of t with those for z in Table 10-3. For each value of P, the z value is smaller than that for t because t is more variable. In our example, $t = 1.86$, with 8 degrees of freedom (d.f.), and t falls at the one-sided 0.05 level. We conclude in our problem that although departure from zero is not as compelling as if we had been dealing with a normal distribution, still it is impressive.

Confidence Limits. At the foot of Table 10-2 we have indicated the calculations for 95 percent confidence limits using the multiplier from the t table. The one-sided 0.025 level corresponds to a two-sided 5 percent level, and this can be used to give a 95 percent confidence interval.

TABLE 10-3

Percentage points for t distribution with 8 degrees of freedom and for the standard normal distribution

	Two-sided probability level, P†					
	0.50	**0.20**	**0.10**	**0.05**	**0.02**	**0.01**
t, with 8 d.f.	0.71	1.40	1.86	2.31	2.90	3.36
z, normal (t with ∞ d.f.)	0.67	1.28	1.64	1.96	2.33	2.58

†P is the probability of a value of t larger than a or smaller than $-a$.

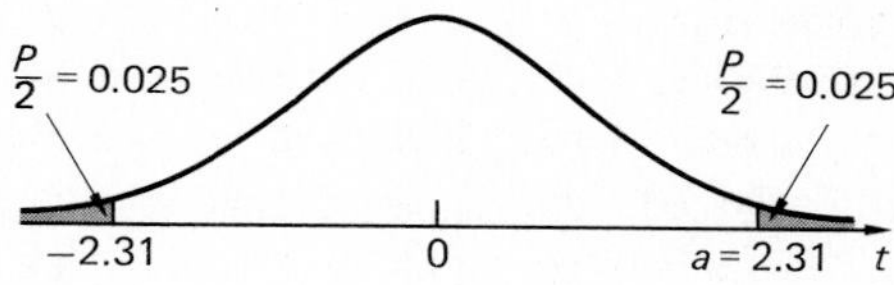

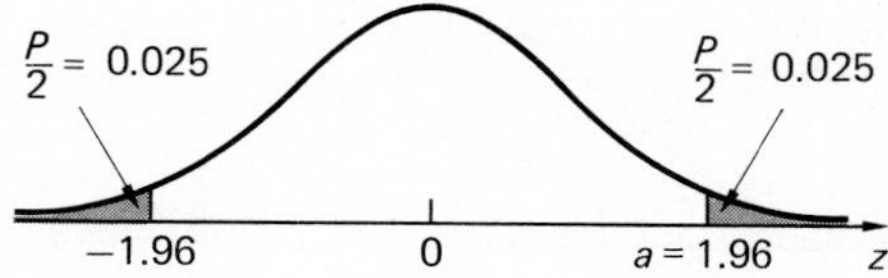

Other Degrees of Freedom. Let us look at a t table for several degrees of freedom. Table 10-4 gives us some comparative values. We get three important impressions:

1. The line for an infinite number of degrees of freedom gives numbers identical with those for a normal distribution.
2. The lines for 1, 2, and 3 degrees of freedom are seriously different from those for the normal.
3. The line for 30 degrees of freedom is fairly close to that for the normal.

The second point shows that data based on very small numbers of degrees of freedom need special treatment.

TABLE 10-4

Values of *t* that give two-sided probability levels for Student *t* distributions

Degrees of freedom	*Probability level, P*†					
	0.50	***0.20***	***0.10***	***0.05***	***0.02***	***0.01***
1	1.00	3.08	6.31	12.71	31.82	63.66
2	0.82	1.89	2.92	4.30	6.96	9.92
3	0.76	1.64	2.35	3.18	4.54	5.84
5	0.73	1.48	2.02	2.57	3.36	4.03
10	0.70	1.37	1.81	2.23	2.76	3.17
20	0.69	1.33	1.72	2.09	2.53	2.85
30	0.68	1.31	1.70	2.04	2.46	2.75
100	0.68	1.29	1.66	1.98	2.36	2.63
1000	0.67	1.28	1.65	1.96	2.33	2.58
∞(normal)	0.67	1.28	1.64	1.96	2.33	2.58

†*P* is the probability of a value of *t* larger than *a* or smaller than $-a$.

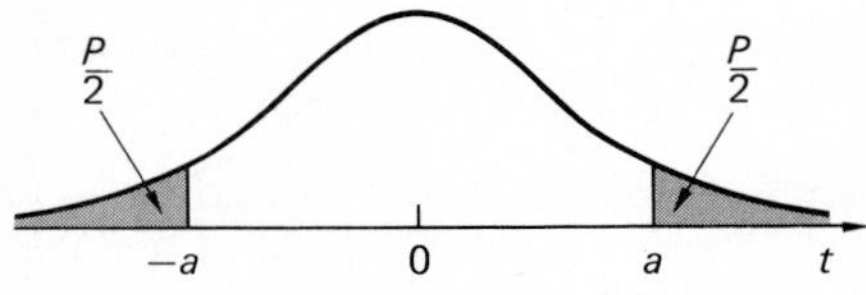

The third point gives some foundation to the popular idea that 30 degrees of freedom represent a good-sized sample. Of course, we have no discontinuity between 29 being "small" and 30 being "large," but it is true that when the degrees of freedom are 30, much of the *t* table looks close to the normal table. Actually, for 20 degrees of freedom, the tabulated values are close to the normal ones, and even for 10 degrees of freedom they are only about 20 percent larger for probabilities 0.02 and 0.01, and even closer for probabilities larger than 0.05. Table A-2 in Appendix III in the back of the book gives a larger *t* table for general use.

Single Samples. An alternative way to look at our marijuana example is that it represents one sample of size 9 from a population of differences. The analysis would be the same as before, and so we have illustrated both analyzing a one-sample problem and analyzing a matched two-sample problem with the same example.

One-sided and Two-sided Tests. To illustrate the general ideas of one- and two-sided tests and of null hypotheses and alternative hypotheses, we first use the population mean μ and its estimate, the sample mean $\bar{x}$ based on a sample of size n. We assume σ is known. We may have special interest in testing whether the population mean μ has the specific value μ_0, called the null hypothesis, as opposed to all other possible values of μ. To form a symmetric, two-sided test at the 5 percent level, we lay off the interval $\mu_0 - 1.96\sigma_{\bar{X}}$ to $\mu_0 + 1.96\sigma_{\bar{X}}$. If $\bar{x}$ falls in the interval, we say we accept the null hypothesis at the 5 percent level. If $\bar{x}$ falls outside the interval, we say we reject the null hypothesis at the 5 percent level. The point is that when $\bar{x}$ is too far in either direction from μ_0, it is hard to believe that μ_0 is the true mean.

In a one-sided test, we may concentrate our attention on the possibility that the true mean μ is larger than μ_0, as opposed to being no larger than μ_0. If μ represented a gain from a new treatment, we might lose interest in the treatment unless we believed that the gain exceeded μ_0 $(\mu > \mu_0)$. Then the null hypothesis would consist of all values of $\mu \leq \mu_0$ and the alternative of all $\mu > \mu_0$. Cut the $\bar{x}$ axis at $\mu_0 + 1.96\sigma_{\bar{X}}$. Using this one-sided test, if $\bar{x} > \mu_0 + 1.96\sigma_{\bar{X}}$, we reject the null hypothesis at the 2½ percent level; if $\bar{x} \leq \mu_0 + 1.96\sigma_{\bar{X}}$, we accept (that is, do not reject) the null hypothesis at the 2½ percent level. Had we wanted a one-sided 5 percent level test, we could have chosen the critical value as $\mu_0 + 1.65\sigma_{\bar{X}}$.

Note that the null and alternative hypotheses address the possible true values of the parameter μ whereas the test addresses the value of the observed statistic $\bar{x}$.

We call the region where the null hypothesis is rejected the critical region. In two-sided tests, the critical region has two parts, as in our example the part $\bar{x} < \mu_0 - 1.96\sigma_{\bar{X}}$ and the part $\bar{x} > \mu_0 + 1.96\sigma_{\bar{X}}$. In the one-sided test, the critical region has only one part, in our example $\bar{x} > \mu_0 + 1.65\sigma_{\bar{X}}$.

More generally, the parameter need not be a mean. It could be the true probability p in a binomial distribution, for example. And the statistic need not be its estimate, but rather a quantity whose value is related to it, such as the number of successes in n trials, in the binomial example.

The choice of null hypothesis often arises rather naturally from the problem. We ask whether a coin is fair. If so, the probability of a head on a toss is $p = ½$, the natural null hypothesis. If it is unfair, $p \neq ½$, the alternative hypothesis. To test the null hypothesis, we may toss the coin n times. If $p \neq ½$, we expect a number of heads further from $½n$ than when $p = ½$. From binomial Table A-5 in Appendix III in the back of the book, we know that when $n = 17$, the probability of 13 or more successes is 0.025, and the probability of 4 or fewer successes is also 0.025. The two-sided 5 percent level test rejects $p = ½$ if the number of successes is $x \leq 4$ or $x \geq 13$.

By subtracting μ_0 from $\overline{x}$ and dividing by $G_{\overline{X}}$, we convert to a test based on z scores.

PROBLEMS FOR SECTION 10-3

1. *Cloud seeding.* In an experiment intended to increase rainfall by seeding clouds, a northern region or a central region was seeded every day at random. The net gains in inches of rain owing to seeding in 6 time periods covering 5 years were as follows:

Period:	1	2	3	4	5	6
Total seeding gain:	1.5	0.4	−0.1	−0.1	2.0	1.0

   Assuming that the 6 periods are independent and have equal weights, use the t distribution to test whether the net gain from seeding is positive against the null hypothesis of zero gain from seeding.

2. Continuation. Set 90 percent confidence limits on the average seeding gain.
3. Random samples of adult males requiring treatment for obesity are assigned to various treatments. Ten subjects assigned to one treatment had weight losses of 9, 1, 5, 11, 7, 6, 7, 4, 5, and 5 pounds in 2 weeks. Set 90 percent confidence limits on the mean loss.
4. Continuation. Set 80 percent limits on the average loss for the treatment reported in Problem 3.
5. The following table shows roughness scores for each person's two hands. For each of 10 people, one hand was treated and the other hand was not. Low numbers indicate less roughness. Use the differences to make a t test for improvement owing to treatment. Save your answer.

Person:	1	2	3	4	5	6	7	8	9	10
Treated:	5	0	3	4	4	5	1	0	2	1
Untreated:	6	2	3	5	4	6	0	2	4	3
Difference:	1	2	0	1	0	1	−1	2	2	2

6. Continuation. Set 95 percent confidence limits on the difference in the means for Problem 5.
7. In response to two different sets of instructions, a random sample of 10 men from a large college population had the following scores on a test of dexterity:

Man	Instruction *A*	Instruction *B*
1	120	128
2	130	131
3	118	127
4	140	132
5	140	141
6	135	137
7	126	118
8	130	132
9	125	130
10	127	135

Set 99 percent confidence limits on the difference between the two population mean scores under the two instructions (do not forget that each man took both tests).

8. Continuation. Test at the one-sided 10 percent level the null hypothesis of no difference in the effects of instructions against the alternative hypothesis that instruction *B* tends to produce higher scores.

9. The following data give the amounts of rain collected by two types of rain gauges for seven storms. Note that each storm was assessed by both gauges.

Storm	Type *A*	Type *B*	Difference
1	1.38	1.44	− .06
2	9.69	10.37	− .68
3	.39	.39	.00
4	1.42	1.46	− .04
5	.54	.55	− .01
6	5.94	6.15	− .21
7	.59	.61	− .02

Test the hypothesis that the mean difference between the two types of gauges is zero.

10. Continuation. Set 80 percent confidence limits on the difference.

11. Continuation. In Problem 9, it would be more reasonable to assess the percentage change, say from type *A* gauge to type *B*. Set 80 percent confidence limits on the percentage change. [The percentage change from 1.38 to 1.44 is 100(.06)/1.38 = 4.]

10-4 TWO INDEPENDENT SAMPLES: MEASUREMENT OUTCOMES, VARIANCES KNOWN

Probably the most common of all group comparisons is based on two independent samples of 0,1 outcomes, as described in Section 9-4, or on two independent measurement samples. For example, instead of analyzing the frequency of zero colds, as we did in Section 9-4, we could, if we had the data, analyze the average number of colds per person and thus deal with measurements.

Two situations must be distinguished:

1. The variances in the population are known.
2. The variances are unknown.

We deal here with the first situation. We shall take up the second situation in Section 10-5.

The Variance of a Difference. If $\sigma^2_{\overline{X}}$ and $\sigma^2_{\overline{Y}}$ are variances of the independent means $\overline{X}$ and $\overline{Y}$, respectively, then the variance of the difference $\overline{X} - \overline{Y}$ for measurements, as for proportions, adds as follows:

$$\text{Var}\,(\overline{X} - \overline{Y}) = \sigma^2_{\overline{X}-\overline{Y}} = \text{Var}\,\overline{X} + \text{Var}\,\overline{Y}.$$

To set 95 percent confidence limits, as described earlier in Section 9-1, we use

$$\overline{x} - \overline{y} \pm 1.96\sigma_{\overline{X}-\overline{Y}},$$

where $\overline{x}$ and $\overline{y}$ are the observed means and 1.96 is the two-sided 95 percent multiplier from the normal table. We use the normal table because we know the variances. The quantity $\sigma_{\overline{X}-\overline{Y}}$ is the standard deviation of the difference between the two means:

$$\sigma_{\overline{X}-\overline{Y}} = \sqrt{\text{Var}\,\overline{X} + \text{Var}\,\overline{Y}} = \sqrt{\frac{\sigma^2_X}{n_X} + \frac{\sigma^2_Y}{n_Y}},$$

where n_X and n_Y are the sample sizes and σ^2_X and σ^2_Y are the variances for individual measurements.

EXAMPLE *Handedness in baseball.* A researcher gathered data to see whether a batter does better when he and the pitcher are opposite-handed or same-handed. He assembled batting averages based on about 50 successive times at bat for many players, as shown in Table 10-5. Test the significance of the difference in batting averages, and set 95 percent confidence limits on it.

TABLE 10-5
Data on batting averages

Batter and pitcher are:	*Number of sets of 50*	*Grand batting average*	*Standard deviation of the sets of batting averages*
X (opposite-handed)	103	0.263	0.06
Y (same-handed)	106	0.231	0.06

SOLUTION. The statistic of interest is

$$\bar{x} - \bar{y} = .263 - .231 = .032.$$

Because the samples are large, we shall assume that the standard deviations are known and apply the formula for $\sigma_{\bar{X}-\bar{Y}}$. The standard deviation of the difference is

$$\sigma_{\bar{X}-\bar{Y}} = \sqrt{\frac{(.06)^2}{103} + \frac{(.06)^2}{106}} = .0083.$$

The critical ratio is

$$z = \frac{.032}{.0083} = 3.86.$$

From Table A-1 in Appendix III in the back of the book, we read the probability of a value of z larger than 3.49 as 0.0002. More extensive tables show the probability of a value of z larger than 3.86 as 0.0001. Thus we confirm the traditional baseball belief in the advantage of opposite-handed batters.

To set 95 percent confidence limits we compute

$$\bar{x} - \bar{y} \pm 1.96\sigma_{\bar{X}-\bar{Y}} = .032 \pm 1.96\,(.0083)$$
$$= .032 \pm .016$$
$$= .016,\ .048.$$

Thus the advantage seems to be between 1.6 and 4.8 percent, with a central estimate of 3.2 percent. Note that the confidence interval does not include zero. Thus at the 5 percent significance level we would reject zero as a possible value for the true advantage. Note that although the difference of 0.032 may seem small, it is 13 percent of the average batting average, which is 0.247.

PROBLEMS FOR SECTION 10-4

1. The observed average yield for a variety of cabbage grown on 10 plots was 120 pounds per plot, with standard deviation of 20 pounds, under one watering treatment; the average yield was 145 pounds per plot on 10 comparable plots, with standard deviation of 25 pounds, under another watering treatment. Set 95 percent confidence limits on the difference in mean pounds per plot.
2. Continuation. Test at the one-sided 10 percent level whether the second watering treatment produces higher yields than the first.
3. Measurements taken under a microscope have a standard deviation of 5. That is, this is the standard deviation of the distribution of sizes when the same object is repeatedly measured. One object is measured as 108 units, another as 98 units. Is it a reasonable possibility that they are the same size? If they are, what is the probability of a greater discrepancy than that observed?
4. Continuation. Set 80 percent confidence limits on the difference in their true means.
5. Washers must fit on rods. The washers have inside diameters W, with mean μ_W and standard deviation $\sigma_W = 0.003$ inch, and the rods have outside diameters R, with mean μ_R and standard deviation $\sigma_R = 0.004$ inch. If $W - R = D > 0$, then the washer fits the rod, otherwise not. Washers and rods are randomly assembled. Estimate the proportion of assemblies that fit if $\mu_D = 0.006$ inch.
6. The scores of high school seniors on a college entrance examination are approximately normally distributed, with standard deviation $\sigma = 100$ and mean $\mu = 500$, for the entire country. For two random samples of size 25 from the seniors in two large school districts, the average scores are observed to be $\bar{x}_1 = 550$ and $\bar{x}_2 = 515$. Assuming that the variability is the same in these districts as in the nation, set 95 percent confidence limits on the true mean difference for these school districts.
7. Continuation. Test the significance of the difference of the mean performances for the seniors in the two school districts using the two-sided 10 percent level.

10-5 TWO INDEPENDENT SAMPLES: MEASUREMENT OUTCOMES, VARIANCES UNKNOWN BUT EQUAL

In the two-sample problem, when the population standard deviations are unknown but are regarded as equal in size, or as a practical matter nearly equal, we replace

$$\sigma^2_{\bar{X}-\bar{Y}} = \frac{\sigma^2}{n_X} + \frac{\sigma^2}{n_Y} = \sigma^2\left(\frac{1}{n_X} + \frac{1}{n_Y}\right)$$

by its sample analogue

$$s^2_{\bar{X}-\bar{Y}} = \frac{\Sigma(x-\bar{x})^2 + \Sigma(y-\bar{y})^2}{n_X + n_Y - 2}\left(\frac{1}{n_X} + \frac{1}{n_Y}\right).$$

The idea is that

$$s^2_X = \frac{\Sigma(x-\bar{x})^2}{n_X - 1} \quad \text{estimates } \sigma^2$$

and

$$s^2_Y = \frac{\Sigma(y-\bar{y})^2}{n_Y - 1} \quad \text{estimates } \sigma^2$$

also. We ought to use both estimates. The numbers of independent measurements they offer for estimating σ^2 are $n_X - 1$ and $n_Y - 1$, respectively. We weight them by their degrees of freedom. Thus the weights are

$$\frac{n_X - 1}{(n_X - 1) + (n_Y - 1)} \quad \text{and} \quad \frac{n_Y - 1}{(n_X - 1) + (n_Y - 1)}.$$

And so the weighted estimate of σ^2 is

$$\begin{aligned} s^2 &= \frac{n_X - 1}{n_X + n_Y - 2} s^2_X + \frac{n_Y - 1}{n_X + n_Y - 2} s^2_Y \\ &= \frac{\Sigma(x-\bar{x})^2 + \Sigma(y-\bar{y})^2}{n_X + n_Y - 2}. \end{aligned}$$

This is a good estimate of σ^2, and it uses the information from both samples.

In addition to replacing σ^2 by s^2 in $\sigma^2_{\bar{X}-\bar{Y}}$ to get $s^2_{\bar{X}-\bar{Y}}$, we also need to allow for the fact that we do not know the true variance but have $n_X + n_Y - 2$ degrees of freedom for estimating it. This recalls the *t* distribution to mind, and, indeed, for normally distributed data, the 95 percent confidence limits on the true difference, $\mu_X - \mu_Y$, between the population means is given by

$$\bar{x} - \bar{y} \pm t_{n_X+n_Y-2}\, s_{\bar{X}-\bar{Y}}.$$

And we can perform tests for differences using the *t* table instead of the normal table.

EXAMPLE *Ambiguity and speech.* To find out if people with less information about a task talk more about it, some researchers gave five groups of people a clear picture of a complicated Tinkertoy to be built, and they gave five groups an ambiguous picture, smaller and shaded. Each group had to tell a silent builder, who had no picture, how to make the object. The groups were scored for the amount of talking done. The groups with the clear picture had a mean speech score $\overline{x} = 22.8$, with $\Sigma(x - \overline{x})^2 = 707.6$; the groups with the ambiguous picture had a mean speech score $\overline{y} = 40.8$, with $\Sigma(y - \overline{y})^2 = 1281.6$. Assuming that the population variances of the speech scores of the two groups are approximately equal (even though the sample variances may not be), set 90 percent confidence limits on $\mu_X - \mu_Y$ and test the hypothesis mentioned.

SOLUTION. The estimated standard deviation of the difference is

$$s^2_{\overline{X}-\overline{Y}} = \frac{707.6 + 1281.6}{8}\left(\frac{1}{5} + \frac{1}{5}\right) = 99.5 \quad \text{or} \quad s_{\overline{X}-\overline{Y}} = 9.97.$$

From Table 10-3, the 90 percent two-sided level for t with 8 degrees of freedom is 1.86:

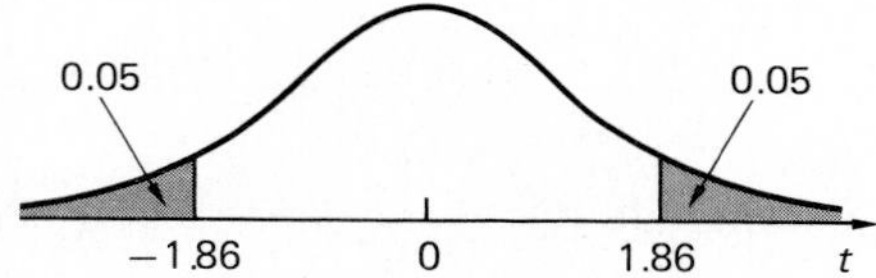

The confidence limits are

$$\overline{x} - \overline{y} \pm 1.86 s_{\overline{X}-\overline{Y}} = 22.8 - 40.8 \pm 1.86\,(9.97) = [-36.5, +.5].$$

Thus the difference in means, although large and in the hypothesized direction, has quite wide confidence limits. Notice that 0 is contained in the confidence interval, and so using this as a test of significance we would not reject the hypothesis of equality at the 10 percent level of significance, even though the direction of the difference encourages us to accept the truth of the hypothesis.

Comment. In experiments on another set of subjects, these same investigators got means of 39.2 and 41.3 for the clear and ambiguous groups, respectively, and so the large difference seen in this example was not replicated.

PROBLEMS FOR SECTION 10-5

1. Sample 1 has observations 6, 9, and 15, and sample 2 has observations 17, 18, 20, and 25. Use the two-sample t approach to set 95 percent confidence limits on the difference between the population means.

2. Continuation. Find the one-sided descriptive level of significance associated with the test that the difference in means (sample 2 minus sample 1) is positive.
3. In a sociological investigation the average satisfaction score for 10 discussion groups was 30, with a sum of squares $\Sigma\,(x - \overline{x})^2 = 100$, and that for 5 other groups using a different procedure was 50, with the sum of squares $\Sigma\,(y - \overline{y})^2 = 56$. Use the t test to assess the difference in means of satisfaction scores.
4. Continuation. Set 80 percent confidence limits on the true difference in means.
5. Treat the data in Problem 5 in Section 10-3 as if it consisted of two independent samples of 10 individuals, and test for significance of the difference in means.
6. Continuation. Set 90 percent confidence limits on the difference in means.
7. Continuation. Compare the result for Problem 5 with that for Problem 5 in Section 10-3.

10-6 TWO INDEPENDENT SAMPLES: UNKNOWN AND UNEQUAL VARIANCES

If both sample sizes are large when the standard deviations are unknown and are believed not to be equal, we estimate $\sigma_{\overline{X}}^2$ and $\sigma_{\overline{Y}}^2$ and substitute into $\sqrt{\sigma_{\overline{X}}^2 + \sigma_{\overline{Y}}^2}$ to estimate $\sigma_{\overline{X}-\overline{Y}}$. We then use normal tables to get the multiplier for setting confidence limits or for testing hypotheses. But when sample sizes are small, careful approximations are complicated, and so we shall suggest a simple rough-and-ready approximation.

If the sample sizes are small, we still estimate $\sigma_{\overline{X}-\overline{Y}}^2$ from

$$s_{\overline{X}-\overline{Y}}^{*2} = \frac{\Sigma\,(x - \overline{x})^2}{n_X\,(n_X - 1)} + \frac{\Sigma\,(y - \overline{y})^2}{n_Y\,(n_Y - 1)}$$

but take the smaller of $n_X - 1$ and $n_Y - 1$ as the degrees of freedom for the multiplier. When $n_X = n_Y$, $s_{\overline{X}-\overline{Y}}^2 = s_{\overline{X}-\overline{Y}}^{*2}$. But our usage will still be conservative for testing because here we use $n_X - 1$ as the degrees of freedom instead of $2(n_X - 1)$.

EXAMPLE *Rate of use of short words.* Mosteller and Wallace studied the distribution of word lengths in writings of Alexander Hamilton and James Madison. Table 10-6 shows the rates of use of one-letter words per 1000 words of text in 8 essays by Hamilton and 7 by Madison. Set 95 percent confidence limits on the difference of the rates. Might their rates be equal?

SOLUTION. From the t table, Table A-2 in Appendix III in the back of the book, for 6 degrees of freedom, the 95 percent multiplier is 2.45. The limits

TABLE 10-6

Rate of use of one-letter words per 1000 words of text

Hamilton papers	*Madison papers*
24	20
21	27
23	19
24	30
33	11
28	17
28	27
37	—
$\Sigma h = 218$	$\Sigma m = 151$

$$\bar{h} = 218/8 = 27.2, \ \bar{m} = 151/7 = 21.6$$

$$s^{*2}_{\bar{H}-\bar{M}} = \frac{\Sigma(h - \bar{h})^2}{n_H(n_H - 1)} + \frac{\Sigma(m - \bar{m})^2}{n_M(n_M - 1)}$$

$$= \frac{208}{8(7)} + \frac{272}{7(6)}$$

$$= 10.19$$

$$s^{*}_{\bar{H}-\bar{M}} = 3.19$$

Source: Adapted from F. Mosteller and D. L. Wallace (1964). *Inference and Disputed Authorship: The Federalist*, p. 258. Addison-Wesley, Reading, Mass.

are

$$\bar{h} - \bar{m} \pm 2.45 s^{*}_{\bar{H}-\bar{M}} = 27.2 - 21.6 \pm 2.45(3.19) = [-2.2, 13.4].$$

The upper and lower confidence limits for the rates fall on both sides of 0, and so equality of rates of use of one-letter words by the two authors is still a reasonable hypothesis at the 5 percent level of significance. Other choices of significance levels would give different confidence limits.

PROBLEMS FOR SECTION 10-6

1. Treat the marijuana data in Table 10-2 as if they were two independent samples of size 9 to test the difference in means.
2. At a local hospital a study was done to examine the effects of group therapy on the aged. The subjects were all elderly outpatients suffering from grief, usually caused by recent trauma (for example, death of a friend). Two kinds of therapy were tried: one that was specifically directed toward grieving and one that was not so directed. The differences in scores on tests designed to measure "grieving" before and after therapy were recorded. The data are in Table 10-7. Assume that the scores for each group are observations from an approximately normal distribution, and carry out a test to determine if the directed therapy gave higher differences than the undirected therapy.

TABLE 10-7
Effects of group therapy for the aged

Treatment 1 (directed)	*Treatment 2 (undirected)*
14.33	2.33
9.33	1.67
22.33	3.00
−6.33	1.67
12.00	4.33
19.00	0.33
12.33	1.33
10.33	2.67
0.67	1.67
	1.67

3. Twenty-seven subjects were randomly divided into two groups, one of size 18 and the other of size 9. The first group was given training in problem solving, with emphasis on general methods and procedure. The other group was given practice in the actual solving of problems. The two groups were then tested, with the following results (high means imply better performance):

	Methods Group	**Practice Group**
n	18	9
$\overline{x}$	19.0	27.4
s^2	14.9	6.0

Carry out a test of the hypothesis that the two groups are from populations with the same mean.

4. A clinical psychologist was trying to assess the effectiveness of a type of behavior therapy to cure bedwetting (enuresis) among 3-year-old children. She had a scale measuring the seriousness of the problem (a high score indicates a serious problem; a score of 7 or less is considered normal for this age group). She selected 12 children, all of whom had scores of 30 prior to treatment. Six were chosen at random and were given the new therapy; the other six were given traditional therapy. Scores for all patients after two months are shown in Table 10-8. Carry out a test to see if the new therapy was superior to the traditional one, assuming that the population variances for the two therapies are equal.

TABLE 10-8
Therapies to cure bedwetting

Traditional therapy (x)	*New therapy* (y)
5	0
10	3
15	6
18	8
20	9
22	10
$\Sigma x_i = 90$	$\Sigma y_i = 36$
$\Sigma x_i^2 = 1558$	$\Sigma y_i^2 = 290$

5. Recompute your test in Problem 4 for the case in which the population variances are unequal.

10-7 ACCEPTING OR REJECTING THE NULL HYPOTHESIS

Why do we reject the null hypothesis, or accept it? As has already been seen, we reject it when the probability of the occurrence of the observed event, or more extreme ones, is small. But that alone is not the reason. We reject it (a) because the data do not support it and (b) because we think that the alternative hypothesis is tenable and that the data do support the alternative. The medical researcher looking for a new medication knows that medications better than the standard ones are found from time to time (after all, the standard was once unknown), but not often. So he will be very cautious about replacing a standard. In most such instances the decisions are not final; new data can overthrow them.

What does it mean to accept the null hypothesis? If a difference between observed means has a z value of 0.8, we do not reject the null hypothesis, but neither do we believe that the difference in population means is exactly zero. Indeed, confidence limits let us tell how far apart we think these means might reasonably be. All we conclude in accepting the null hypothesis is that the data do not give strong evidence against equality of means. Clearly, we cannot expect means of two distinct populations to be alike to infinitely many decimals, and so we are sure from nondata considerations that the means are not exactly equal. Again, the issues are these: What are we to believe? What is the source of the evidence? We note that "accepting" and "rejecting" are labels we attach to various beliefs and that their use may be complicated.

10-8 SUMMARY OF CHAPTER 10

1. This chapter on measurements explains how to set confidence limits or test against a conjectured value in several cases, using results related to the normal distribution.
2. The use of the normal approximation for measurements is justified by the central limit theorem:

   The Central Limit Theorem. If a population of measurements has mean μ and variance σ^2, then in a random sample of n measurements drawn with replacement, as n grows large, the probability that the sample mean $\overline{X}$ is within a distance $z\sigma_{\overline{X}}$ of μ is given by $P(-z \leq Z \leq z)$ from the standard normal distribution, $z > 0$.
3. For the one-sample case, the unknown population variance (equivalently, a two-sample matched-pairs case, with unknown variance of the difference), we use the statistic

$$t = \frac{\overline{D}}{s_{\overline{D}}},$$

   where

$$s_{\overline{D}} = \sqrt{\frac{\Sigma(D - \overline{D})^2}{n(n-1)}}.$$

   to see if μ is far from zero. The quantity t is referred to a t table with $n - 1$ degrees of freedom.
4. For two independent samples with known population variances, 95 percent confidence limits for the difference in the population means, $\mu_X - \mu_Y$, are

$$\overline{x} - \overline{y} \pm 1.96\sigma_{\overline{X}-\overline{Y}},$$

   where

$$\sigma_{\overline{X}-\overline{Y}} = \sqrt{\frac{\sigma_X^2}{n_X} + \frac{\sigma_Y^2}{n_Y}}.$$

5. If in the two-sample case the variances are unknown, but equal, the 95 percent confidence limits for $\mu_X - \mu_Y$ are

$$\overline{x} - \overline{y} \pm t_{n_X+n_Y-2}\, s_{\overline{X}-\overline{Y}},$$

where

$$s_{\overline{X}-\overline{Y}} = \sqrt{\frac{\Sigma(x-\overline{x})^2 + \Sigma(y-\overline{y})^2}{n_X + n_Y - 2}\left(\frac{1}{n_X} + \frac{1}{n_Y}\right)}$$

6. For two independent samples with unknown and unequal variances, we use the conservative 95 percent limits

$$\overline{x} - \overline{y} \pm t^* \, s^*_{\overline{X}-\overline{Y}},$$

where

$$s^*_{\overline{X}-\overline{Y}} = \sqrt{\frac{\Sigma(x-\overline{x})^2}{n_X(n_X-1)} + \frac{\Sigma(y-\overline{y})^2}{n_Y(n_Y-1)}}$$

and t^* is taken from the t table, with the smaller of $n_X - 1$ and $n_Y - 1$ as the degrees of freedom.

7. We attach labels "accepting" and "rejecting" to our beliefs about parameters (such as a difference of means). The interpretation of these labels is complicated.

SUMMARY PROBLEMS FOR CHAPTER 10

1. In this chapter we have treated sample means and differences of sample means as if they were normally distributed. What justification have we for this?
2. Distinguish between $\overline{x}$ and μ_X.
3. Distinguish between s and σ.
4. The quantity that might be expected to be distributed approximately according to a standard normal distribution is

$$Z = \frac{\overline{X} - \mu_X}{\sigma_{\overline{X}}},$$

but we have generally computed $\overline{x}/\sigma_{\overline{X}}$. What practical feature of our questions makes this reasonable?

5. In this chapter, sometimes we have used a standard normal table, sometimes a t table. What determines which we choose?
6. Carry out the weighting of s^2_X and s^2_Y of Section 10-5 and thus show that the weighted sum is

$$\frac{\Sigma(x-\overline{x})^2 + \Sigma(y-\overline{y})^2}{n_X + n_Y - 2}.$$

7. Compare the variance of the difference in Section 10-4 with the estimated variance in Section 10-5. Specifically point out the similarities in the two formulas.
8. Why do we make tests of significance or set confidence limits?
9. Suppose that a null hypothesis is that $\mu = 0$. If we accept the null hypothesis after reviewing the data with a significance test, does that mean that we believe that the value of μ is zero? Why or why not?
10. Why do we reject null hypotheses?

REFERENCE

F. Mosteller, R. E. K. Rourke, and G. B. Thomas, Jr. (1970). *Probability with Statistical Applications*, second edition. Addison-Wesley, Reading, Mass.

Fitting Straight Lines Using Least Squares

Learning Objectives

1. Predicting the values of new observations using linear relations
2. Reviewing the aims of fitting straight lines to data
3. Choosing a criterion for fitting straight lines
4. Fitting straight lines to data using the method of least squares to minimize the sum of the squares of the residuals

11-1 PLOTTING DATA AND PREDICTING

Were predicting the outcomes or values of future events not such a difficult job, we would all be rich. To make prediction a reasonable statistical task, we need information on approximate relationships between the variable whose value we would like to predict and other variables whose values we can observe or control, such as the relationship between the breaking strength of pottery and its temperature at firing.

Today, high-speed computers can do most of the computational work for plotting data and fitting linear relations. In this chapter we give, without proof, formulas for fitting straight lines to data. By using such formulas, many people have written computer programs to help us compute. We need to learn the logic behind the formulas and then let the computer do the work of calculating and printing out the information we request.

In Chapter 3 we explored in an informal manner how to find and summarize approximate linear relationships. We now offer formal methods that are widely used in practice for fitting and examining linear relations. We turn to an example that uses these methods for prediction purposes.

EXAMPLE 1 *Winning speeds at the Indianapolis 500.* The most famous American automobile race, the Indianapolis 500, is held each Memorial Day at the Indianapolis Speedway. Owing to advancing technology, each year the racers tend to go faster than in previous years. We can assess this phenomenon by examining the speed of the winning drivers over a period of years. Table 11-1 gives the winning speeds in miles per hour for 1961 through 1970. The years have been coded by subtracting 1960 from each.

Because graphs are essential to good statistical analysis, first we plot winning speed against year in Fig. 11-1. By and large, the winning speeds do go up from one year to the next, giving an approximate linear relationship between y and x, speed and year. In Chapter 3 we used the black-thread method to fit straight lines. As we saw, one difficulty with the black-thread method is that placing the thread involves subjective judgments. Different people applying the method get different answers. Later in this chapter we learn about an objective method called least squares, which we used to determine the line drawn in Fig. 11-1. An equation of this fitted line is

$$y = 137.60 + 1.93x.$$

(Objectivity and repeatability are not always the first considerations; doing a good job is important too. For example, we could fit a straight line by passing a horizontal line through the highest observed point in a graph, but the fit, while objective and repeatable, would ordinarily be poor.)

TABLE 11-1

Winning speeds at the Indianapolis 500 auto race for 1961–1970

Year	x = *Year* − *1960*	y = *Winning speed*
1961	1	139.130
1962	2	140.293
1963	3	143.137
1964	4	147.350
1965	5	151.388
1966	6	144.317
1967	7	151.207
1968	8	152.882
1969	9	156.867
1970	10	155.749

Summary statistics: $n = 10$, $\Sigma x = 55$, $\Sigma x^2 = 385$, $\Sigma y = 1482.320$, $\Sigma y^2 = 220{,}086.699$, $\Sigma xy = 8312.167$.

Source: *The World Almanac and Book of Facts 1980*, p. 910. Newspaper Enterprise Association, New York.

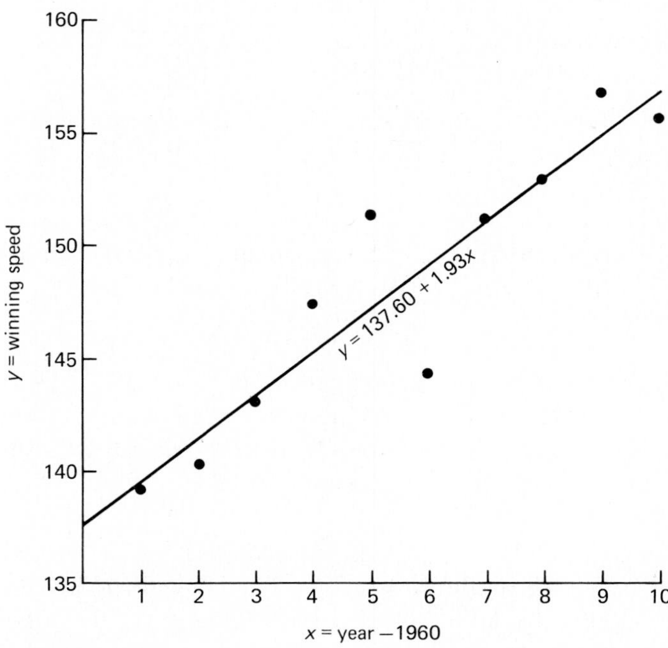

Figure 11-1 Graph of winning speeds in miles per hour against year for the Indianapolis 500 auto races, 1961–1970.

How well the fitted line describes the relationship between y and x depends on how closely the points follow the line. The vertical distances from the line measure the departure in this problem. To understand this approximate linear relationship, we should try to learn why the two points for 1965 and 1966 lie so far from the fitted line. For example, was there an accident in 1966 that slowed the race for a substantial period? It is more difficult to guess why 1965 should be high; news stories of the event might help us.

Suppose that we are satisfied with the notion of an approximate linear relationship between winning speed and year. We can then use this relationship to predict the winning speeds, say, for 1971, 1972, and 1973. We make our predictions by substituting the values of x for these three years into our linear equation:

$$1971: \quad y = 137.60 + (1.93)(11) = 158.83$$

$$1972: \quad y = 137.60 + (1.93)(12) = 160.76$$

$$1973: \quad y = 137.60 + (1.93)(13) = 162.69$$

We can even look back a ways, and predict, or should we say "postdict," what the winning speed should have been in 1950:

$$1950: \quad y = 137.60 + 1.93(-10) = 118.30$$

How good are these predictions? Although usually we cannot immediately compare predictions with actual outcomes, here we can:

Year	Predicted Speed	Actual Speed	Actual − Predicted
1971	158.83	157.74	−1.09
1972	160.76	163.47	2.71
1973	162.69	159.01	−3.68
1950	118.30	124.00	5.70

(Source: *The World Almanac and Book of Facts 1980*, p. 910.)

The difference Actual minus Predicted speed is −3.68 for 1973. Other differences are −1.09 for 1971 and 2.71 for 1972, and for the 10-year gap going backward in time, 5.70. This 1950 deviation is only a bit larger than the 4.14 for 1965.

To further examine our approximate linear relationship in Fig. 11-1, we can ask if the actual values for 1950, 1971, 1972, and 1973 lie farther away from the fitted line than the 10 original points. Two of the 10 points in Fig. 11-1 lie even farther from the line than do the actual points for 1972 and 1973, whereas 6 of the points lie farther away than the actual value for 1971. We can tentatively conclude that the predictions based on our fitted straight line are fairly good, because the variability of the actual values about the predictions is similar to the observed variability of the points about the line in Fig. 11-1. In the next chapter we shall make a more formal assessment of our ability to predict in this example.

At the same time that we can be pleased about the fit of our line, we know that the equation has basic flaws. For example, more than the Civil War was interfering with the Indianapolis 500 in 1861. Cars were not running backward then, as the equation predicts (they weren't running at all), but the result illustrates the danger of extrapolation from a rough-and-ready equation. This equation is designed for empirically estimating points near the 1961–1970 interval. Nothing in the equation provides knowledge of the invention of cars or of the creation of an annual race or of changes in rules. Nevertheless, linear approximations often work well over small intervals, the meaning of "small" depending on the problem.

PROBLEMS FOR SECTION 11-1

1. Find a news story covering the Indianapolis 500 for 1966 and see if any explanation is offered for the low winning speed.
2. Rewrite the fitted equation in Example 1 so that winning speed is a linear function of year rather than of the coded variable x (= year minus 1960).
3. Using the fitted line $y = 137.60 + 1.93x$ from Example 1, predict the winning speed in 1974. Compare the prediction with the actual winning speed, 158.59 miles per hour.
4. Use the equation in Problem 3 to estimate the speed in 1861.
5. Using the equation in Problem 3, for what year does the estimated speed equal zero? It must be difficult to go 500 miles at that pace!
6. Fit a straight line to the data in Table 11-1 using the black-thread method, and compare your equation with $y = 137.60 + 1.93x$.
7. *Project.* Use an almanac to find the winning speeds for each year 1971 through 1980. Plot these data on a graph, and fit a straight line to the data using the black-thread method.
8. Continuation. Compare your equations in Problems 6 and 7. Do they differ appreciably? If so, suggest a reason.

11-2 AIMS OF FITTING LINEAR RELATIONS

Before we describe formal methods for fitting straight lines to data, let us briefly review some reasons for using straight lines.

Aim 1: To Get a Formula Relating Two Variables. A formula is a compact way to describe the relationship between the variables, especially when the relationship is only approximate.

EXAMPLE 2 *Ohm's law.* Physics offers a variety of laws relating physical variables to one another, and many of these laws can be described in terms of straight lines. All electrical devices possess a property known as resistance. If we place such a device into a simple electrical circuit hooked up to batteries, we get what is called a voltage (electrical pressure) across the device. By adding or taking away batteries, we change the voltage. What is the relationship between the voltage and the flow of electrical charge, known as the current, passing through the electrical device? How does the property of resistance link up to the relationship between current and voltage?

Table 11-2 gives the result of an experiment with a particular electrical device. The device offers resistance when placed into an electrical circuit

TABLE 11-2
Relating electrical current to voltage

x *Voltage* *(volts)*	y *Current* *(amperes)*
2	.0015
4	.0030
6	.0048
8	.0069
10	.0081
12	.0100

Summary statistics: $n = 5$, $\Sigma x = 42$, $\Sigma y = .0343$, $\Sigma x^2 = 364$, $\Sigma y^2 = .000247510$, $\Sigma xy = .3000$.

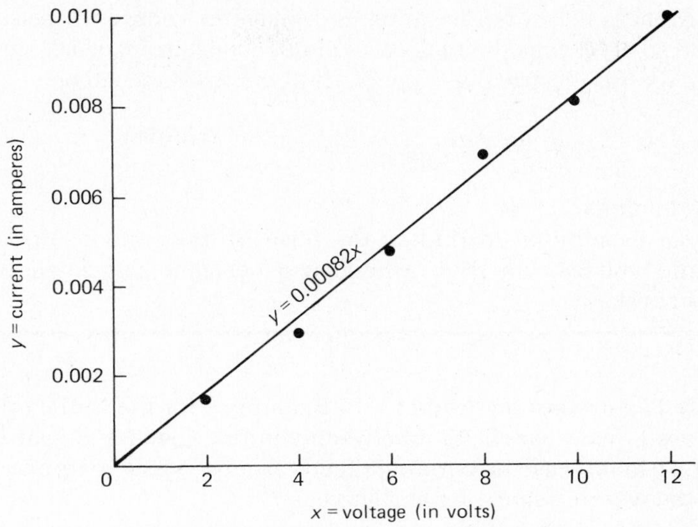

Figure 11-2 Relating current to voltage across an electrical device.

with batteries hooked up, and the flow of current is measured. The voltage was changed by adding or taking away batteries, and each time the current was measured. The resistance was not changed. What is the relationship between the voltage and the flow of current through the device?

We begin by graphing the six points in Fig. 11-2. The relationship between current, y, and voltage, x, looks very much like a straight line. If the voltage is zero, corresponding to no battery, the current is also zero, and so the line passes through the origin. We used the black-thread method to draw the line. Its equation is

$$y = .00082x.$$

We note that the form of this straight line is

$$\text{current} = \text{constant} \times \text{voltage}.$$

For this circuit and electrical device, we have discovered what physicists call Ohm's law. They write the law as

$$\text{current} = \frac{\text{voltage}}{\text{resistance}},$$

where current is measured in amperes, voltage in volts, and resistance in ohms. And so, if we take the reciprocal of our "constant" 0.00082, we get the resistance for this electrical device. We find the resistance to be

$$\frac{1}{.00082} = 1220,$$

measured in ohms.

And so, in addition to finding the form of the relationship between current and voltage, we have estimated a constant associated with the circuit, its resistance.

Aim 2: To Forecast or Predict. In Example 1 for the Indianapolis 500 data, we used the linear relation between winning speed and year for 1961 through 1970 to forecast the winning speed for subsequent years and even to retrocast the winning speed for earlier years.

Aim 3: To Assess Predictions. To determine how well one variable serves as a predictor of another, we fit a straight line to a set of observations and then we see how close the observed points lie to the line. Both the Indianapolis 500 and the Ohm's law examples illustrate this idea, but let us add another.

EXAMPLE 3 *Minnesota snowfalls.* In Minneapolis–St. Paul, Minnesota, a television weatherman in 1974 predicted a small amount of snow for the remainder of the winter, based on the amount of snow that fell in November. He said that the greater the November snowfall, the more snow could be expected for the rest of the winter. In Fig. 11-3 we give a scatterplot of snowfall data from 1920–1921 to 1973–1974, where x = November snowfall (in inches) and y = snowfall for the remainder of the year (also in inches). The graph shows no evidence of a rising relationship between November snowfall and that for the remainder, and so we conclude that the weatherman's prediction is not well founded in the data.

Aim 4: To Check on a Theory. Many theories can be described in terms of straight-line relations, and to verify a theory we can check on the sign or even the value of the slope or y intercept. In a way, the preceding example checked the weatherman's theory, even though we rejected the theory. His statement implied that the slope was positive, but we found no evidence for this.

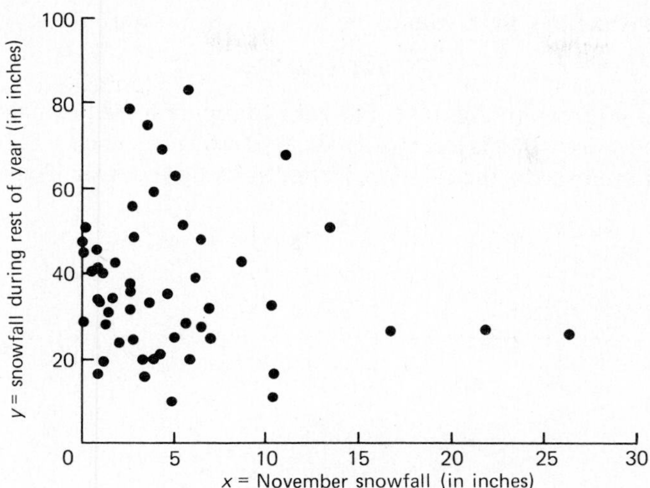

Figure 11-3 Snowfall data for the Twin Cities, Minnesota, 1920–1974.
Source: Constructed from data in B. F. Watson (1974). *Minnesota and Environs Weather Almanac 1975.* Freshwater Biological Research Foundation, Navarre, Minn.

EXAMPLE 4 *Metabolic rate and body weight.* Food nutrition experts have claimed that the metabolic rate of an animal is proportional to its weight raised to the ¾ power, that is, if R = metabolic rate and W = weight, then

$$R = C \times W^{0.75},$$

where C is a constant. This equation looks formidable, but we can simplify it considerably by using logarithms (see the discussion in Chapter 2). First, we recall that the logarithm of a product of several quantities is the sum of their logarithms. Then, taking logarithms of both sides of the equation, we get

$$\log R = \log C + 0.75 \log W.$$

If we let

$$r = \log R, \quad c = \log C, \quad \text{and} \quad w = \log W,$$

then our new equation can be written in an easier-to-read form as

$$r = c + 0.75w.$$

To check this theory, a sample of 100 Wyoming University women was taken, their weights and metabolic rates were measured, and the following straight line was fitted to the data:

$$r = 2.548 + 0.334w.$$

These data are shown in Fig. 11-4. The results appear to be inconsistent with the theory because 0.334 is not close to 0.75. If 0.334 is a good estimate, then the ⅓ power seems to be the relation, rather than the ¾ power. To pursue this

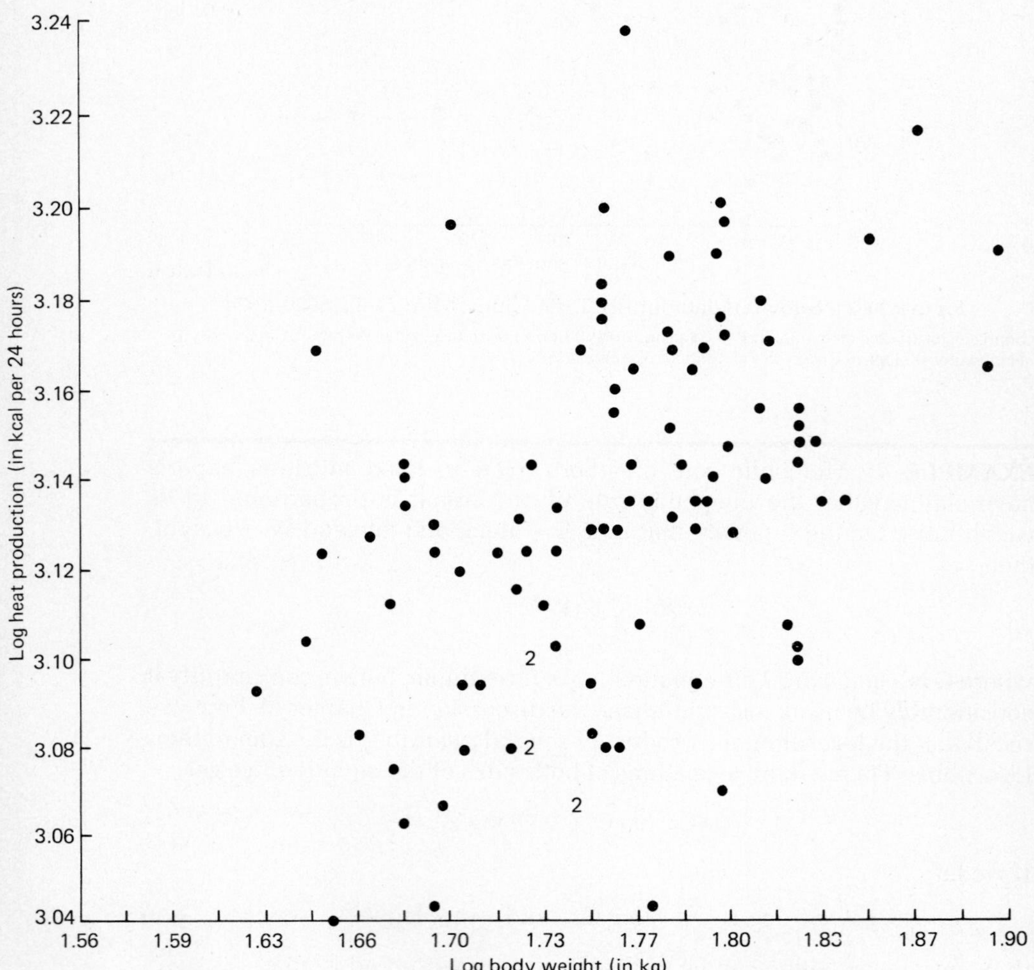

Figure 11-4 Scatterplot of log of metabolic rate versus log of weight for 100 Wyoming University women.

Source: Based on data reported in E. J. McKittrick (1936). Basal metabolism of Wyoming University women, *Journal of Nutrition*, Volume II, table on p. 319.

further, we need to know more about the reliability of the estimate of the slope.

Aim 5: To Evaluate a Constant. A theory often suggests that we fit a particular curve to a set of data in order to assess the value of some unknown constant that is part of the theory. In Example 4, we assessed the value of the exponent.

Aim 6: To Take Out Effects. Suppose that a variable such as the strength of a metal wire is known to depend on at least one variable such as the diameter of the wire, but may also be related to several other variables. We could fit a straight line to the strength–diameter data, take residuals (the observed strength value minus that predicted by the line of strength versus diameter), and then plot these residuals against the other variables. This sort of procedure takes out the linear effect of the diameter of the wire, so that it does not disguise the relationship between the strength of the wire and the other variables.

WHY STRAIGHT LINES?

Most of the aims we have just described apply not only to straight lines but also to curves in general. Some reasons for fitting straight lines as opposed to other curves are the following:

Simplest Function. Of all the functions relating two variables, the straight line is the easiest to describe, understand, and report. For convenience in interpolation and prediction, the straight line has no equal.

Actual Linearity. Many relationships are linear or are at least very nearly so.

Transformations to Linearity. Even when approximate relationships are not linear, we often can get a linear relationship by transforming one or both variables as we did in Chapter 3. In Example 4 we transformed a product relationship into a linear one by taking logarithms of both variables.

Limited Range. Even though a functional relation may not be linear, if we restrict ourselves to a limited range of the variables, we often find that straight lines give good approximations, as in the Indianapolis 500 example. For a mathematical example, in Fig. 11-5 we have plotted the quadratic curve $y = x + x^2$, for x between 0 and 4. Yet if we restrict our attention to the values of x between 1.5 and 2.5, say, the curve is nearly straight. A close

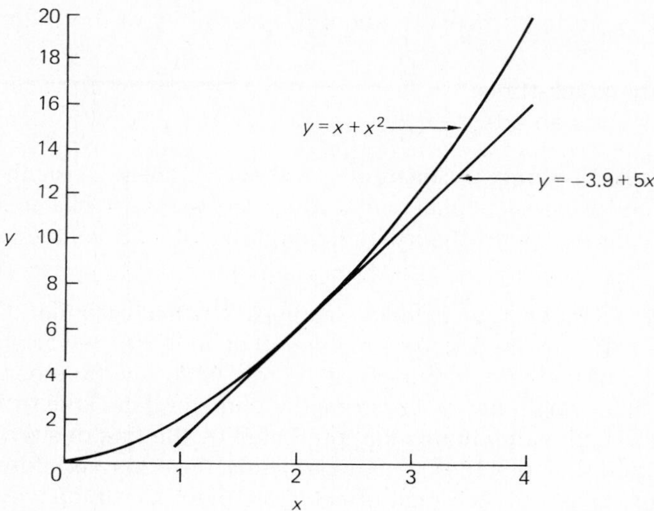

Figure 11-5 Plot of $y = x + x^2$ for x between 0 and 4, and a linear approximation for the range of x between 1.5 and 2.5.

approximation is given in that interval by

$$y = -3.9 + 5x.$$

Statistical Theory. For a variety of theoretical and computational reasons it is easier to fit straight lines to data than to fit more complicated curvilinear functions.

None of these reasons is meant to exclude fitting curves other than straight lines, but the reader may well want these explanations of why the straight line is so often used.

PROBLEMS FOR SECTION 11-2

1. List five reasons for fitting straight lines to data.

2. Explain why the fitted straight line in Example 4 might not provide a good fit for very large values of *w*.

3. Write down an equation that is used to describe a well-known theory in a subject you have studied. Is the equation a linear one? If not, can you use transformations to make it linear? How?

4. Make a graph of the equation $y = x^2 - 3x + 4$ for values of x between 3 and 7, and approximate the resulting curve with a straight line.

11-3 CRITERIA FOR FITTING STRAIGHT LINES

Although the black-thread method of fitting straight lines to a set of (x, y) points often does an adequate job, we need a method that is both *reliable* and *objective*. For the Indianapolis 500 data in Example 1, the black-thread method would have yielded a result much like the line plotted in Fig. 11-1. Nevertheless, when we have a swarm of points as in Fig. 11-4 relating metabolic rate and weight, we may have trouble placing the line by eye. We also want a method that is repeatable, so that two different people will fit exactly the same line to the same set of points. If we do not have these properties, investigators' lines may be biased to favor their own preferences. We all need a little protection from our own biases.

We have a variety of objective methods or criteria for fitting straight lines, and we must make a choice among them. But before we discuss these methods we must emphasize the importance of looking at our data before we try to fit a straight line. Before we do anything else, we should plot the y values against the x values to see how the variables relate. We often have a computer make such plots for us, as was done for Figs. 11-3 and 11-4. Some of the things we might see are the following:

1. The points all lie quite close to a straight line.
2. The points all lie quite close to a smooth curve, but not a straight line.
3. The points appear to be scattered all over the place, with no regular pattern relating y values to x values.
4. Most of the points lie quite close to a straight line (or some other type of curve), but a few are far away.

Situation 1 is ideal, and often we can handle situation 2 by the use of transformations such as logarithms or square roots. Situation 4 presents the most trouble, and we must decide if and how we should let the extreme points influence the line we fit. This situation is especially troublesome if we are relying on the black-thread method or some other subjective method. We should try to check to see if the extreme departures represent copying errors or are faulty in some other way.

EXAMPLE 5 *Heights and weights.* The heights and weights of 15 high school sophomore boys are given in Table 11-3. The plot of height against weight, given in Fig. 11-6 (or even by a detailed look at the values themselves), reveals that height and weight are roughly linearly related, but that one point (subject 7) lies far from the line. It appears that the height and weight for subject 7 have been interchanged (few boys are over 14 feet tall).

TABLE 11-3

Heights and weights of 15 high school sophomore boys

Subject	*Weight (pounds)*	*Height (inches)*
1	135	68
2	105	62
3	189	74
4	128	68
5	142	68
6	150	70
7	74	173
8	110	62
9	130	66
10	148	71
11	98	60
12	183	72
13	155	67
14	115	63
15	132	68

Choosing a Criterion. To focus our discussion for the remainder of this section, we consider fitting a straight line to the following five points:

x:	1	2	3	4	5
y:	29	28	36	45	46

It might be reasonable to require our fitted line to pass through the point that represents the average value of x and the average value of y. We use the bar notation for averages, so that $\bar{x} = 3$, $\bar{y} = 36.8$, where $\bar{x}$ (read: x-bar) is the average of the x values and $\bar{y}$ (read: y-bar) is the average of the y values. Then the point representing these averages is $(\bar{x}, \bar{y}) = (3, 36.8)$. Infinitely many lines pass through (3, 36.8), so that we need some additional way to tie down our fitted line. The five original points and the point $(\bar{x}, \bar{y}) = (3, 36.8)$ are plotted in Fig. 11-7.

Because we are trying to predict y from x, we usually measure the closeness of a point to a fitted line by the vertical deviation or residual; that

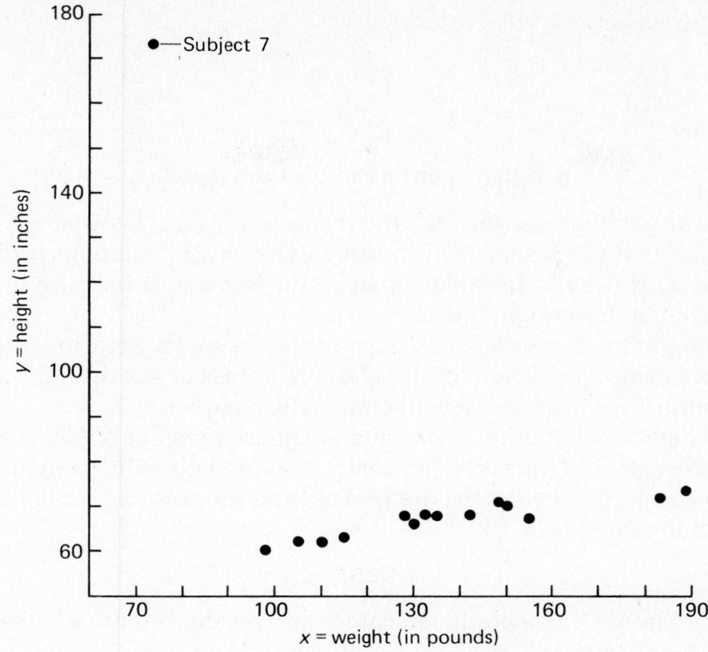

Figure 11-6 Heights and weights of high school boys.

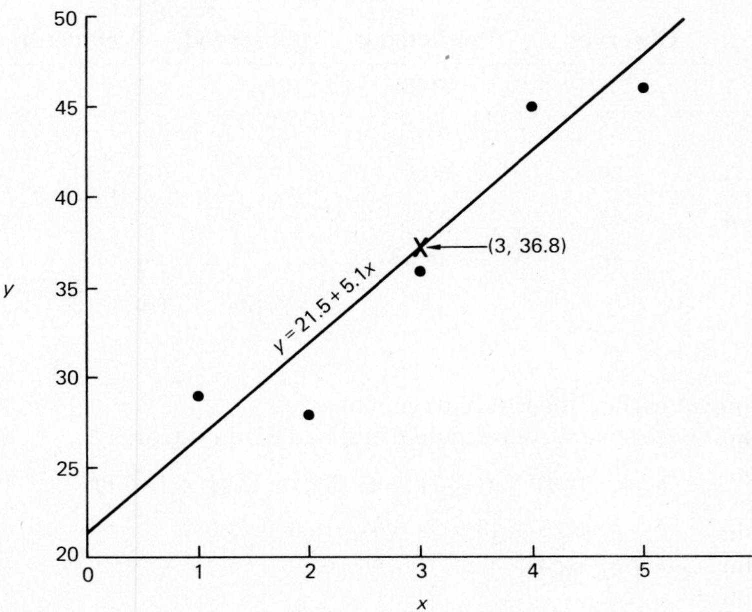

Figure 11-7 Least-squares line for five-point example. (Based on data in Table 11-4).

is given by

$$\text{residual} = \text{observed } y - \text{predicted } y.$$

For any line that passes through the mean point, $(\bar{x}, \bar{y})$, it is a fact that the sum of the residuals is zero. (This result will be illustrated numerically later.) Thus we must use the residuals in some further way if they are to help us choose among the straight lines.

We might try to minimize the sum of the residuals, ignoring their signs; that is, we compute the sum of the absolute values of the residuals. To find this minimum turns out to be mathematically complex.

We might try to minimize the sum of squared residuals. Not only is this task mathematically tractable, but also it is computationally straightforward. This approach is known as the **method of least squares** and for our example results in the fitted line

$$y = 21.5 + 5.1x$$

plotted in Fig. 11-7. The predicted values and residuals for this least-squares line are as follows:

x	Observed y	Predicted y	Residual (Observed − Predicted)
1	29	26.6	2.4
2	28	31.7	−3.7
3	36	36.8	−0.8
4	45	41.9	3.1
5	46	47.0	−1.0
		Total	0.0

As we noted earlier, the residuals sum to zero.

Finally, we square each residual and add up the squares:

$$\begin{aligned} \text{sum} &= (2.4)^2 + (-3.7)^2 + (-0.8)^2 + (3.1)^2 + (-1.0)^2 \\ &= 5.76 + 13.69 + 0.64 + 9.61 + 1.00 \\ &= 30.70. \end{aligned}$$

This value of 30.7 is the minimum possible value for the sum of squared residuals. Any line other than the least-squares line gives a larger sum of squared residuals.

PROBLEMS FOR SECTION 11-3

Let us reassure ourselves that the least-squares line

$$y = 21.5 + 5.1x$$

really does minimize the sum of squared residuals for our five-point example. We do not give the mathematical proof that the least-squares line gives the minimum value. Rather, we explore part of the problem numerically to get a better feel for what is going on.

First, we vary the slope of the line, but we use the same y intercept as for the least-squares line, that is, 21.5. Here are the sums of squared residuals that go with four different values for the slope:

Slope	Sum of Squared Residuals
4.0	97.25
4.5	50.50
5.5	39.50
6.0	75.25

1. Compute the sums of squared residuals for the lines with slopes 5.0 and 5.2.
2. Continuation. Which of the seven lines having different slopes (4.0, 4.5, 5.5, 6.0, 5.0, 5.2, and 5.1) but having the same y intercept 21.5 has the smallest sum of squared residuals?
3. Continuation. For these seven lines, make a graph of the sums of squared residuals versus slopes.

Next we vary the y intercept of the line, but we use the least-squares slope value, 5.1. Here are the sums of squared residuals that go with four different y intercepts:

y Intercept	Sum of Squared Residuals
19	61.95
20	41.95
22	31.95
23	41.95

4. Compute the sum of squared residuals for the line with y intercept 21 and slope 5.1.
5. Continuation. For the six lines with slope 5.1 and intercepts 19, 20, 21, 22, 23, and 21.5, make a graph of the sums of squared residuals versus y intercept.

6. Continuation. Which of the six lines has the smallest sum of squared residuals?
7. Continuation. Explain why the calculations in Problems 1 through 6 do not prove that least squares give the minimum value of the sum of squared deviations.
8. Find some other line that passes through $(\bar{x}, \bar{y}) = (3, 36.8)$, and compute the sum of the residuals and the sum of their squares for its fit to the five points. Does the sum of the residuals equal zero? Is the sum of squares larger than 30.7?
9. Why is the method of least squares for fitting straight lines preferable to the black-thread method?

11-4 LEAST SQUARES

The method of least squares has much to recommend it. First, it is part of a general method that can be used to fit more complicated linear and curvilinear relationships. Second, the least-squares line possesses many nice statistical properties. Third, it results in relatively simple formulas for the slope and intercept, so that we need not carry out the minimization by trial and error or by some more sophisticated computational method.

Although the method of least squares does have strengths, it also has weaknesses. For example, a least-squares line is highly sensitive to outliers, that is, residuals that are much larger in size than all the others. This is one reason why we always make a graph of our points first: to look for wild values that are bound to influence our answers.

Next, we present the general situation in which we apply the least-squares formulas. Suppose that we have a set of n pairs of (x, y) values and that we wish to fit a linear relationship of the form

$$y = b + mx$$

to the data. Using subscripts to distinguish the n points, we have

$$P_1 = (x_1, y_1), P_2 = (x_2, y_2), \ldots, P_n = (x_n, y_n).$$

The subscripts number the points arbitrarily. For example, we might have $n = 3$ points, where $P_1 = (2, 3)$, $P_2 = (3, 4)$, $P_3 = (4, 1)$. For specified values of b and m and a particular x value, say x_i, the predicted value of y_i is $b + mx_i$, and the sum of squared residuals is

$$S = \Sigma(y_i - b - mx_i)^2. \tag{1}$$

The method of least squares chooses b and m to minimize the sum of squares (1). We label those minimizing values as $\hat{m}$ and $\hat{b}$ (read: m-hat and b-hat):

SLOPE:

$$\hat{m} = \frac{\Sigma(x_i - \overline{x})(y_i - \overline{y})}{\Sigma(x_i - \overline{x})^2} \tag{2}$$

and

INTERCEPT:

$$\hat{b} = \overline{y} - \hat{m}\overline{x} \tag{3}$$

Here $\overline{x}$ is the mean of the x values, that is,

$$\overline{x} = \frac{1}{n}\Sigma x_i\,,$$

and $\overline{y}$ is the mean of the y values, that is,

$$\overline{y} = \frac{1}{n}\Sigma y_i\,.$$

We use the hats on *m* and *b* in formulas (2) and (3) to remind ourselves that these are the **minimizing values** of *m* and *b* for the method of least squares. Unless there is some possibility of ambiguity, we do not normally include the limits for the summation. Thus in formula (2) it is understood that the summations in both the numerator and denominator run from 1 to *n*.

We present these formulas for the least-squares line without proof. For a proof that expressions (2) and (3) minimize the sum of squared residuals in expression (1), the reader may consult Chapter 11 of *Probability with Statistical Applications* (Mosteller et al., 1970). The proof there is based only on facts about quadratics.

An alternative but equivalent expression for the least-squares slope $\hat{m}$ is useful for computation:

$$\hat{m} = \frac{\Sigma x_i y_i - n\overline{x}\,\overline{y}}{\Sigma x_i^2 - n\overline{x}^2} \tag{4}$$

If we have a pocket calculator available for computing sums, sums of squares, and sums of products, expression (4) will be especially helpful. And if we have to work by hand, it may also help.

Because the formula for $\hat{b}$ in (3) involves $\bar{x}$, $\bar{y}$, and $\hat{m}$, we can write the least-squares line as follows:

$$y = \bar{y} + \hat{m}(x - \bar{x}) \quad (5)$$

It is now easy to see from (5) that this line does pass through $(\bar{x}, \bar{y})$, as we claimed earlier. If we substitute $\bar{y}$ for y and $\bar{x}$ for x, the equation is satisfied, which means that the point $(\bar{x}, \bar{y})$ falls on the line. For the ith data point, $P_i = (x_i, y_i)$, the line predicts a y value

$$\hat{y}_i = \bar{y} + \hat{m}(x_i - \bar{x}),$$

and the residual for the observation (x_i, y_i) is

$$\begin{aligned} i\text{th residual} &= y_i - \hat{y}_i \\ &= (y_i - \bar{y}) - \hat{m}(x_i - \bar{x}). \end{aligned} \quad (6)$$

To get the minimum sum of squared residuals, we can either plug the values of $\hat{m}$ and $\hat{b}$ from (2) and (3) into expression (1) or simply sum the squared values of the least-squares residuals in (6). In either approach, algebraic manipulation yields as the minimum sum of squares

$$\hat{S} = \Sigma(y_i - \bar{y})^2 - \frac{[\Sigma(y_i - \bar{y})(x_i - \bar{x})]^2}{\Sigma(x_i - \bar{x})^2},$$

or

$$\hat{S} = \Sigma(y_i - \bar{y})^2 - \hat{m}^2\Sigma(x_i - \bar{x})^2. \quad (7)$$

A convenient computational form for expression (7) is the following:

$$\hat{S} = \Sigma y_i^2 - n\bar{y}^2 - \hat{m}^2(\Sigma x_i^2 - n\bar{x}^2) \quad (8)$$

We use this formula for the minimum sum of squares to measure variability, and we practice its computation in the following examples. The primary uses of $\hat{S}$ come in Chapter 12.

TABLE 11-4
Least-squares calculations for Example 5

Point	*(a)* x	*(b)* y	*(c)* x^2	*(d)* y^2	*(e)* xy
P_1	1	29	1	841	29
P_2	2	28	4	784	56
P_3	3	36	9	1296	108
P_4	4	45	16	2025	180
P_5	5	46	25	2116	230
Totals	15	184	55	7062	603

We start by reviewing the least-squares computations for our five-point example from the last section. Table 11-4 reproduces in columns (a) and (b) the x and y values for the five points. Then column (c) gives the squares of the x values, and at the bottom of the column we find the sum of squares:

$$\Sigma x_i^2 = 55.$$

In columns (d) and (e) we have the squares of y values and the products of the x values and y values. The totals for these columns are

$$\Sigma y_i^2 = 7062 \quad \text{and} \quad \Sigma x_i y_i = 603.$$

Finally, we compute $\overline{x}$ and $\overline{y}$ by dividing the totals for columns (a) and (b) by 5:

$$\overline{x} = \frac{15}{5} = 3 \quad \text{and} \quad \overline{y} = \frac{184}{5} = 36.8.$$

Now we are all set to use the least-squares formulas. From expression (4) we have the slope

$$\hat{m} = \frac{\Sigma x_i\, y_i - n\overline{x}\overline{y}}{\Sigma x_i^2 - n^2} = \frac{603 - (5)(3)(36.8)}{55 - (5)(3^2)}$$

$$= \frac{603 - 552}{55 - 45} = \frac{51}{10}$$

$$= 5.1,$$

and from expression (3) the y intercept

$$\hat{b} = \overline{y} - \hat{m}\overline{x} = 36.8 - (5.1)(3) = 36.8 - 15.3$$

$$= 21.5.$$

Thus we have checked that the fitted line is

$$y = 21.5 + 5.1x.$$

Next we compute, using formula (8), the minimum sum of squared residuals as

$$\begin{aligned}\hat{S} &= \Sigma y_i^2 - n\bar{y}^2 - \hat{m}^2(\Sigma x_i^2 - n\bar{x}^2) \\ &= 7062 - 5(36.8)^2 - (5.1)^2(55 - 45) \\ &= 7062 - 6771.2 - 260.1 \\ &= 30.70.\end{aligned}$$

We computed $\hat{S}$ directly in the last section by squaring the residuals:

$$\begin{aligned}\hat{S} &= (2.4)^2 + (-3.7)^2 + (-0.8)^2 + (3.1)^2 + (-1.0)^2 \\ &= 5.76 + 13.69 + 0.64 + 9.61 + 1.00 \\ &= 30.70.\end{aligned}$$

This calculation provides a check on the computational form.

EXAMPLE 6 *Least squares for the Indianapolis 500 problem.* In Example 1 we considered fitting a straight line to predict the winning speeds of the Indianapolis 500 auto race. Using the data in Table 11-1, find the least-squares line and the minimum sum of squared residuals.

SOLUTION. Using the data from Table 11-1, we have

$$\begin{aligned}&n = 10, \\ &\Sigma x_i = 55, \quad \Sigma y_i = 1482.320, \\ &\Sigma x_i^2 = 385, \quad \Sigma y_i^2 = 220086.699 \quad \text{(to 3 decimals)}, \\ &\qquad \Sigma x_i y_i = 8312.167.\end{aligned}$$

The least-squares coefficients are then

$$\hat{m} = \frac{\Sigma x_i y_i - n\bar{x}\bar{y}}{\Sigma x_i^2 - n\bar{x}^2} = \frac{8312.167 - 10(55/10)(1482.320/10)}{385 - 10(55/10)^2} = 1.932,$$

$$\hat{b} = \bar{y} - \hat{m}\bar{x} = (1482.320/10) - 1.932(55/10) = 137.606.$$

Thus the fitted line is

$$y = 137.606 + 1.932x,$$

and the minimum sum of squared residuals is, by formula (8),

$$\hat{S} = \Sigma y_i^2 - n\bar{y}^2 - \hat{m}^2\,(\Sigma x_i^2 - n\bar{x}^2)$$

$$= 220086.699 - 219727.258 - (1.932)^2(385 - 302.5)$$

$$= 51.500.$$

We started our calculations in the example using speeds reported to three decimal places. This is the way the speeds are reported in *The World Almanac*. We often keep more decimals in our calculations than we appear to need in our final answer. We do this so that our answers are not made inaccurate by rounding errors. It may at first blush seem bizarre to keep three decimal places in a number as large as 220086.699, but when we subtract off 219727.258, and then 307.942, the result is 51.499. Then the three decimal places do not look quite so excessive, but one or two decimal places might easily do here. If we do our calculations on a pocket calculator, it requires little effort to carry an extra decimal or two just in case we need them later (this is the spare-tire principle). If our calculations are done using a standard least-squares computer program, we need not worry too much about how many decimal places to use for computations.

It is not easy to give a rule for the number of decimal places to use in calculations. The best advice is to keep as many decimal places as possible and round the answer at the end. We might still run into trouble following this advice, but by and large it should stand us in good stead.

EXAMPLE 7 *Submarine sinkings.* In Chapter 3 (Example 10) we looked informally at the results of a historical study of the actual and reported numbers of German submarines sunk each month by the U.S. Navy in World War II. The original data are given in Table 3-3. Find the least-squares line for predicting the actual number of sinkings from the navy's reported numbers.

SOLUTION. Using the data in Table 3-3, we have

$$n = 16,$$

$$\Sigma x_i = 123, \qquad \Sigma y_i = 140,$$

$$\bar{x} = 7.688, \qquad \bar{y} = 8.750,$$

$$\Sigma x_i^2 = 1287, \qquad \Sigma y_i^2 = 1682,$$

$$\Sigma x_i\, y_i = 1431.$$

The data points are plotted in Fig. 11-8. The least-squares coefficients are then

$$\hat{m} = \frac{\Sigma x_i y_i - n\overline{x}\overline{y}}{\Sigma x_i^2 - n\overline{x}^2} = \frac{1431 - 16(7.688)(8.750)}{1287 - 16(7.688)^2} = 1.039,$$

$$\hat{b} = \overline{y} - \hat{m}\overline{x} = 8.750 - 1.039(7.688) = .762,$$

and the fitted line (rounding the coefficients to two decimal places) is

$$y = .76 + 1.04x.$$

Note that we needed to keep the third decimal place through all the calculations in order to make sure our final answer is correct to two decimal places. Because every predicted value slightly exceeds the navy's reported numbers, this equation implies that the navy underestimated the actual number of sinkings, but not by very much. We have included the least-squares line in Fig. 11-8.

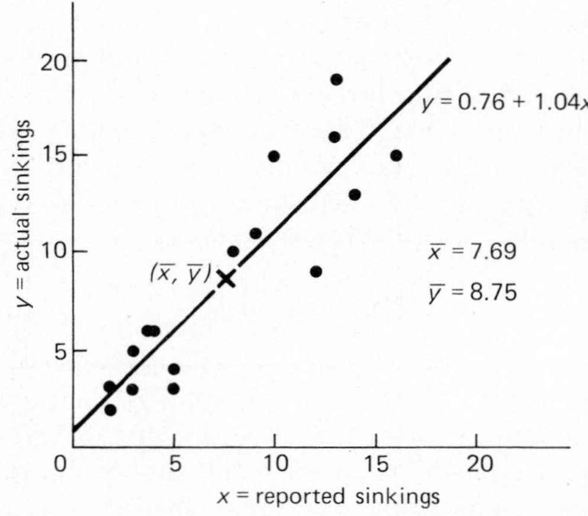

Figure 11-8 Actual and reported sinkings of German submarines in World War II.

The minimum sum of squared deviations is, by formula (8),

$$\begin{aligned}\hat{S} &= \Sigma y_i^2 - n\overline{y}^2 - \hat{m}^2(\Sigma x_i^2 - n\overline{x}^2) \\ &= 1682 - 16(8.75)^2 - (1.039)^2[1287 - 16(7.688)^2] \\ &= 457 - (1.039)^2(341.314) = 88.544.\end{aligned}$$

Remark. In Example 10 of Chapter 3, we fit a different straight line to these data, one that goes through $(\overline{x}, \overline{y})$ and the origin, (0, 0):

$$y = 1.14x.$$

Because we multiply the x value by a number slightly larger than 1, we consistently increase the reported values, and the fitted line again implies that the navy slightly underestimated the actual number of sinkings. This line does almost as good a job of minimizing the sum of squared residuals as does the least-squares line:

$$\Sigma(y_i - 1.14x_i)^2 = 91.905.$$

We might ask if the difference between the two fitted lines is important enough statistically for us to worry about. To answer this question, we require the formal model described in the next chapter.

PROBLEMS FOR SECTION 11-4

1. Given the data

y:	1	2	10	11
x:	−2	−1	+1	+2

find the least-squares line for predicting y from x. Be sure to plot the points on a graph of y versus x.

2. Continuation. Compute the residuals for the least-squares line in Problem 1, and sum their squares.

3. Continuation. Compare your answer in Problem 2 with the minimum sum of squared residuals you get using computational formula (8).

A calculator or computer will be of considerable help in working the following set of problems.

An economist studied the demand for feed corn in the United States following World War II as it related to the prices and quantities of livestock, hogs, poultry, and eggs being produced. Some of his data are given in Table 11-5. First he calculated the aggregate value in billions of dollars of the livestock, hogs, poultry, and eggs produced in a year. He wanted to relate demand to aggregate value in order to study the implications of price changes. Let the aggregate value be variable x and the demand be variable y. The data in Table 11-5 are used in Problems 4 through 7. The following summary may be useful.

$$n = 27,$$

$$\Sigma x_i = 496.94, \quad \Sigma y_i = 4611.88, \quad \bar{x} = 18.4052, \quad \bar{y} = 170.8104,$$

$$\Sigma x_i^2 = 9288.9440, \quad \Sigma y_i^2 = 818462.9624, \quad \Sigma x_i y_i = 86728.0601.$$

TABLE 11-5
Feed corn demand in the U.S., 1948–49 to 1974–75

Crop year (beginning Oct. 1)	x *Aggregate value (billions of dollars)*	y *Demand (billions of lbs)*
1948–49	14.42	126.28
1949–50	14.73	142.52
1950–51	14.68	138.99
1951–52	14.97	143.08
1952–53	15.77	129.53
1953–54	16.23	133.67
1954–55	16.77	125.55
1955–56	17.44	132.50
1956–57	17.35	133.11
1957–58	16.96	141.96
1958–59	17.26	155.79
1959–60	17.64	170.41
1960–61	18.08	173.15
1961–62	18.25	179.87
1962–63	18.66	176.79
1963–64	19.47	168.50
1964–65	19.45	165.54
1965–66	19.65	188.22
1966–67	20.28	186.42
1967–68	20.55	196.39
1968–69	20.63	200.48
1969–70	21.03	212.52
1970–71	21.60	200.54
1971–72	21.67	222.77
1972–73	20.87	241.36
1973–74	21.23	234.98
1974–75	21.30	190.96

Source: A. W. Womack (1975). "Domestic Demand for U.S. Feed Grains—An Econometric Analysis," Ph.D. dissertation, Department of Agricultural and Applied Economics, University of Minnesota.

4. Using the data in Table 11-5, make a graph of demand versus aggregate value (y versus x).

5. Find the least-squares line for demand as a linear function of aggregate value.

6. Continuation. Plot the least-squares line on your graph from Problem 4. How well does the least-squares line in Problem 5 seem to fit the data?

7. Compute the minimum sum of squared residuals for the least-squares line from Problem 5.

8. Find the least-squares line for winning times for the Olympic 1500-meter run as a linear function of year using the data in Table 3-6.

9. Continuation. Compute the minimum sum of squared residuals for the least-squares line from Problem 8, and plot the residuals versus year.

11-5 SUMMARY OF CHAPTER 11

Formula numbers correspond to those used earlier in the chapter.

1. We fit straight lines to data for a variety of reasons:

a) to get a formula relating two variables,
b) to forecast or predict,
c) to assess predictions,
d) to check a theory,
e) to evaluate a constant,
f) to take out effects.

2. Straight lines are the simplest of functions relating variables, and they are the easiest to report. The relationships between many different pairs of variables can be described approximately by the use of straight lines. The formula for a straight line is used to summarize the information in data sets, just as a mean might be used for data involving only one variable.

3. Although we have many different ways to fit a regression line to a set of (x, y) points, we often use the method of least squares. This method minimizes the sum of squared residuals; that is, it minimizes the sum of squares of the differences between the observed y values and the predicted y values.

4. The least-squares regression line has a simple form:

$$y = \hat{b} + \hat{m}x,$$

where

$$\hat{m} = \frac{\Sigma x_i y_i - n\bar{x}\bar{y}}{\Sigma x_i^2 - n\bar{x}^2} \tag{4}$$

and the intercept is

$$\hat{b} = \bar{y} - \hat{m}\bar{x}. \tag{3}$$

5. The minimum sum of squared residuals achieved by the least-squares regression line is

$$\hat{S} = \Sigma y_i^2 - n\bar{y}^2 - \hat{m}^2 (\Sigma x_i^2 - n\bar{x}^2). \tag{8}$$

SUMMARY PROBLEMS FOR CHAPTER 11

The summary problems for Chapter 3 contain data on how fast people forget. The first six problems here deal with a least-squares analysis of these data.

1. Find the least-squares line for predicting average retention score, R, from the time until testing, T.
2. Continuation. Compute the residuals from your line in Problem 1, and plot them against T. Does there appear to be a systematic pattern?
3. Continuation. Compute the minimum sum of squared residuals for the least-squares line in Problem 1.
4. Transform the values of T to $t = \log T$ (as in Problem 5 of the summary problems for Chapter 3), and compute the least-squares line for predicting R from t.
5. Continuation. Compute the residuals from your line in Problem 4, and plot them against t. Has the systematic pattern disappeared?
6. Continuation. Compute the minimum sum of squared residuals for the least-squares line in Problem 4. Is it larger or smaller than the value from Problem 3?

The remaining problems deal with least-squares analyses of data on liver weight, heart weight, and body weight of 10-month-old male mice, given in Table 3-4.

7. Using the data for the normal parental strain A body weight, compute the least-squares line for predicting liver weight.
8. Compute the least-squares line for predicting liver weight using the data for the normal parental strain B.
9. Compute the least-squares line for predicting liver weight using the data for the diabetic offspring.
10. Continuation. Compute the minimum sum of squared deviations for the fitted line in Problem 7.

11. Continuation. Compute the minimum sum of squared deviations for the fitted line in Problem 8.

12. Continuation. Compute the minimum sum of squared deviations for the fitted line in Problem 9.

13. Continuation. Pool the data from the three sources in Problems 7, 8, and 9, using the method of least squares, and fit a single line.

14. Continuation. Compare the minimum sum of squared residuals from Problem 13 with the sum of the three separate sums of squares in Problems 10, 11, and 12. Does one line fit all three kinds of data equally well?

15 through 22. Repeat Problems 7 to 14 using heart weight instead of liver weight.

REFERENCE

F. Mosteller, R. E. K. Rourke, and G. B. Thomas, Jr. (1970). *Probability with Statistical Applications*, second edition, Chapter 11. Addison-Wesley, Reading, Mass.

12

The Linear Regression Model and Its Use

Learning Objectives

1. Linking the fitting of straight lines and least squares to more formal aspects of statistics through the linear regression model
2. Estimating the proportion of variability explained by a least-squares regression line
3. Making inferences about the true regression line from least-squares estimates

★ 4. Assessing the variability of a predicted y value for a new x value using the linear regression model and the least-squares estimates

12–1 FITTING STRAIGHT LINES TO DATA

Do election turnout percentages go up with increasing city voter registration? Is the low-level radiation generated by nuclear power plants related to increased cancer mortality? Is the amount of fluoride in a city's drinking water related to the amount of tooth decay among children? Researchers often investigate such questions by fitting straight lines to data.

Fitting a straight line to a set of points leads to a simple summary involving two quantities: the slope m and the y intercept b. Once we fit a straight line, we look at the slope and the intercept. We ask such questions as these: Is the slope positive or negative? The sign tells us whether the relationship between the two variables is an increasing one or a decreasing one. If the slope is zero, we have essentially no linear predictive relation between the variables. Is the size of the slope reasonable? Is the y intercept zero?

Some problems suggest specific values of m and b that demand attention. If we want to predict change in the gross national product from changes in the money supply using a linear relation, then economic theory might suggest specific values of m and b for us to check out, such as $b = 0$ and $m > 0$. (The value $b = 0$ implies that no change in money supply goes with no change in gross national product, and $m > 0$ means the more change in money supply, the larger the change in gross national product.) Similarly, if we wish to predict the total expenditures of an ice cream plant in terms of the number of gallons of ice cream produced, we expect a value of b greater than zero corresponding to the fixed costs of production (such as rent) and a positive slope (roughly corresponding to the production cost per gallon).

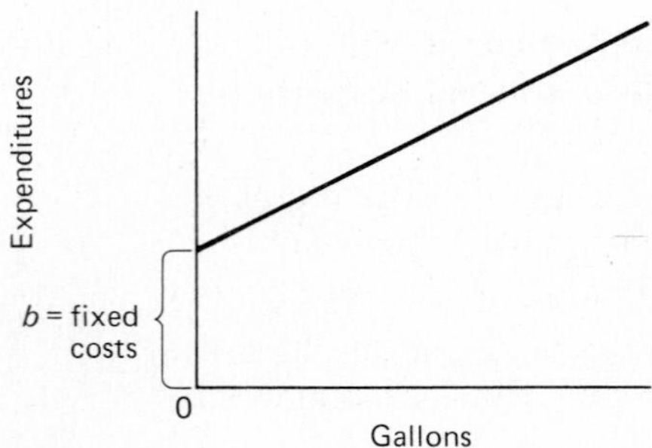

To check on specific contemplated values of the slope and intercept, or to make statements about the accuracy of predictions from fitted straight lines, we need to introduce the idea of variability about a straight-line relation. This chapter presents a model for data fitted by straight lines and then uses the model to do such things as analyze the differences between fitted and comtemplated slopes or y intercepts. The model and the methods used throughout draw heavily on the ideas for the analysis of measurement data in Chapter 10.

12-2 THE LINEAR REGRESSION MODEL

In this section we discuss three different ways that fitting straight lines ties in with the more formal features of statistics. First, we explain why we use the term *regression* when fitting curves to data. Second, we introduce a structure for measurements, called a model, that helps us analyze the reliability of estimates. Third, we provide the basic ideas of (a) testing hypotheses and of (b) confidence intervals in regression problems.

LINEAR REGRESSION

The fitting of straight lines often is referred to as linear regression theory, due to the pioneering work in the late 1800s of Sir Francis Galton, a British biometrician. Galton wanted to discover scientific laws governing the inheritance of physical traits, such as height, and he studied the heights of fathers (x) and the heights of their adult sons (y). He found that tall fathers tended to have tall sons and short fathers to have short sons. Nevertheless, the sons of tall fathers were, on the average, shorter than their fathers, and the sons of short fathers were, on the average, taller than their fathers. Galton called this phenomenon a "regression toward mediocrity." He called the line that he fitted predicting a son's height y from a father's height x the regression of the son's height on the father's height, or the regression of y on x. As a result of Galton's work on inheritance, we now refer to the theory of curve fitting as "regression theory," no matter what field we apply it to. We call the straight line for predicting y from x the **linear regression** of y on x. The coefficients $\hat{b}$ and $\hat{m}$ we used in the preceding chapter are the **least-squares regression coefficients.**

USING A MODEL

Although using least squares to fit a linear regression line makes perfectly good sense, we need to introduce a more careful mathematical description of the situation so that we can

a) assess how good a job the linear regression does in describing the relationship between y and x,

b) compare the least-squares regression coefficients with some contemplated values, and

c) decide how accurately we can predict a new value of y from a given value of x.

The mathematical or theoretical description we use is called a model. Our model for regression will specify how pairs of (x, y) values relate.

EXAMPLE 1 *Stopping distances.* Suppose that we measure the stopping distances for a car traveling on a specific road at various speeds and that we want to predict the distance, d, from the speed, x. Actually, we prefer to predict the square root of the distance

$$y = \sqrt{d}$$

from the speed, instead of *d* itself. Part of the idea of using the square root of *d* comes from a physical law relating to the dissipation of inertial forces. Even without the use of physics, the idea of taking the square root of *d* will be suggested on statistical grounds later in our discussion.

If we stop the same car traveling at 20 miles per hour three different times on a standard surface, we might observe three different values for the square root of the distance it travels: 3.5, 4.1, 3.8. Similarly, if we stop the same car traveling at a speed of 30 miles per hour four times, we might observe four different values: 5.2, 5.9, 5.6, 6.2. For any chosen speed, x, there are many possible values of y, the square root of the stopping distance. The actual value will vary from trial to trial, and for a given speed x it will have an average value or mean μ_x. The average values corresponding to different x's may have no special relationship, or they may follow a straight line,

$$\mu_x = b + mx,$$

which we call the "true regression of y on x." We do not get the exact value $b + mx$ every time we stop the car going at x mph. Some causes are variations in reaction time of the driver, in the mechanical action of the brakes, in road conditions, and so on.

Instead of observing the true regression of y on x, we observe

$$\begin{aligned} y &= \text{true regression} + \text{random error} \\ &= b + mx + e. \end{aligned}$$

Suppose the true regression of y on x in the car-stopping example is $y = -1.0 + .23x$. By subtracting the values of the true regression from the observed values of y,

$$e = y - (-1.0 + .23x),$$

we find that the random errors for $x = 20$ are -0.1, 0.5, and 0.2 and for $x = 30$ are -0.7, 0.0, -0.3, and 0.3. In practice, we do not know whether the true regression follows a straight line, let alone the exact values of b and m. Because we observe only the values of y, we do not know the actual values of the errors themselves. We fit a straight line to the data to get estimates of b and m. If the errors are small and we have more than two x values, then examining the residuals (observed y values minus fitted y values) from a fitted straight line can often tell us if the straight-line assumption is reasonable.

This example illustrates the linear regression model with random error. We turn now to a more general development.

THE NORMAL MODEL

Our basic model underlying the fitting of straight lines assumes that data are collected by selecting n values of x $(x_1, x_2, \ldots, x_n)$ and then observing the corresponding values of y $(y_1, y_2, \ldots, y_n)$. The value y_i is the sum of the true regression of y on x, namely $b + mx_i$, and a random error e_i:

$$y_i = b + mx_i + e_i.$$

But e_i is not directly observable unless we have the unusual situation of knowing b and m.

To complete the model, we must make some assumptions about the properties of the errors $e_1, e_2, \ldots, e_n$. The widely used model assumes that

a) the errors are independent random values and are generated according to the same probability distribution, and that

b) this probability distribution is the normal distribution with a mean of zero and variance σ^2 (the value being ordinarily unknown).

Because the errors have a mean of zero, the value of y for a given value of x is, on the average, $b + mx$; that for x_1 is $b + mx_1$, which equals the average value of y_1, and so on.

We begin by acting as if the model and the properties of the error terms are true, but we usually do not know whether such a description is correct. In the course of our analysis we must critically examine the model for contrary evidence. We can use the residuals from the least-squares regression line to help check the description of the model and its properties, and perhaps change them if circumstances warrant. Assumption (b), about normality of the error terms, is the one we believe the least. Although it holds for few practical examples, it is often a satisfactory approximation, and it allows us to make approximate inferences.

The basic regression model and the related statistical methods are used in two different situations:

In the **control-knob** or **experimental** situation the x variable is, more or less, under the control of the investigator. Thus, specific values of x can be chosen, with the resulting y values being the outcome of an experiment. For example, when we measure the stopping distance of a car, we can control the speed at which it travels, and we can get repeated measurements of the stopping distance for a given speed.

In the **observational** situation the x variable is not controlled by the investigator. Rather, a sample of individuals or items is drawn at random from some population, and the values of x and y are measured. Now the x values, as well as the y values, vary, even though the regression model treats the x values as fixed. For example, in Galton's problem he could not control the heights of the fathers he observed. They were what they were.

Before studying the relationship of the square root of stopping distances to speed, we relate stopping distances themselves to speed. The findings help explain our later use of $\sqrt{d}$.

EXAMPLE 2A Driving speed and stopping distance. In the stopping-distance problem of Example 1, we had observations for speeds of 20 and 30 miles per hour. Table 12-1 contains these observations along with additional ones for speeds of 40, 50, and 60 miles per hour. (Note that the second column of the table gives stopping distances, d, not their square roots, $y = \sqrt{d}$.) The plot of stopping distance versus speed in Fig. 12-1 also shows the least-squares line of d on x. Although some readers will be able to detect a noticeable curvature of the points relative to the fitted line, it can be seen more clearly if we plot the residuals from the line, $d - \hat{d}$ (listed in Table 12-1), against the x values, as in Fig. 12-2. Thus the true regression of d on x appears to be a curve, not a straight line. In addition, the spread of the distances for high speeds is greater than that for low speeds. This increase in the variability of the errors violates our first assumption that the errors come from the same distribution.

Thus, from regressing distance on speed, we found (a) lack of linearity and (b) unequal variability. These failures of assumptions suggest trying some other regression model. Physics tells us that the inertia of a car, and possibly the stopping distance, is proportional to the square of speed. Thus we might find that the square root of the stopping distance is more nearly linearly related to speed than is distance itself.

TABLE 12-1

Stopping distances for car traveling at various speeds

x *Speed (mph)*	d *Stopping distance (ft)*	$\hat{d}$†	$d - \hat{d}$ *Residuals for predicted stopping distance*	$y = \sqrt{d}$	$\hat{y}$††	$y - \hat{y}$ *Residuals for predicted root distance*
20	12.3	1.4	10.9	3.51	3.65	−0.14
20	16.8	1.4	15.4	4.10	3.65	0.45
20	14.5	1.4	13.1	3.81	3.65	0.16
30	27.0	41.6	−14.6	5.20	6.02	−0.82
30	34.8	41.6	− 6.8	5.90	6.02	−0.12
30	31.4	41.6	−10.2	5.60	6.02	−0.42
30	38.4	41.6	− 3.2	6.20	6.02	0.18
40	70.6	81.8	−11.2	8.40	8.39	0.01
40	75.5	81.8	− 6.3	8.69	8.39	0.30
40	68.9	81.8	−12.9	8.30	8.39	−0.09
50	121.0	122.0	− 1.0	11.00	10.76	0.24
50	130.0	122.0	8.0	11.40	10.76	0.64
50	123.2	122.0	1.2	11.10	10.76	0.34
50	116.6	122.0	− 5.4	10.80	10.76	0.04
60	179.2	162.2	17.0	13.39	13.13	0.26
60	171.6	162.2	9.4	13.10	13.13	−0.03
60	151.3	162.2	−10.9	12.30	13.13	−0.83
60	166.4	162.2	4.2	12.90	13.13	−0.23

Summary data: $n = 18$, $\Sigma x = 740$, $\Sigma y = 155.70$, $\Sigma x^2 = 34{,}000$, $\Sigma y^2 = 1549.67$, $\Sigma xy = 7247.40$.

† $\hat{d}$ is predicted stopping distance from regression of d on x: $\hat{d} = -79.0 + 4.02x$.
†† $\hat{y}$ is prediction from regression of y on x: $\hat{y} = -1.09 + 0.237x$.

EXAMPLE 2B *Driving speed and square root of stopping distance.* The idea of using a transformation to make variabilities comparable is one we encountered in Chapter 3, Section 3-2. Here the multiple observations at each speed form a group, and we wish to pool residuals.

In Fig. 12-3 the square roots of the stopping distances, y, are plotted against speed, x, and the points no longer deviate in a recognizable pattern

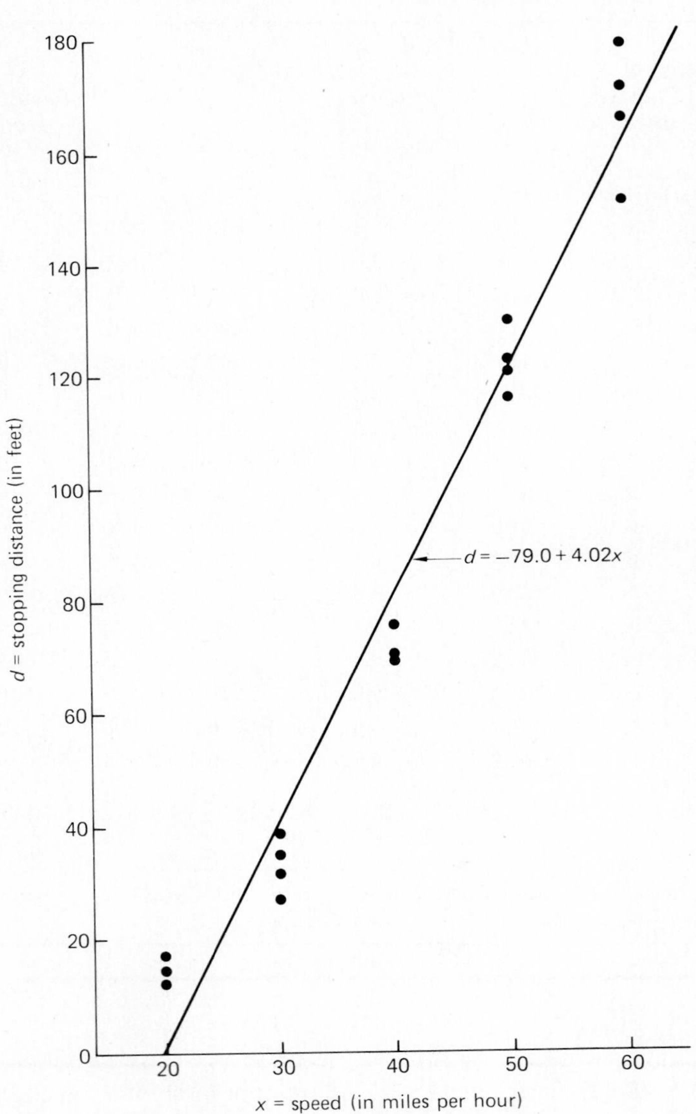

Figure 12-1 Stopping distance versus speed. (Based on data in Table 12-1.)

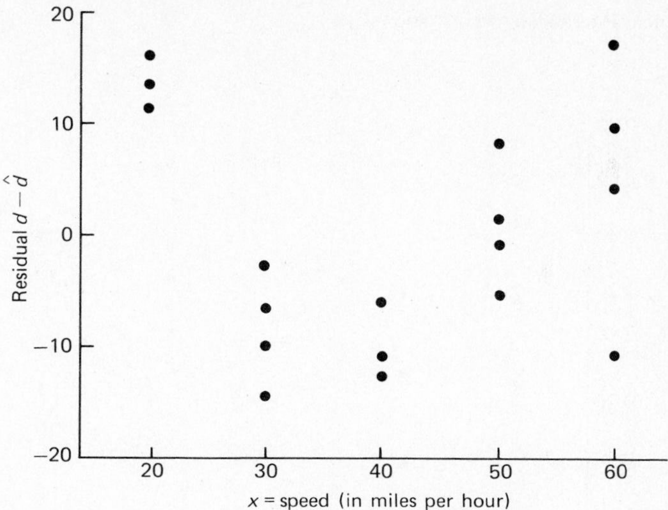

Figure 12-2 Residuals of least-squares fit, for predicting stopping distance, versus speed.

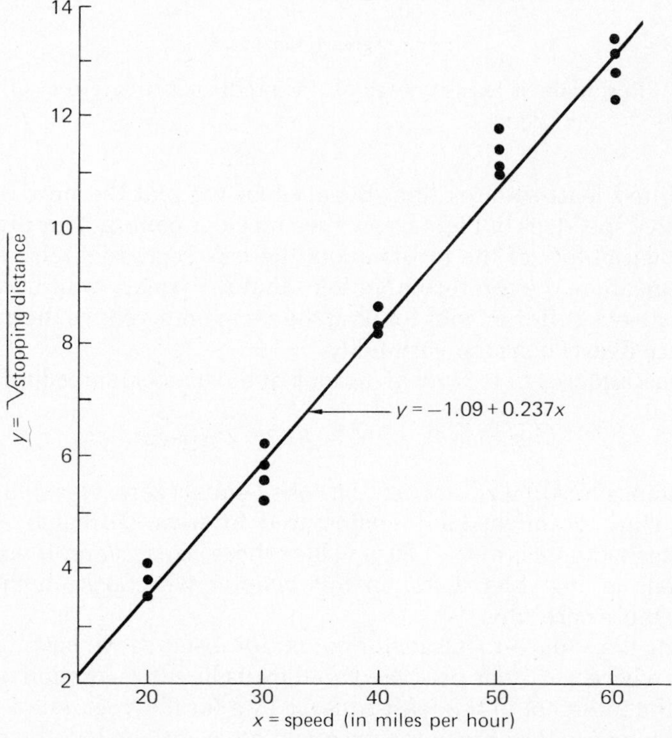

Figure 12-3 Square root of stopping distance versus speed. (Based on data in Table 12-1.)

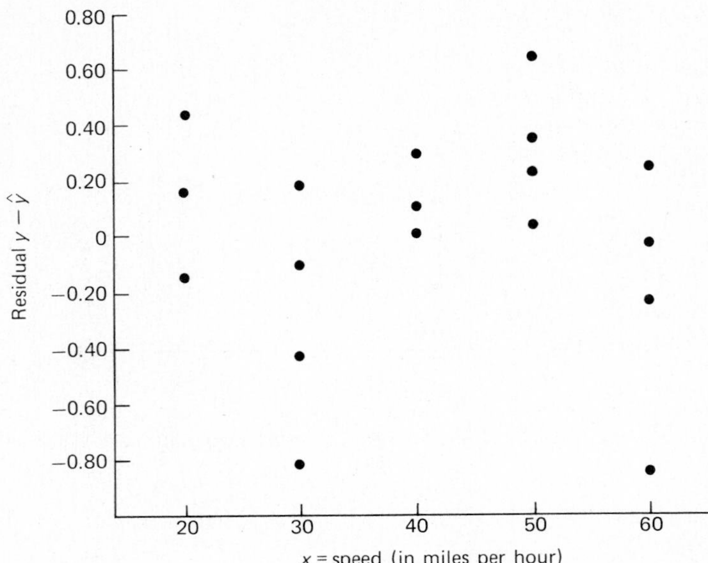

Figure 12-4 Residuals of least-squares fit, for predicting square root of distance, versus speed.

from the fitted least-squares line. Even when we plot the new residuals, $y - \hat{y}$, against speed, as in Fig. 12-4, we see no clear pattern. Our problem of increasing variability of the errors about the true regression also seems to have disappeared. We are fortunate here that the square-root transformation has done two different jobs for us at the same time: reduce the curvature and produce nearly constant variability.

The least-squares line. Now let us look at the least-squares line:

$$\sqrt{\text{distance}} = -1.09 + 0.237 \times (\text{speed}).$$

For the distance to equal zero when the speed equals zero, we need a zero y intercept. Thus we might have conjectured in advance that $b = 0$. The least-squares estimate is $\hat{b} = -1.09$, a value whose size is larger than those of the residuals in Fig. 12-4. Later in this chapter we check on the $b = 0$ conjecture more carefully.

By using the square-root transformation for distance we have produced a situation where $y = \sqrt{d}$ is predicted well by its linear regression on speed. Although the intercept in the least-squares line for this regression does not quite correspond to the zero value we might have anticipated, the residuals from the line present no recognizable pattern.

TESTS AND INTERVALS FOR REGRESSION COEFFICIENTS

In many situations some values of the regression coefficients, b and m, have meaning for us, and we would like to check how plausible these values are. For example, if we are predicting a son's height from the height of his father, then we might want to know whether the slope of the true regression is 1 and the y intercept is 0; that is, $m = 1$ and $b = 0$. These values imply that for fathers of a given height, their sons, on the average, will have the same height. If we had a large enough sample, and if Galton's finding is still valid, we would find that $m < 1$. And likely b would not turn out to be zero, because successive generations are growing taller.

Throughout this chapter we use the same general approach that we described in Section 9-1 to test the plausibility of contemplated values, when the sample size is large. It is based on the idea that

$$\text{critical ratio} = \frac{\text{observed statistic} - \text{contemplated value}}{\text{standard deviation of the statistic}}$$

should be a number near zero (say between -2 and $+2$) when the contemplated value equals the true value of the parameter—the true intercept or the slope. When the contemplated value departs from the true value, the ratio may be far from zero. In Chapter 9 we used ratios of this form to carry out tests involving parameters from the binomial, and in Chapter 10 we used them for parameters from normal distributions. The linear regression model with independent errors drawn from a normal distribution allows us to perform similar tests here involving the parameters (coefficients) of the regression model. The key to this approach, as before, is to measure the distance between the observed statistic and the contemplated value in units of the variability of the statistic.

For most of the tests in this chapter we do not know the standard deviation of the statistic, but we do know how to estimate it. If we modify the ratio by using an estimate in place of the true standard deviation, the probability distribution for the ratio changes to reflect the resulting uncertainty. Then, instead of comparing the critical ratio to values from the normal distribution with mean 0 and variance 1, we compare it to values from the t distribution with degrees of freedom determined by the sample size (see Chapter 10 for a discussion of the use of the t distribution). If the sample size is large, say over 30, the two approaches give nearly identical results.

The same basic approach can be used to get confidence intervals for the true parameters. The intervals we use have these limits:

Confidence limits =

$$\text{observed statistic} \pm \text{multiple of standard deviation of the statistic.}$$

What multiple of the standard deviation do we use? If we actually know the value of the standard deviation, then we choose a multiple from tables of the normal distribution with mean 0 and variance 1 (often 1.96, a choice that corresponds to a 95 percent confidence interval). When we do not know the standard deviation and estimate it from the data, we choose a multiple from tables of the *t* distribution, as we did in Chapter 10.

PROBLEMS FOR SECTION 12-2

In Chapters 7 and 10 we learned about the normal probability distribution and the *t* distribution; we make use of these ideas again in this chapter because of the linear regression model with normally distributed errors. The following problems deal with the use of Tables 7-2 and 7-3 for the normal distribution and Table 10-4 for the *t* distribution, or Tables A-1 and A-2 in Appendix III in the back of the book.

1. What is the probability that a measurement taken from a normal distribution lies within 1 standard deviation of its mean? What is the probability it lies within 2 standard deviations?
2. A measurement comes from a normal distribution with mean $\mu = 0$ and standard deviation $\sigma = 5$. What is the probability that the measurement is less than (a) 0.0, (b) -5.0, (c) 10, (d) 15, (e) 17.5?

Recall from Chapter 10 that for *n* degress of freedom we find a two-sided multiplier, t_n, from the *t* table, Table A-2 in Appendix III in the back of the book. For example, the two-sided 95 percent multiplier for 8 degrees of freedom is 2.31.

3. From Table A-2 find the following two-sided multipliers, where the subscript on *t* gives the number of degrees of freedom: (a) t_1, 95 percent, (b) t_2, 90 percent, (c) t_2, 99 percent, (d) t_5, 95 percent, (e) t_∞, 98 percent, (f) t_∞, 99 percent.

Chatterjee and Price (1977) studied data for 27 industrial establishments of varying sizes and recorded the numbers of supervised workers (x) and the numbers of supervisors (y). Their data are reproduced here in Table 12-2 for use in Problems 4 and 5.

4. Make a plot of y versus x, and plot on it the least-squares regression line

$$y = 14.4 + 0.105x.$$

Is the linear regression a reasonable model?

5. Continuation. Plot the residuals $y - \hat{y}$ from the least-squares line in Problem 4 versus x. Which assumption for the linear regression model with normal error terms appears to be violated?
6. Using the data on the distances of planets from the sun in Table 3-2, check whether the least-squares regression line is

$$\log d_n = 0.332 + 0.233n.$$

TABLE 12-2

Numbers of supervised workers and numbers of supervisors in 27 industrial establishments

Plant number	x *Number of workers*	y *Number of supervisors*
1	294	30
2	247	32
3	267	37
4	358	44
5	423	47
6	311	49
7	450	56
8	534	62
9	438	68
10	697	78
11	688	80
12	630	84
13	709	88
14	627	97
15	615	100
16	999	109
17	1022	114
18	1015	117
19	700	106
20	850	128
21	980	130
22	1025	160
23	1021	97
24	1200	180
25	1250	112
26	1500	210
27	1650	135

Summary data: $n = 27$, $\Sigma x = 20{,}500$, $\Sigma y = 2550$, $\Sigma x^2 = 19{,}245{,}812$, $\Sigma y^2 = 293{,}500$, $\Sigma xy = 2{,}323{,}945$.

Source: *Regression Analysis by Example*, by S. Chatterjee and B. Price, p. 44. Copyright 1977 by John Wiley & Sons, New York. Reprinted by permission of John Wiley & Sons, Inc.

7. Continuation. Which situation best describes the data analyzed: control knob or observational?

12-3 ESTIMATING VARIABILITY

We introduce some extra summation notation that simplifies writing about sums and products of deviations. Then we use it in formulas for variability of the residuals.

In Chapter 11 we worked with the least-squares regression coefficients for a sample of n points, $(x_1, y_1), (x_2, y_2), \ldots, (x_n, y_n)$. Certain quantities used to compute these coefficients come up over and over in this chapter, and therefore we introduce the following three symbols:

Sum of squares of deviations from the mean:

$$SSx = \Sigma (x_i - \overline{x})^2, \text{ definitional form (note two } S\text{'s)},$$
$$= \Sigma x_i^2 - n\overline{x}^2, \quad \text{hand-computation form},$$
$$SSy = \Sigma (y_i - \overline{y})^2 = \Sigma y_i^2 - n\overline{y}^2.$$

Sum of product of deviations from the means:

$$Sxy = \Sigma (x_i - \overline{x})(y_i - \overline{y}), \quad \text{definitional form (note one } S\text{)},$$
$$= \Sigma x_i y_i - n\overline{x}\overline{y}, \quad \text{hand-computation form}.$$

Using this new notation, we write the least-squares coefficients given in Chapter 10 as follows. Estimates of m and b:

LEAST-SQUARES ESTIMATES:

$$\hat{m} = \frac{Sxy}{SSx}$$
$$\hat{b} = \overline{y} - \hat{m}\overline{x} \tag{1}$$

These coefficients are our estimates of the unknown quantities m and b, respectively.

Estimating the Variance. The vertical scatter about the line is measured by the variance, σ^2, for the errors in the normal-error model. This unknown variance is also a parameter, and we shall ordinarily need to estimate it for several reasons. We may just want to know how variable the data are, or we may need the estimate to help us deal with the uncertainty of other quantities, such as $\hat{b}$ and $\hat{m}$. To estimate σ^2 we divide the minimum sum of squared deviation $\hat{S}$ (in our new notation):

RESIDUAL SUM OF SQUARES:

$$\hat{S} = SSy - \hat{m}^2 (SSx) \tag{2}$$

by $n - 2$ to get s_e^2, the estimate of σ^2:

ESTIMATE OF THE ERROR VARIANCE:

$$s_e^2 = \frac{\hat{S}}{n - 2} \tag{3}$$

The estimate s_e^2 is like the average squared deviation from the estimated regression line, but we divide $\hat{S}$ by $n - 2$ instead of by n because we have used up 2 of our n pieces of information in estimating b and m, leaving $n - 2$. We say that there are $n - 2$ degrees of freedom associated with our estimate s_e^2. (See the earlier discussion of degrees of freedom in Section 10-1.)

EXAMPLE 3 *Four-point example.* Fit a straight line to the following four points:

x:	1	2	3	4
y:	1	2	3	5

Estimate σ^2, the unknown error variance.

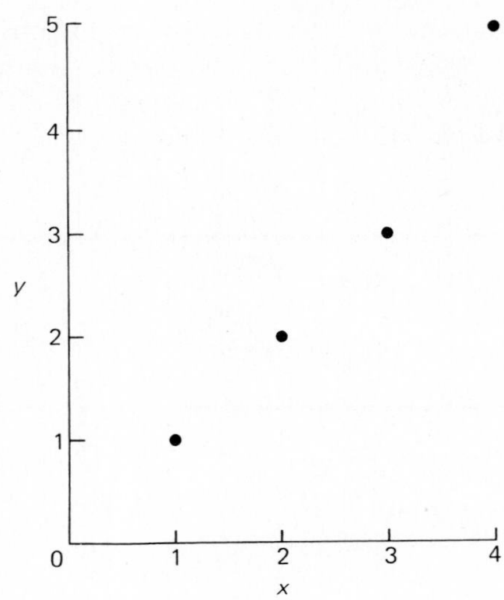

SOLUTION. For this example we have these summary statistics:

$$\bar{x} = \frac{1 + 2 + 3 + 4}{4} = \frac{10}{4} = 2.5,$$

$$\bar{y} = \frac{1 + 2 + 3 + 5}{4} = \frac{11}{4} = 2.75,$$

$$\Sigma x^2 = 1 + 4 + 9 + 16 = 30,$$

$$\Sigma y^2 = 1 + 4 + 9 + 25 = 39,$$

and

$$\Sigma xy = (1 \times 1) + (2 \times 2) + (3 \times 3) + (4 \times 5) = 34.$$

Thus

$$SSx = \Sigma x^2 - n\bar{x}^2 = 30 - 4(2.5)^2 = 30 - 25 = 5,$$

$$SSy = \Sigma y^2 - n\bar{y}^2 = 39 - 4(2.75)^2 = 39 - 30.25 = 8.75,$$

and

$$Sxy = \Sigma xy - n\bar{x}\bar{y} = 34 - 4(2.5)(2.75) = 34 - 27.5 = 6.5.$$

The least-squares coefficients are

$$\hat{m} = \frac{Sxy}{SSx} = \frac{6.5}{5} = 1.3$$

and

$$\hat{b} = \bar{y} - \hat{m}\bar{x} = 2.75 - (1.3)(2.5) = 2.75 - 3.25 = -0.5.$$

Thus the straight line we have fitted to the four points is

$$y = -0.5 + 1.3x.$$

The minimum sum of squared deviations about this line is

$$\hat{S} = SSy - \hat{m}^2\,(SSx) = 8.75 - (1.3)^2 5$$

$$= 8.75 - 8.45 = 0.3.$$

Our estimate of σ^2 is

$$s_e^2 = \frac{\hat{S}}{n-2} = \frac{0.3}{2} = 0.15.$$

Thus the estimated standard deviation of the unknown error term is

$$s_e = \sqrt{0.15} \approx 0.39$$

$$\approx 0.4.$$

This illustrates the calculation of the least-squares line and of the estimated variance and standard deviation of the errors.

EXAMPLE 4 *Indianapolis 500 prediction variability.* In Example 6 in Chapter 11 we estimated the winning speed for the Indianapolis 500 auto race in miles per hour (y) from the year of the race (x = year minus 1960). We learned that the least-squares regression line based on the 1961–1970 data has as an equation, when coefficients are rounded,

$$y = 137.60 + 1.93x,$$

and the minimum sum of squared residuals is, from Chapter 11,

$$\hat{S} = 51.50.$$

Because there are $n = 10$ observations, our estimate for the unknown error variance is

$$s_e^2 = \frac{51.50}{8} = 6.44.$$

The square root of this quantity, $s_e = \sqrt{6.44} = 2.54$, is the estimated standard deviation of the unknown normal-error term. We use this value of s_e in some further calculations later in this chapter.

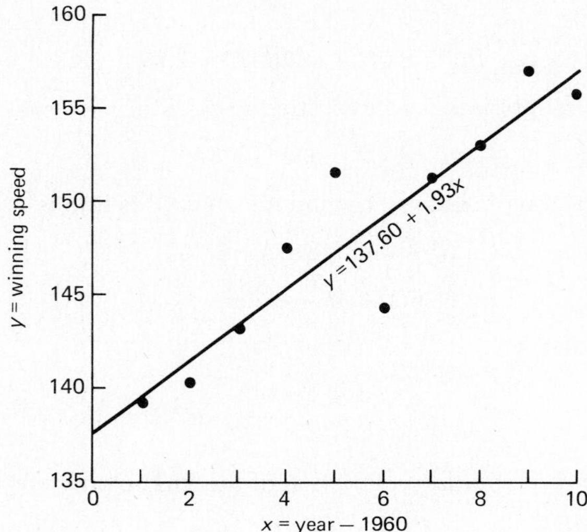

The value of the estimated standard deviation of the errors, $s_e = 2.54$, is small when compared with the magnitudes of the y values we are trying to predict. This means that our least-squares line is doing a good job of prediction, with relatively small error, in this example.

Illustrative Examples. Throughout the rest of this chapter, the topics usually are illustrated numerically with the data of Example 3 to firm up notation and illustrate the calculations. Then a second example illustrates an interpretation in an applied setting.

PROBLEMS FOR SECTION 12-3

1. Find the slope and the y intercept of the least-squares regression line fitted to these points:

x:	−1	−1	0	0	1	1
y:	0	1	0	1	2	1

2. Continuation. Compute the values of $\hat{S}$ and s_e^2.

3. Continuation. Plot the points and the line from Problem 1, and find out what fraction of the y values are within s_e of the line (as measured vertically—up and down). What fraction are within $2s_e$?

In Example 3 in Chapter 11 we showed in Fig. 11-3 a plot of snowfall data for Minneapolis–St. Paul (1920–1974), where x = amount of November snowfall and y = amount of snowfall for the remainder of the winter. In Table 12-3 we have an

TABLE 12-3
Snowfall data for Minneapolis–St. Paul, 1950–1970

Year	x *November snowfall (inches)*	y *December–May snowfall (inches)*
1950–51	5.6	83.3
1951–52	10.8	67.4
1952–53	10.1	32.8
1953–54	1.9	23.8
1954–55	6.4	27.1
1955–56	6.0	36.7
1956–57	6.8	32.3
1957–58	10.3	10.9
1958–59	3.3	15.8
1959–60	6.9	21.2
1960–61	2.4	37.8
1961–62	2.5	78.8
1962–63	5.6	28.9
1963–64	0.0	28.9
1964–65	4.3	69.4
1965–66	1.6	34.5
1966–67	3.4	74.8
1967–68	0.8	16.4
1968–69	4.9	63.2
1969–70	3.8	57.0

Summary data: $n = 20$, $\Sigma x_i = 97.4$, $\Sigma y_i = 841.0$,
$\Sigma x_i^2 = 656.48$,
$\Sigma y_i^2 = 45{,}488.96$, $\Sigma x_i y_i = 4091.77$.

Source: Data reprinted by permission of the Freshwater Biological Research Foundation, Navarre, Minnesota, from the *1975 Minnesota and Environs Weather Almanac*, by Bruce Watson, Copyright 1974.

excerpt of these data running from 1950–51 to 1969–70. Use the data in this table to solve Problems 4 through 6.

4. Find the least-squares line for predicting December–May snowfall from November snowfall. Be sure to plot the data. (Hint: Begin by computing SSx, SSy, and Sxy.)
5. Continuation. Compute the values of $\hat{S}$ and s_e^2 for the fitted regression line in Problem 4.
6. Continuation. What fraction of the y values in Problem 4 are within s_e of the line, vertically? Within $2s_e$?
7. Use the summary statistics data in Table 12-2 to compute $\hat{S}$ and s_e.

12-4 THE PROPORTION OF VARIABILITY EXPLAINED

When we fit a least-squares line to a set of points, the vertical scatter of the points about the line is smaller than the scatter of the y values about their mean $\bar{y}$, unless $\hat{m} = 0$. In this section we relate these two forms of variability by means of a quantity that, as we shall show, gives the proportion of variability of y "explained" by x.

The magnitude of s_e^2, our estimate of the error variance σ^2, tells us the variability of the data points about the least-squares regression line. The larger s_e^2, the greater the variability. If the variability is large, we may find that the regression line does not help us much in predicting y values. If we did not have the x values available to help in the prediction of the y values, we might predict the y_i's by the sample average. This average, $\bar{y}$, is also the value of $\hat{b}$ in expression (1) of Section 12-3 that results from setting the estimated slope $\hat{m} = 0$. The value of the sum of squared deviations when $\hat{m} = 0$ is

$$SSy = \Sigma\,(y_i - \bar{y})^2.$$

If we use the x values, we predict the value

$$\hat{y}_i = b + mx_i$$

when the actual value is y_i, and so our minimum sum of squared deviations is the residual sum of squares:

MINIMUM SUM OF SQUARED DEVIATIONS:

$$\hat{S} = \Sigma\,(y_i - \hat{y}_i)^2 \qquad (4)$$

This is the same minimum sum of squared deviations $\hat{S}$ that we compute using expression (2):

$$\hat{S} = SSy - \hat{m}^2 SSx. \tag{2}$$

Note that $\hat{S}$ is nonnegative, because it is a sum of squares in expression (4). So are SSx and SSy and $\hat{m}^2$. In fact, $\hat{S}$ is positive unless all predicted values equal the observed, and then $\hat{S}$ is zero. Also, $\hat{S}$ can never be larger than SSy.

EXAMPLE 5 For the four points of Example 3, find the predicted values, $\hat{y}_i$, and show that expression (4) yields the same minimum sum of squared deviations, $\hat{S}$, as before.

SOLUTION. The predicted y values, $\hat{y} = \hat{b} + \hat{m}x$, for the four points are

$$x_1 = 1, \quad \hat{y}_1 = -0.5 + (1.3)(1) = 0.8,$$

$$x_2 = 2, \quad \hat{y}_2 = -0.5 + (1.3)(2) = 2.1,$$

$$x_3 = 3, \quad \hat{y}_3 = -0.5 + (1.3)(3) = 3.4,$$

$$x_4 = 4, \quad \hat{y}_4 = -0.5 + (1.3)(4) = 4.7.$$

Thus,

$$\begin{aligned}
\hat{S} &= \Sigma (y_i - \hat{y}_i)^2 \\
&= (1 - 0.8)^2 + (2 - 2.1)^2 + (3 - 3.4)^2 + (5 - 4.7)^2 \\
&= (0.2)^2 + (-0.1)^2 + (-0.4)^2 + (0.3)^2 \\
&= 0.04 + 0.01 + 0.16 + 0.09 \\
&= 0.30,
\end{aligned}$$

which agrees with the value of $\hat{S}$ from Example 3, computed using expression (2). Thus, in this example, we verify that $\hat{S}$ is the sum of the squares of the residuals or vertical deviations from the fitted line.

PROPORTION OF VARIABILITY EXPLAINED

The value of $\hat{S}$ when m is set equal to zero is SSy. So the difference $SSy - \hat{S}$ is the amount of variability in the y's accounted for by using the x values and the regression line. Thus the fraction of SSy that is removed by the fitted regression line is as follows.

Proportion of y variance explained:

$$r^2 = \frac{SSy - \hat{S}}{SSy}, \tag{5a}$$

or

$$r^2 = \frac{\hat{m}^2 SSx}{SSy}. \tag{5b}$$

Another version of formula (5) that is valuable for computing purposes is the following:

COMPUTING FORMULA FOR r^2:

$$r^2 = \frac{(Sxy)^2}{(SSx)(SSy)} \tag{6}$$

We get this formula by substituting the value of $\hat{m}$ from equation (1) into equation (5b) and then doing some algebra. It is common practice to say that r^2 is the fraction or proportion of the variability in the y values accounted for or explained by the regression line. Here the term *explained* is used in a technical sense; it does not necessarily mean that the variable x is causally related to the variable y. Thus r^2 measures the proportional reduction in sum of squares we get when we use x's to estimate y's instead of estimating them all as $\bar{y}$.

If we recall from expression (2) that $\hat{S}$ can never exceed SSy, then expression (5a) implies that $r^2 \geq 0$. Moreover, because $\hat{S}$ is a sum of squares, the least it can be is 0, and so we have 1 as an upper limit to r^2. Thus, r^2 must lie between 0 and 1:

$$0 \leq r^2 \leq 1.$$

The value of r^2 is 1 when and only when a set of points is perfectly fitted by the regression line based on the x values. The value of r^2 is 0 when the regression line is of no help at all in predicting the y values.

EXAMPLE 6 Compute r^2 for the regression equation in the four-point problem of Examples 3 and 5.

SOLUTION. Recall that $SSy = 8.75$ and that $\hat{S} = 0.30$. Thus from expression (5a), we have

$$r^2 = \frac{8.75 - 0.30}{8.75}$$

$$= \frac{8.45}{8.75} = 0.966.$$

Alternatively, we can compute r^2 in expression (6):

$$r^2 = \frac{(6.5)^2}{(5)(8.75)}$$

$$= \frac{42.25}{43.75} = 0.966.$$

Thus, the percentage of the variance of y explained by x is 96.6 percent.

Figure 12-5 shows several plots of sets of (x, y) values, the fitted least-squares regression line, and the value of r^2. Note that in plot (f) the values of r^2 and $\hat{m}$ are zero, but y and x are perfectly related by a simple curve. Thus r^2 can tell us that a line is not a good predictor, but it can fail to tell us that a curve other than a line may be a great help in predicting. This finding reemphasizes the need to plot the data.

CORRELATION COEFFICIENT

We used the notation r^2 in expressions (5) and (6), instead of just r. The correlation coefficient

$$r = \frac{Sxy}{\sqrt{(SSx)(SSy)}} \qquad (7)$$

is the sample **correlation coefficient** between x and y. The possible values of r satisfy the inequality

$$-1 \leq r \leq 1.$$

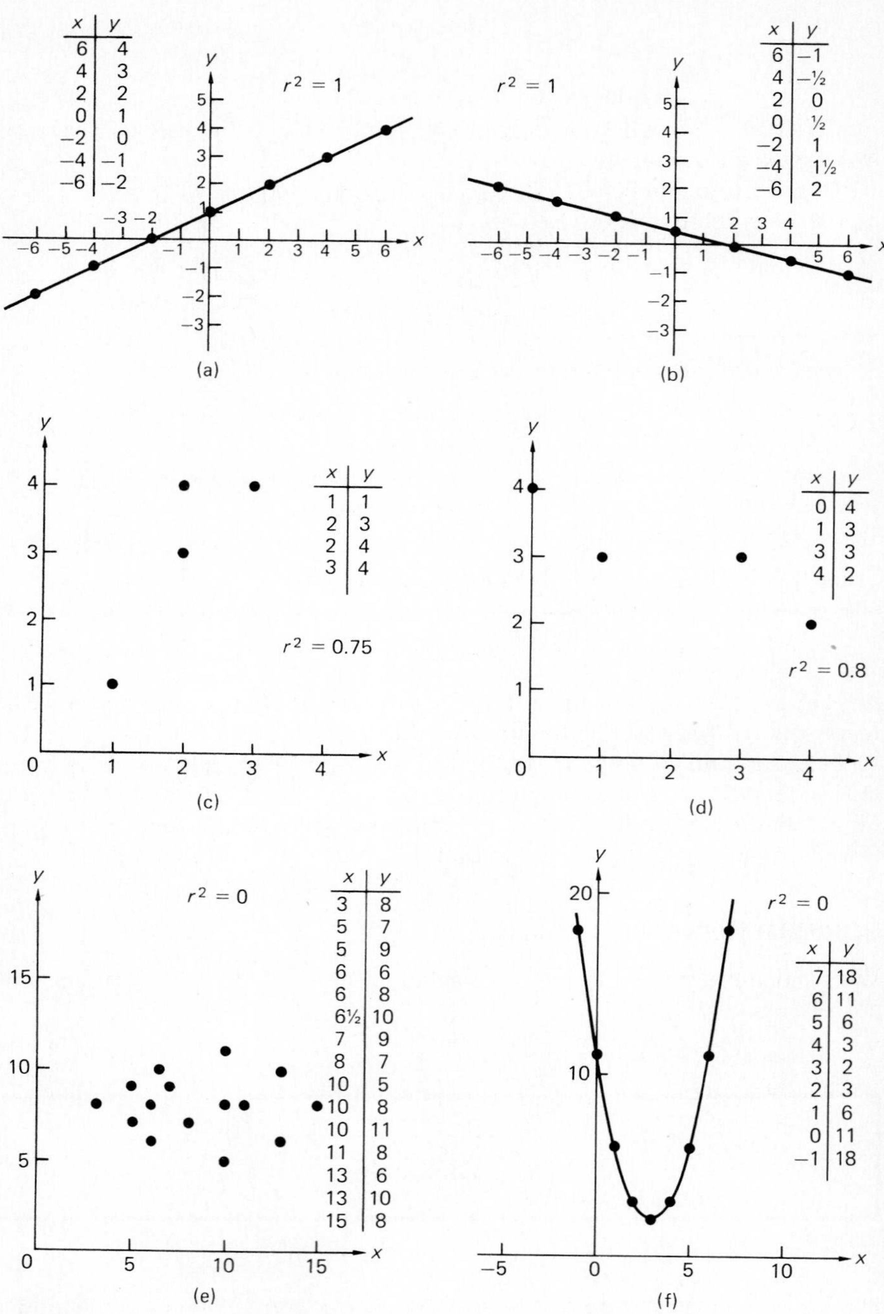

x	y
6	4
4	3
2	2
0	1
−2	0
−4	−1
−6	−2

x	y
6	−1
4	−½
2	0
0	½
−2	1
−4	1½
−6	2

x	y
1	1
2	3
2	4
3	4

x	y
0	4
1	3
3	3
4	2

x	y
3	8
5	7
5	9
6	6
6	8
6½	10
7	9
8	7
10	5
10	8
10	11
11	8
13	6
13	10
15	8

x	y
7	18
6	11
5	6
4	3
3	2
2	3
1	6
0	11
−1	18

Figure 12-5 Illustrations of amount of variance explained by a linear regression from 100 percent when the points fall on a line, as in (a) and (b), to zero percent when there is no linear predictability, as in (e) and (f).

Source: Mosteller *et al.* (1970, p. 412).

Note that formula (7) is the same if we interchange x and y. This is why we speak and write of the "co-relation" or correlation between x and y. Thus r^2 has the same value when we are predicting x from y as it does when we are predicting y from x.

PROBLEMS FOR SECTION 12-4

1. From the data of Fig. 12-5(a) calculate, from formula (6) directly, that $r^2 = 1$. Is $r = +1$ or $r = -1$?

2. From the data of Fig. 12-5(c), calculate, from formula (6), that $r^2 = 0.75$. What is the value of the sample correlation coefficient r?

3. From the data of Fig. 12-5(f), calculate

a) Σx_i^2 and SSx,
b) Σy_i^2 and SSy, and
c) $\Sigma x_i y_i$ and Sxy.

It is a fact that the pair of quantities in (a) are the same only when $\bar{x} = 0$, in (b) only when $\bar{y} = 0$, and in (c) only when $\bar{x} = 0$ and/or $\bar{y} = 0$.

4. Continuation. From the quantities in Problem 3, calculate directly that $r^2 = 0$ for the data in Fig. 12-5(f).

5. If $r = -0.3$, what percentage of variability is explained by the least-squares regression line? What if $r = 0.3$?

6. If $SSy = 10$, what percentage of the variability is explained by the fitted regression line when $\hat{S} = 5$? What if $\hat{S} = 2.5$?

7. What percentage of the variability is explained by the fitted regression line for the snowfall data in Table 12-3? What does this value say about the value of November snowfall as a predictor?

8. For the Indianapolis 500 data in Example 6 in Section 11-4, we have these summary data:

$$n = 10,$$

$$\Sigma x_i = 55, \qquad \Sigma y_i = 1482.320,$$

$$\Sigma x_i^2 = 385, \qquad \Sigma y_i^2 = 220086.699 \quad \text{(to 3 decimals)},$$

$$\Sigma x_i y_i = 8312.167.$$

Compute the value of r^2.

9. For the submarine-sinkings data of Example 7 in Section 11-4, we have these summary data:

$$n = 16,$$

$$\Sigma x_i = 123, \qquad \Sigma y_i = 140,$$

$$\Sigma x_i^2 = 1287, \qquad \Sigma y_i^2 = 1682,$$

$$\Sigma x_i y_i = 1431.$$

Compute the value of r^2.

12-5 TESTS AND CONFIDENCE LIMITS FOR *m*

When we fit straight lines to a set of data we ordinarily are working with a sample rather than a population. The parameters b and m of the true linear regression of y on x are population values, and the least-squares estimates $\hat{b}$ and $\hat{m}$ are sample values. Thus $\hat{b}$ and $\hat{m}$ will vary from one sample to the next. In this section we introduce methods for testing the plausibility of contemplated values of m and for setting confidence limits on m. For large samples, we lean on the theory of the normal distribution, but for small samples, say $n \leq 30$, we use the t distribution.

The linear regression model with normally distributed errors described in Section 12-2 is the sampling model for the estimation of the true coefficients b and m. If $\hat{m}$ could be computed for all possible samples of size n that have the same fixed x values, $x_1, x_2, \ldots, x_n$, then the least-squares slope $\hat{m}$ would have a distribution with mean m and variance

$$\sigma_{\hat{m}}^2 = \frac{\sigma^2}{SSx}.$$

In words, the estimated slope $\hat{m}$ is, on the average, the true slope m, and so it is an unbiased estimate of m. When the "errors" in the linear regression model come from a normal distribution, the distribution of $\hat{m}$ about m is also normal, and we can use this result to make inferences about the true but unknown regression slope m. More realistically, if the errors come from a distribution that is approximately normal, we can make approximate inferences.

We estimate the variance of $\hat{m}$, $\sigma^2_{\hat{m}}$, by substituting s_e^2 for σ^2:

ESTIMATED VARIANCE OF $\hat{m}$:

$$s^2_{\hat{m}} = \frac{s_e^2}{SSx} \tag{8}$$

How does the estimated variability of the estimated slope, $s^2_{\hat{m}}$, respond to the data? The numerator of the right-hand side of formula (8) shows that the smaller the natural variability of the data about the line, the smaller $s^2_{\hat{m}}$ is, on the average, and thus the better determined is the slope. Similarly, the denominator in formula (8) shows that the more spread out the points are in the x direction, the more leverage they have on the slope, and so the larger SSx, the better determined is the slope. This means that if we wish to make $s^2_{\hat{m}}$ small, and if we can actually pick the values of x ourselves, we would be wise to spread them out so that SSx is large. The price for doing this is that we may not find out the shape of the true curve between the extremes; that is, we may miss departures from linearity.

CHECKING A CONTEMPLATED VALUE OF *m*

Suppose that we wish to check whether or not a particular number, say 5, is a plausible value of the true regression slope m. To do this we use the critical-ratio approach outlined in Section 12-2. To begin, we take the observed value of $\hat{m}$, subtract the particular value of interest (in this case 5), and then divide by the estimated standard deviation of $\hat{m}$. This yields the statistic

$$t_{\hat{m}} = \frac{\hat{m} - 5}{s_{\hat{m}}}.$$

Once we plug in the values of $\hat{m}$ and $s_{\hat{m}}$, $t_{\hat{m}}$ is simply a number. The correct procedure uses a t table with $n - 2$ d.f. However, if n is large, as an approximate procedure, we compare the value of $t_{\hat{m}}$ with those in Table 7-2 or 7-3 or Table A-1 in Appendix III in the back of the book for the normal distribution with mean 0 and variance 1. If $t_{\hat{m}}$ (with either a positive or negative sign) is large (for example, larger than 1.96), we have evidence to suggest that the true slope m is different from 5.

The general formula for the t statistic used to check the plausibility of the contemplated value m is the following:

t STATISTIC FOR CONTEMPLATED VALUE OF m:

$$t_{\hat{m}} = \frac{\hat{m} - m}{s_{\hat{m}}} \tag{9}$$

If n is large, we can also set approximate confidence limits for m using values from the normal table. The 95 percent confidence limits on the true slope m are given by

$$\hat{m} \pm 1.96\, s_{\hat{m}},$$

where 1.96 is the two-sided 95 percent multiplier from the normal table.

If n is small (say $n \leq 30$), the correct procedure is to compare the value of $t_{\hat{m}}$ with the values in the t table for $n - 2$ degrees of freedom, Table A-2 in Appendix III in the back of the book. Similarly, we can set confidence limits for m using values from the t table for $n - 2$ degrees of freedom.

We illustrate with examples.

EXAMPLE 7 *Testing the slope in the four-point problem.* Compute the value of $s_{\hat{m}}$ for the four-point problem of Example 3. Suppose we contemplate $m = 0$ as a value for the slope. Is this a plausible value for m?

SOLUTION. From Example 3 we recall that $s_e^2 = 0.15$ and $SSx = 5$. Thus

$$s_{\hat{m}} = \sqrt{\frac{0.15}{5}} = \sqrt{0.03} = 0.17.$$

To check the plausibility of the contemplated value $m = 0$, we compute

$$t_{\hat{m}} = \frac{\hat{m} - 0}{s_{\hat{m}}} = \frac{1.3}{0.17} = 7.6.$$

The probability of observing a value as large as 7.6 from a t distribution for $n - 2 = 4 - 2$ degrees of freedom is less than 0.01. Thus we doubt (a) that the true slope is zero and (b) that it is close to zero.

EXAMPLE 8 *Submarine sinkings reexamined.* In Example 7 in Chapter 11 we fitted the least-squares regression line

$$y = 0.76 + 1.04x$$

to the data for predicting the actual number of submarine sinkings y (by month) from the navy's reports of submarine sinkings x. A possible model for the data would postulate that, aside from error, the navy systematically undercounted the number of actual sinkings by subtracting 1; that is, $m = 1$ and $b = 1$. How plausible is this model?

SOLUTION. We begin by exploring the plausibility of $m = 1$. We need $s_{\hat{m}}$. From Example 7 in Chapter 11 we have that

$$n = 16, \quad \hat{m} = 1.04, \quad \hat{S} = 88.54,$$

and

$$SSx = 1287 - 16(7.688)^2 = 341.31.$$

Thus

$$s_e^2 = \frac{\hat{S}}{n-2} = \frac{88.54}{14} = 6.324,$$

and the estimated variance of the slope is

$$s_{\hat{m}}^2 = \frac{s_e^2}{SSx} = \frac{6.324}{341.31} = 0.0185.$$

Taking the square root, we have $s_{\hat{m}} = 0.136$. To test whether or not m might be 1, we compute

$$t_{\hat{m}} = \frac{\hat{m} - 1}{s_{\hat{m}}} = \frac{1.04 - 1}{0.136} = \frac{0.04}{0.136} = 0.29.$$

Next we look up the value 0.29 in Table A-2 in Appendix III in the back of the book for $n - 2 = 16 - 2 = 14$ degrees of freedom. We find that the probability of a value of 0.29 or one at least as extreme (larger than 0.29 or smaller than -0.29) is approximately 0.78. Thus we conclude that the contemplated value of $m = 1$ is plausible, because larger deviations from $m = 1$ than that observed will occur in 78 out of 100 investigations where m actually is 1.

Suppose now that $m = 1$. Let us look at the fitted line with this slope and passing through $(\bar{x}, \bar{y})$. Our new estimated intercept is

$$\bar{y} - \bar{x} = 8.75 - 7.69 = 1.06,$$

and our fitted regression line is approximately

$$y = 1 + x.$$

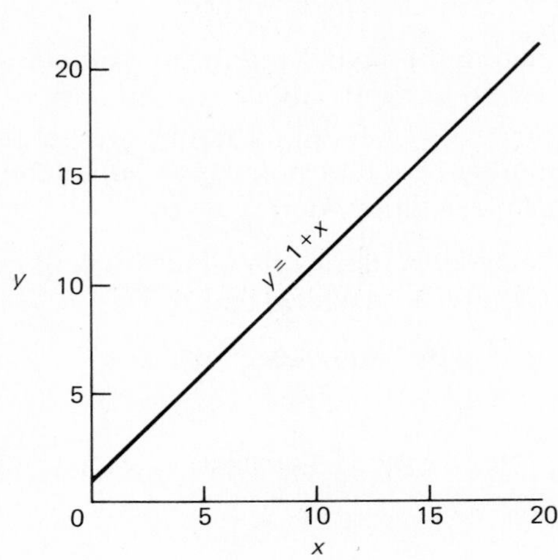

One possible interpretation of this line is that, on the average, the navy missed one submarine per month in its calculation. These estimates were made by the navy for its own use, not as propaganda devices, and so they may have been deliberately conservative to prevent unjustified optimism.

EXAMPLE 9 Find 99 percent confidence limits for the slope in Example 8 on submarine sinkings.

SOLUTION. For this problem, $n = 16$, $\hat{m} = 1.04$, and $s_{\hat{m}} = 0.136$. To set 99 percent limits we need the value t_{14} from Table A-2 in Appendix III in the back of the book:

$$t_{14} = 2.98.$$

Thus the 99 percent confidence limits for m are

$$\hat{m} \pm 2.98 s_{\hat{m}} = 1.04 \pm 2.98(0.136) = [0.63, 1.45].$$

The plausible value of $m = 1$ discussed in Example 8 lies close to the middle of the interval between the two limits.

RELATIONSHIP BETWEEN r^2 AND $t_{\hat{m}}^2$

The value $m = 0$ is a special one. It means that the x values are of no use in linearly predicting the y values, and we have no linear relationship between x and y. In many examples it is natural to want to check out the possibility that $m = 0$. To do this, we use the t statistic from equation (9), with $m = 0$:

$$t_{\hat{m}} = \frac{\hat{m}}{s_{\hat{m}}} \tag{10}$$

We compare its value with those of the t distribution with $n - 2$ degrees of freedom.

This test statistic for checking the plausibility of $m = 0$ is directly related to r^2, the fraction of variability explained by the regression line, by the following formula:

$$t_{\hat{m}}^2 = \frac{(n - 2)r^2}{1 - r^2} \quad \text{(testing } m = 0\text{)} \tag{11}$$

This result means that if we have already computed r^2, then by taking the square root in equation (11), we have the statistic for testing $m = 0$.

Illustrative Check. Let us check the agreement of formulas (10) and (11) through an example. For the data in Fig. 12-5(d), $SSx = 10$, $SSy = 2$, and $Sxy = -4$;

$$r^2 = \frac{(-4)^2}{10(2)} = 0.8,$$

as indicated, and

$$\frac{(n - 2)r^2}{1 - r^2} = \frac{2(0.8)}{1 - 0.8} = \frac{1.6}{0.2} = 8.$$

Computing formula (11) directly, we get

$$\hat{m} = \frac{Sxy}{SSx} = -0.4,$$

$$\hat{S} = SSy - \hat{m}^2 (SSx) = 2 - 0.16(10) = 0.4,$$

$$s_e^2 = \frac{\hat{S}}{n-2} = 0.2,$$

$$s_{\hat{m}}^2 = \frac{s_e^2}{SSx} = \frac{0.2}{10} = 0.02.$$

After squaring formula (10), we have

$$\frac{\hat{m}^2}{s_{\hat{m}}^2} = \frac{0.16}{0.02} = 8.$$

Thus we check that the two methods of calculation give the same result in this example.

EXAMPLE 10 *Minnesota snowfalls.* For the estimated regression line using the November snowfall in Minneapolis–St. Paul to predict the amount of snowfall for the remainder of the winter (December–May), check the plausibility of $m = 0$.

SOLUTION. Using the summary data in Table 12-3, we find

$$\hat{m} = \frac{Sxy}{SSx} = -3.90, \quad SSx = 182.14, \quad \text{and} \quad SSy = 10{,}124.91.$$

We compute

$$r^2 = \frac{(Sxy)^2}{(SSx)(SSy)} = \frac{(-3.90)^2}{(182.14)(10{,}124.91)} = 0.00001.$$

The square of the statistic for testing the plausibility of $m = 0$ is

$$t_{\hat{m}}^2 = \frac{18(0.00001)}{0.99999} = 0.00018,$$

and thus

$$t_{\hat{m}} = \pm\sqrt{0.00018} = \pm 0.013.$$

(In testing whether or not m could reasonably be 0, we use the negative value of $t_{\hat{m}}$ here because $\hat{m}$ is negative.) This value of $t_{\hat{m}}$ is so close to zero that we need not look it up in our table of the t distribution. The value $m = 0$ is indeed plausible, and this test suggests that the December–May snowfall amount is scarcely related to the November snowfall amount. This, of course, is the conclusion we had reached earlier.

PROBLEMS FOR SECTION 12-5

Use the data in Fig. 12-5(c) to solve Problems 1, 2, and 3.

1. Confirm that $SSx = 2$, $SSy = 6$, and $Sxy = 3$.
2. Continuation. Compute $\hat{m}$ and $s_{\hat{m}}$, and thus show that if our contemplated value is $m = 0$, then $t_{\hat{m}} = 2.45$.
3. Continuation. Using the relation between r^2 and $t_{\hat{m}}$, equation (11), and the value $r^2 = 0.75$, compute $t^2_{\hat{m}}$. Compare your answer with the square of $t_{\hat{m}}$ found in Problem 2.

In Example 2 in Section 11-2 we looked at a small experiment relating the current passing through an electrical device to the voltage across it. Summary statistics of the data in Table 11-2 are as follows:

$$n = 6, \quad \Sigma x = 42, \quad \Sigma y = 0.0343, \quad \Sigma x^2 = 364,$$

$$\Sigma y^2 = 0.00024751, \quad \Sigma xy = 0.3000.$$

Use this information to solve Problems 4, 5, and 6.

4. For this experiment, fit a least-squares regression line to the data.
5. What is the value of the estimated standard error, $s_{\hat{m}}$, of the least-squares slope in Problem 4?
6. Continuation. Because the electrical device under study is made primarily with a particular metal, a physicist claims the resistance of the device should be 1.3 ohms. This translates into a slope, m, for our straight line equal to $1/1.3 = 0.77$, because m is the inverse of the resistance measured in ohms. In light of the data in Problems 4 and 5, how plausible is this value of m?

In Example 4 in Section 11-2 we considered data relating metabolic rate and body weight. The data are displayed in Fig. 11-4. Some summary statistics are as follows:

$$n = 100, \quad SSx = 0.3521, \quad SSy = 0.1705, \quad \text{and} \quad Sxy = 0.1177.$$

The fitted least-squares regression line is

$$y = 2.55 + 0.33x.$$

Use this information to solve Problems 7 and 8.

7. One of our reasons for looking at these data is to check whether or not the slope of the true regression line is approximately 0.75. How plausible is such a value of m?

8. Continuation. Construct 95 percent confidence limits for the slope m, and comment on the plausibility of the contemplated value $m = 0.75$.

12-6 TESTS AND CONFIDENCE LIMITS FOR *b*

Now that we have learned how to use the general approach of Section 12-2 to test for contemplated values of the slope of the true regression of y on x, we can do other related tests in a similar manner. For example, suppose that we wish to check the plausibility of a particular value of the regression line's y intercept b. To do this we need the estimated variance of the least-squares intercept, $\hat{b}$, which is as follows:

ESTIMATED VARIANCE OF $\hat{b}$:

$$s_{\hat{b}}^2 = \frac{\Sigma x_i^2}{n(SSx)} s_e^2 \qquad (12)$$

In thinking about the variability of $\hat{b}$, it helps to recall that if $\hat{m} = 0$, $\hat{b}$ estimates $\bar{y}$, because

$$\hat{b} = \bar{y} - \hat{m}\bar{x}.$$

The variability of $\bar{y}$ is estimated by s_e^2/n, which is a factor in formula (12). When we estimate a value of y for a value of x near $\bar{x}$, the variability is approximately s_e^2/n. When we estimate values of y for x's distant from $\bar{x}$, we must use the value of $\hat{m}$ to interpolate or extrapolate, and $\hat{m}$'s uncertainty must be taken into account. The coefficient or multiplier

$$\frac{\Sigma x_i^2}{SSx}$$

for s_e^2/n does that. When $\bar{x} = 0$, this coefficient is 1, because in that special case, $\hat{b}$ is the value of y at $\bar{x}$. But when $\bar{x}$ is far from 0, Σx_i^2 is much larger than SSx, and so then the uncertainty in $\hat{b}$ is much larger than that of $\bar{y}$.

To check on the plausibility of a contemplated value of b, we subtract it from $\hat{b}$ and divide by $s_{\hat{b}}$:

t STATISTIC FOR CONTEMPLATED VALUE OF b:

$$t_{\hat{b}} = \frac{\hat{b} - b}{s_{\hat{b}}} \tag{13}$$

Suppose that we want to check the plausibility of $b = 10$. Our test statistic will be

$$t_{\hat{b}} = \frac{\hat{b} - 10}{s_{\hat{b}}},$$

and we refer the value of the statistic to the t distribution with $n - 2$ degrees of freedom, as we did in Section 12-5 with $t_{\hat{m}}$.

Similarly, we construct confidence limits for b using the t distribution. For example, 90 percent limits are given by

$$\hat{b} \pm t_{n-2}\, s_{\hat{b}}. \tag{14}$$

As before, when n is large (say > 30), we can use as an approximation values from the standard normal probability table in place of the values from the t distribution with $n - 2$ degrees of freedom.

EXAMPLE 11 For the data in Example 3, check on the plausibility of the contemplated value $b = 0$.

SOLUTION. From Example 3 we recall that

$$n = 4, \quad \Sigma x^2 = 30, \quad SSx = 5, \quad \text{and} \quad s_e^2 = 0.15.$$

Thus the estimated variance of the slope $\hat{b} = -0.5$ is

$$s_{\hat{b}}^2 = \frac{30}{(4)(5)} 0.15 = 0.225.$$

To check on the plausibility of $b = 0$, we compute

$$t_{\hat{b}} = \frac{-0.05 - 0}{\sqrt{0.225}} = \frac{-0.05}{0.474} = -0.105.$$

Because this value of $t_{\hat{b}}$ is near zero, which is the middle of the t distribution, with 2 degrees of freedom, we conclude that the contemplated value $b = 0$ is plausible.

EXAMPLE 12 *Temperatures below the earth's surface.* In Example 5 in Chapter 3 we looked at data on temperatures taken at various depths in an artesian well at Grenoble, France, using a point 28 meters below the earth's surface as the origin for both temperature and depth. We add the point $x = 0$ and $y = 0$ to the four points studied earlier, because we know that the temperature difference at the origin is, in fact, zero:

number of meters below origin, x:	0	40	150	220	270
number of degrees above the temperature at the origin, y:	0	1.2	4.7	9.3	10.5

From these data we compute the following summary statistics:

$$n = 5, \quad \Sigma x_i^2 = 145{,}400, \quad \Sigma x_i = 680, \quad \bar{x} = 136,$$

$$\Sigma y_i^2 = 220.27, \quad \Sigma y_i = 25.7, \quad \bar{y} = 5.14,$$

$$\Sigma x_i y_i = 5634,$$

and

$$SSx = 145{,}400 - 5(136)^2 = 52{,}920,$$

$$SSy = 220.27 - 5(5.14)^2 = 88.172,$$

$$Sxy = 5634 - 5(136)(5.14) = 2138.8.$$

The coefficients for the least-squares line for predicting temperature from depth are

$$\hat{m} = \frac{2138.8}{52920} = 0.0404,$$

and

$$\hat{b} = 5.14 - (0.0404)(136) = -0.354,$$

so that the least-squares prediction line is

$$y = -0.354 + 0.0404x.$$

In our earlier look at these data we did not include the point (0, 0), but we did make sure that our line passed through the origin. By fitting the line without forcing it through the origin, we can check whether $b = 0$ is plausible. The minimizing value of the sum of squared deviations is

$$\hat{S} = 88.172 - (0.0404)^2 (52920) = 1.798,$$

and, using formula (12), we get

$$s_{\hat{b}}^2 = \frac{\Sigma x_i^2}{n(SSx)} s_e^2 = \frac{145{,}400}{5(52920)} \times \frac{1.798}{3} = 0.329.$$

Taking the square root, we have $s_{\hat{b}} = 0.574$. To get 95 percent confidence limits for the true y intercept b, we compute

$$\hat{b} \pm t_3 s_{\hat{b}} = -0.354 \pm 3.18(0.574) = [-2.18, 1.47].$$

The value $b = 0$ falls near the middle of the interval formed by these limits. Thus zero is indeed a plausible value for the y intercept.

To give a quantitative measure to the plausibility, we can compute

$$t_{\hat{b}} = \frac{-0.354 - 0}{0.574} = -0.62.$$

Interpolation in the t table with 3 degrees of freedom suggests that the chance of a deviation from zero larger than 0.62 is about 0.58 when the true value of b is zero. Thus, again, zero is a reasonable value for b.

Whether one prefers the general least-squares line just given or the line through the origin depends on how firmly one believes in the linearity. If that has no solid foundation, the general line is probably the better choice over the region of measurement.

PROBLEMS FOR SECTION 12-6

In Example 2B we gave the equation

$$y = -1.09 + 0.237x$$

for predicting y, the square root of the distance required to stop a car traveling at a speed of x miles per hour. Use the following summary statistics of the data to solve Problems 1 through 4:

$$n = 18, \quad \bar{x} = 41.11, \quad \bar{y} = 8.65, \quad SSx = 3577.78,$$

$$SSy = 202.865, \quad Sxy = 846.40.$$

1. Verify the values of the least-squares coefficients. Show your work.
2. Continuation. Find a 99 percent confidence interval for b using the two-sided multiplier from Table A-2 in Appendix III in the back of the book for the t distribution.
3. Continuation. How different would your interval be if you used the approximate multiplier from the standard normal table?
4. In Example 2B it was argued that $b = 0$ should be the value for the y intercept if the true regression of y on x really is linear. How plausible is this value of b?

★ 12-7 PREDICTING NEW y VALUES

In addition to constructing tests and confidence limits for slope and intercept, we often want to assess the variability of a new point or of an estimate of a value of y based on a value of x. Even though the variabilities of the y's around the line may be the same for each x, the departure of a new value from the estimate based on the fitted line does depend on the distance the new x is from the center of the previous measurements, $\bar{x}$. We explain here how to take this into account.

Making predictions from an estimated regression line is straightforward. We take a new x value, say x_0, and our predicted y value is $\hat{y}_0 = \hat{b} + \hat{m}x_0$. When we actually do get to observe the y value corresponding to x_0, say y_0, it will not be equal to $\hat{y}_0$, because we had to estimate the true regression parameters b and m by $\hat{b}$ and $\hat{m}$. Also, y_0 equals the true regression of y on x plus an error term, which we write in the reverse of the usual order:

$$y_0 = e_0 + mx_0 + b.$$

The difference between our prediction $\hat{y}_0$ and the observed value y_0 is

$$\begin{aligned} y_0 - \hat{y}_0 &= e_0 + mx_0 + b - \hat{b} - \hat{m}x_0 \\ &= e_0 - (\hat{m} - m)x_0 - (\hat{b} - b). \end{aligned}$$

The variance of this difference is based on contributions from each of the three parts on the right-hand side and takes into account the relationship between $\hat{b}$ and $\hat{m}$. We give without proof the formula

$$\begin{aligned} \mathrm{Var}(y_0 - \hat{y}_0) &= \sigma^2 + \frac{\sigma^2}{n} + \frac{\sigma^2 (x_0 - \bar{x})^2}{SSx} \\ &= \sigma^2\left(1 + \frac{1}{n} + \frac{(x_0 - \bar{x})^2}{SSx}\right). \end{aligned} \tag{15}$$

To appreciate this formula we look at its three parts:

1. The first term σ^2 is the contribution to variability because we are estimating the result for a single observation.
2. The second term σ^2/n is the contribution from the uncertainty of the level of the line, which is essentially the variability of $\bar{y}$.
3. The third part is the extra uncertainty because we are not sure of the slope of the line, and the farther from $\bar{x}$ we go, the more this matters. That is reflected in $(x_0 - \bar{x})^2$. When $x_0 = \bar{x}$, this third term vanishes.

If we substitute s_e^2 for σ^2 in equation (15), our estimate of the variance of $y_0 - \hat{y}_0$ is as follows:

ESTIMATED VARIANCE OF $y_0 - \hat{y}_0$:

$$s_{\hat{y}_0}^2 = s_e^2\left(1 + \frac{1}{n} + \frac{(x_0 - \bar{x})^2}{SSx}\right) \tag{16}$$

The value of $s_{\hat{y}_0}^2$ is a measure of how accurately we can predict a new y value, y_0, from a given x value, x_0.

Let us note two special things about expression (16). First, the estimated variance can never be smaller than $s_e^2(1 + 1/n)$, the value it takes when $x_0 = \bar{x}$. When our original sample is large, we can ignore the $1/n$ term, and this minimum value is approximately s_e^2. Second, the farther x_0 is from $\bar{x}$, the larger $s_{\hat{m}_0}^2$ becomes. This allows us to be properly cautious about predictions far beyond the range of x values in the sample used to calculate the least-squares coefficients when the relation is actually linear. It does not allow for the possibility that the true relation is nonlinear. Finally, when $x_0 = 0$, $\hat{y}_0 = \hat{b}$, the last two terms of (16) become

$$\begin{aligned} s_e^2\left(\frac{1}{n} + \frac{\bar{x}^2}{SSx}\right) &= s_e^2\left(\frac{SSx + n\bar{x}^2}{n(SSx)}\right) \\ &= s_e^2\left(\frac{\Sigma x^2 - n\bar{x}^2 + n\bar{x}^2}{n(SSx)}\right) \\ &= s_e^2\left(\frac{\Sigma x^2}{n(SSx)}\right). \end{aligned}$$

This is just the value of $s_{\hat{b}}^2$ from formula (12).

Our general method for getting confidence limits is applicable for the predicted value of a new observation. For 95 percent limits we take 95 percent confidence limits for y_0:

$$\hat{y}_0 \pm t_{n-2}\, s_{\hat{y}_0} \qquad (17)$$

Because the farther x_0 is from $\bar{x}$, the larger $s_{\hat{y}_0}$ is, confidence limits for an x value near $\bar{x}$ are closer together than those for x values farther away.

The confidence limits in (17) give us a way to check whether a newly observed y value, y_0, is consistent with predictions based on a previously computed least-squares regression line. For a value x_0, if the 95 percent confidence limits do not contain the new y value, y_0, then we have grounds for doubting that the new point is consistent with the least-squares line.

EXAMPLE 13 For the data in Example 3, find the estimated variance of $y_0 - \hat{y}_0$ when $x_0 = 2$.

SOLUTION. Recall that $n = 4$, $\bar{x} = 2.5$, $SSx = 5$, and $s_e^2 = 0.15$. Then, substituting in formula (16), we have

$$\begin{aligned} s_{\hat{y}_0}^2 &= 0.15\left(1 + \tfrac{1}{4} + \frac{(2 - 2.5)^2}{5}\right) \\ &= 0.15\left(1 + 0.25 + \frac{0.25}{5}\right) \\ &= 0.15\,(1.25 + 0.05) \\ &= 0.195. \end{aligned}$$

(The contribution of the third term is almost negligible, because $x_0 = 2$ is so close to $\bar{x} = 2.5$.) Thus the estimated standard error of $y_0 - \hat{y}_0$, at $x_0 = 2$, is $s_{\hat{y}_0} = \sqrt{0.195} = 0.44$.

EXAMPLE 14 *Indianapolis 500 predictions.* In Example 1 in Chapter 11, in our calculations with the winning speeds from 1961 through 1970 for the

Indianapolis 500 auto race, we used the least-squares regression line to make predictions for years beyond 1970. Is the winning speed of 158.59 miles per hour for 1974 consistent with the least-squares prediction?

SOLUTION. The least-squares prediction for 1974 is

$$\begin{aligned}\hat{y}_0 &= 137.60 + 1.93(1974 - 1960)\\ &= 137.60 + 1.93(14)\\ &= 164.62.\end{aligned}$$

Useful summary statistics from previous calculations based on the data in Table 11-1 are

$$n = 10, \quad \bar{x} = 5.5, \quad s_e = 2.54, \quad \text{and} \quad SSx = 82.5.$$

And here we have $x_0 = 1974 - 1960 = 14$. Substituting these values into equation (16) and taking the square root of both sides, we get

$$\begin{aligned}s_{\hat{y}_0} &= 2.54\sqrt{1 + \frac{1}{10} + \frac{(14 - 5.5)^2}{82.5}}\\ &= 2.54\sqrt{1 + 0.1 + 0.88} = (2.54)(1.41)\\ &= 3.58.\end{aligned}$$

To get 95 percent limits for the new y value (the winning speed for 1974) we need the value

$$t_8 = 2.31$$

from Table A-2 in Appendix III in the back of the book. Then the 95 percent confidence limits are

$$164.62 \pm 2.31(3.58) = 164.62 \pm 8.27 = [156.35, 172.89].$$

These limits are plotted in Fig. 12-6. The actual winning speed for 1974 was 158.59 miles per hour, and this lies within our interval, although not by much. Thus the new point seems reasonably consistent with the estimated regression based on data from 1961–1970.

To get a more quantitative assessment, we compute

$$\begin{aligned}t_{\hat{y}_0} &= \frac{y_0 - \hat{y}_0}{s_{\hat{y}_0}} = \frac{158.59 - 164.62}{3.58}\\ &= -1.68.\end{aligned}$$

Interpolating in the t table with 8 degrees of freedom gives a probability of a larger departure than this of about 0.14.

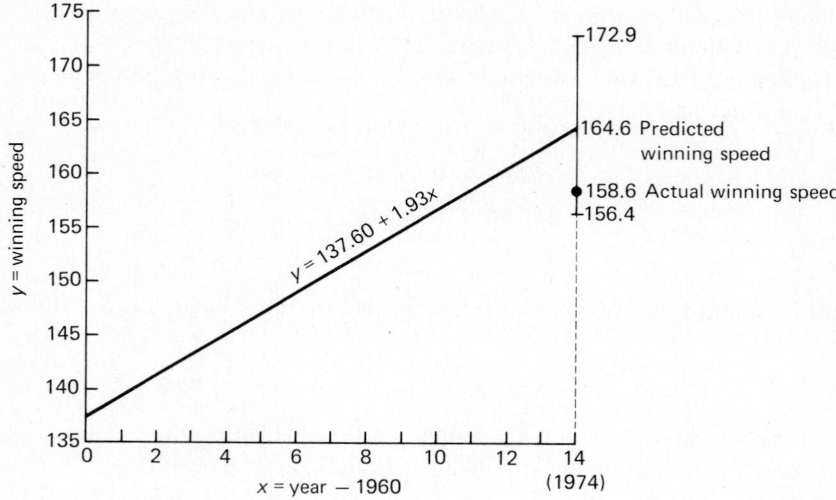

Figure 12-6 Predicting Indianapolis 500 winning speed for 1974.

We should, however, pause before trying to extrapolate with the estimated regression line in order to make predictions far from 1961–1970. Although the $(x_0 - \overline{x})^2$ term is intended to allow for uncertainty in the slope as we extrapolate, that is not what worries us. The possibility exists that a whole new automotive technology might be invented and that new machines with different engines and different fuel would be allowed to race. Our formula does not allow for such systematic changes in the process. This is a major danger in such extrapolations.

PROBLEMS FOR SECTION 12-7

Use the Indianapolis 500 data in Table 11-1 to solve Problems 1, 2, and 3. Useful summary statistics are

$$n = 10, \quad \overline{x} = 5.5, \quad s_e = 2.54, \quad \text{and} \quad SSx = 82.5.$$

The least-squares regression line is

$$y = 137.60 + 1.93x.$$

1. Based on the least-squares regression line, the retrospective prediction for the winning speed in 1950 is 118.30 miles per hour. Is this prediction consistent with the actual winning speed of 124.00 miles per hour?

2. Continuation. Construct 95 percent confidence limits for the winning speed at the Indianapolis 500 auto race in 1979.
3. Continuation. Look up the actual winning speed for the 1979 Indianapolis 500 in an almanac and see whether or not it falls within your interval in Problem 2.

Problems 4 through 6 deal with the data on heights (x) and weights (y) of 15 high school sophomore boys given in Table 11-3. Before doing any calculations, be sure to reverse the height and weight values for subject 7. Some summary statistics for the resulting data are

$$n = 15, \quad \Sigma x_i = 1013, \quad \Sigma y_i = 2093,$$

$$SSx = 263.73, \quad SSy = 10{,}575.73, \quad \text{and} \quad Sxy = 1560.73.$$

4. Compute the least-squares regression coefficients for predicting weight (y) from height (x), and find s_e, the estimated standard deviation of the regression error term.
5. Continuation. A new boy whose height is 65 inches is about to enter the sophomore class. What is your prediction of his weight?
6. Continuation. Construct 95 percent confidence limits to go with your prediction in Problem 5.

The data on the distances of the planets from the sun in Table 3-2 are used to solve Problems 7, 8, and 9. Useful summary statistics are

$$\text{number of planets} = 10, \quad \Sigma n = 55, \quad \Sigma \log d_n = 16.146,$$

$$\Sigma n^2 = 385, \quad \Sigma(\log d_n)^2 = 30.5826, \quad \Sigma n\,(\log d_n) = 108.040,$$

$$\text{least-squares line:} \quad \log d_n = 0.332 + 0.233n.$$

7. Predict the logarithm of the distance from the sun for a new 11th planet.
8. Continuation. Construct 99 percent confidence limits to go with the prediction in Problem 7.
9. Continuation. What values of d_{11} correspond to these limits?

Table 3-6 gives data on winning times for the 1500-meter run for men in the Olympics. Let

$$y = \text{time in seconds minus } 200$$

and

$$x = \text{year minus } 1900.$$

Summary statistics are

$$n = 16, \quad \Sigma x = 584, \quad \Sigma y = 480, \quad \Sigma xy = 13{,}748,$$

$$\Sigma x^2 = 29{,}952, \quad \Sigma y^2 = 16{,}140,$$

$$SSx = 8636, \quad SSy = 1740, \quad Sxy = -3772,$$

and the least-squares line is

$$y = 45.94 - 0.437x.$$

Use this information to solve Problems 10, 11, and 12.

10. Predict the winning time for the 1500-meter run for the 1976 Olympics.

11. Continuation. Construct 95 percent confidence limits for your prediction in Problem 10.

12. Continuation. Look up the actual winning time for the 1976 Olympic 1500-meter race in an almanac and see whether or not it falls within your interval in Problem 11.

12-8 SUMMARY OF CHAPTER 12

Equation numbers match those used earlier in the chapter.

1. The linear regression model for n pairs of (x, y) quantities (x_1, y_1), $(x_2, y_2), \ldots, (x_n, y_n)$, is

$$y_i = b + mx_i + e_i,$$

where b and m are the unknown regression coefficients. The random errors $e_1, e_2, \ldots, e_n$ are assumed to be independent and distributed according to the normal probability distribution with mean 0 and variance σ^2. This normality assumption is rarely met in practice, but its use allows us to make approximate tests and construct approximate confidence limits.

2. In terms of the special notation for sums of squared deviations, the least-squares estimates for b and m are

$$\hat{m} = \frac{Sxy}{SSx},$$

$$\hat{b} = \bar{y} - \hat{m}\bar{x}, \tag{1}$$

and the minimum sum of squared residuals is

$$\hat{S} = SSy - \hat{m}^2(SSx). \tag{2}$$

3. The estimated variance of the random error terms is

$$s_e^2 = \frac{\hat{S}}{n-2}. \tag{3}$$

4. The proportion of the variability SSy explained by the linear regression line is

$$r^2 = \frac{SSy - \hat{S}}{SSy}, \tag{5a}$$

$$r^2 = \frac{\hat{m}^2\, SSx}{SSy}, \tag{5b}$$

$$r^2 = \frac{(Sxy)^2}{(SSx)\,(SSy)}. \tag{6}$$

The quantity r^2 is the square of the sample correlation coefficient,

$$r = \frac{Sxy}{\sqrt{(SSx)(SSy)}}, \tag{7}$$

which lies between -1 and $+1$.

5. The estimated variances of the least-squares coefficients are

$$s^2_{\hat{m}} = \frac{s^2_e}{SSx} \tag{8}$$

$$s^2_{\hat{b}} = \frac{\Sigma x^2}{n(SSx)}\, s^2_e \tag{12}$$

6. The estimated variance for predicting a new y value at $x = x_0$ is

$$s^2_{\hat{y}_0} = s^2_e\left(1 + \frac{1}{n} + \frac{(x_0 - \bar{x})^2}{SSx}\right). \tag{16}$$

7. To test the plausibility of contemplated values of unknown parameters on new y values, we compute the critical ratio

$$\frac{\text{observed statistic} - \text{contemplated value}}{\text{estimated standard deviation of the statistic}},$$

and we compare it to values from the t distribution with $n - 2$ degrees of freedom. Confidence intervals for these unknown values take the form

observed statistic $\pm$ multiple of estimated standard deviation of the statistic,

where again the multiplier comes from the t distribution with $n - 2$ degrees of freedom.

SUMMARY PROBLEMS FOR CHAPTER 12

Table 11-5 gives data on demand for corn feed. Summary statistics for these data are

$$n = 27, \quad \Sigma x = 496.94, \quad \Sigma y = 4611.88,$$

$$\Sigma x^2 = 9288.9440, \quad \Sigma y^2 = 818{,}462.9624, \quad \Sigma xy = 86{,}728.0601.$$

Use this information to solve Problems 1 through 5.

1. Compute a 95 percent confidence interval for the slope m in the regression demand (y) on aggregate value (x).

2. Continuation. Comment on the plausibility of the value $m = 0$.

3. Continuation. What proportion of variability of demand is explained by the fitted least-squares line?

4. Continuation. An examination of the data in Table 11-5 shows a general increase in demand over time that may be related to factors other than aggregate value. Compute the regression of demand on year (minus 1948), using the starting date of each crop year.

5. Continuation. What proportion of variability of demand is explained by the fitted regression line in Problem 4? Compare this value with the one you computed in Problem 3.

Table 12-1 gives data on stopping distances for a car traveling at various speeds. Summary statistics are given at the bottom of the table. The least-squares line for predicting the square root of the distance is

$$y = -1.09 + 0.237x,$$

where $y = \sqrt{\text{distance}}$ and x = speed. Use this information to solve Problems 6 through 10.

6. Predict the value of $y = \sqrt{d}$ for a new observation for a car traveling at 35 miles per hour.

7. Continuation. Construct 90 percent limits for your predicted value in Problem 6.

8. Predict the value of $y = \sqrt{d}$ if the new observation is for a car traveling at 70 miles per hour.

9. Continuation. Construct 90 percent limits for your predicted value in Problem 8.

10. Continuation. Compare the sizes of the intervals formed by the limits in Problems 7 and 9, and explain why one interval is larger than the other.

REFERENCES

S. Chatterjee and B. Price (1977). *Regression Analysis by Example*, Chapters 1 and 2. Wiley, New York.

F. Mosteller, R. E. K. Rourke, and G. B. Thomas, Jr. (1970). *Probability with Statistical Applications*, second edition, Chapter 11. Addison-Wesley, Reading, Mass.

S. Weisberg (1980). *Applied Linear Regression*, Chapter 1. Wiley, New York.

13

Regression with Two Predictors

Learning Objectives

1. Developing the linear regression model for two predictors
2. Extending the method of least squares to fit linear regression equations with two predictors
3. Comparing the effectiveness of one versus two predictors
4. Applying linear regression ideas to multiplicative models involving two predictors

13-1 PREDICTING WITH TWO VARIABLES

Once we have used regression methods to get a linear relation

$$y = b + mx$$

between a dependent variable, y, and one independent variable or predictor, x, then the idea of using more than one predictor suggests itself. The next thing to try is two predictors.

EXAMPLE 1 *Picture frames.* We all know the formula for the circumference of a rectangle:

$$C = 2W + 2L,$$

where C is the circumference and W and L are the width and length.

L

W

Picture frames have a rectangular shape, and the employees in a frame shop often use a crude measuring stick and this formula to estimate the circumference. The stick measures distances only to the nearest inch.

To see how close he comes to the measured circumference, the shop manager performed an experiment with four frames. First, the width and length were measured to the nearest inch with the measuring stick, and then the circumference was measured to the nearest 1/10 of an inch using a tape measure. Because these measured quantities differ from the true quantities, we label them with the corresponding lower-case letters, w, l, and c. The results of the experiment are as follows:

Measurements to Nearest Inch			Observed Measurement to Nearest 1/10 Inch	Residuals
w	l	$2w + 2l$	c	$c - 2w - 2l$
10	15	50	50.8	0.8
15	25	80	80.1	0.1
20	30	100	99.6	−0.4
15	30	90	89.5	−0.5

The difference between the measured circumference c and the estimated one $2w + 2l$ is a **residual.** The results in the final column show us that even in a simple situation, prediction equations are subject to error.

In this example of a formula with two predictors, the equation $C = 2W + 2L$ represents a geometric law where we know exactly the form of the relation. In the social and biological sciences, we are likely to have less assured forms for our relationships. We may read that some variable y, such as a child's performance on a standardized test, is a "function of" other variables, u and w, such as parents' education and the child's age. (Now w is a general variable, no longer the width measurement in the example.) The expression "function of" means the same in everyday life as in mathematics; that is, it means "depends on in an as yet unspecified way." In statistics we usually expect our models to include some uncertainty, fuzz, or deviations that imply that although there may be some general trend, the forecasts are not exact. Thus we have as a mathematical model a regression equation:

$$y = b_0 + b_1u + b_2w + e \qquad (1)$$

Here e is taken to be a random error, just as in linear regression with one predictor. Let us take another example where our experience is a good guide.

EXAMPLE 2 *Ski trip expenses.* Suppose that we go on a ski trip. The total cost of the trip will have three components: (a) the preparation costs, which include purchases of maps, waxes, etc., (b) the daily expenditure (such as tow fees, lodging, and meals) times the number of days for the trip, and (c) the automobile expenses, the number of miles times the cost per mile. The preparation costs, daily expenditures, and automobile costs per mile are roughly constant from trip to trip, but the number of days and number of miles driven will vary if the trips are to different ski areas and at different times. Table 13-1 contains the information for five different trips.

We wish to estimate the coefficients b_0, b_1, and b_2 in equation (1) for this problem. Because we have already suggested an interpretation for each, we

TABLE 13-1
Expenses for five different ski trips

u *Number of days*	w *Number of miles driven*	y *Total expenses in dollars*	$\tilde{y}$ *Naive prediction†*	$y - \tilde{y}$ *Residual*	$\hat{y}$ *Least-squares prediction††*	$y - \hat{y}$ *Residual*
2	120	180	183	−3	173.7	6.3
4	60	315	324	−9	321.0	−6.0
6	40	480	471	9	475.2	4.8
5	180	420	417	3	421.0	−1.0
3	100	245	255	−10	249.1	−4.1
				−10		0.0
				$\Sigma(y - \tilde{y})^2 = 280$		$\Sigma(y - \hat{y})^2 = 116.54$

† $\tilde{y} = 15 + 75u + 0.15w$.
†† $\hat{y} = -5.2 + 78.9u + 0.176w$.

can make some guesses as to their values. For example:

$$\text{preparation costs} \quad \tilde{b}_0 = \$15,$$

$$\text{daily expenditures} \quad \tilde{b}_1 = \$75,$$

$$\text{cost per mile} \quad \tilde{b}_2 = \$0.15.$$

Using these, we can get predictions for the total expenditures using the linear relation

$$\begin{aligned} \tilde{y} &= \tilde{b}_0 + \tilde{b}_1 u + \tilde{b}_2 w \\ &= 15 + 75u + 0.15w. \end{aligned}$$

The five values of $\tilde{y}$ are listed in Table 13-1, along with the residuals, $y - \tilde{y}$. No residual exceeds \$10, and thus it appears that we have a reasonable prediction equation. The sum of squared residuals is

$$\Sigma(y - \tilde{y})^2 = 280.$$

In most problems we do not know what the model actually involves—what the variables are, whether the equation is linear, whether there are only two predictors, and so on. Even if we know that formula (1) gives the correct model, we do not know the values of b_0, b_1, and b_2 or even the reasonable guesses that we used in this example. As in the case of one predictor, we need to estimate the unknown regression coefficients. One way to estimate these uses an extension of the method of least squares that we used in Chapters 11 and 12 for the case of one predictor. Using a least-squares computer program for the data in Table 13-1, we get the following estimates:

$$\hat{b}_0 = -5.2, \ \hat{b}_1 = 78.9, \quad \text{and} \quad \hat{b}_2 = 0.176.$$

The predicted values using the least-squares equation are

$$\hat{y} = \hat{b}_0 + \hat{b}_1 u + \hat{b}_2 w$$

$$= -5.2 + 78.9u + 0.176w,$$

and the values of $\hat{y}$ are listed in Table 13-1 along with the new residuals $y - \hat{y}$.

Notice that the residuals from the least-squares line add to zero (the other residuals do not) and that the sum of squared residuals for the least-squares regression line

$$\Sigma (y - \hat{y})^2 = 116.54$$

is smaller than the sum of squares for the line based on our guesses, 280. Both of these results follow from properties of the method of least squares.

The least-squares result does not look just as we expected. We had tentatively labeled b_0 as the preparation costs, but our estimate $\hat{b}_0 = -5.2$ is negative! This may mean that our original interpretation was not quite correct, or more likely that we need more than five observations to get a more accurate estimate of b_0. The estimates of the other two coefficients make more sense, and they are reasonably close to our expectations.

PROBLEMS FOR SECTION 13-1

Draper and Smith (1980) gave an example of steam consumption in a factory, with

y = number of units of steam used per month,
u = average atmospheric temperature for month in degrees Fahrenheit,
w = number of operation days per month.

(Their regression coefficients are given on their p. 196; data on their pp. 615–616.)

1. Which variable is the dependent variable, and which are the independent ones?

2. Continuation. If we have a prediction equation for this problem of the form

$$y = b_0 + b_1 u + b_2 w,$$

what signs do you expect b_1 and b_2 to have? Explain your answer.

3. Continuation. For a set of 25 observations, Draper and Smith computed the fitted equation

$$y = 9.1 - 0.0724u + 0.2029w.$$

Do the coefficients agree in sign with your predictions from Problem 2?

4. Continuation. Using the equation in Problem 3, estimate the value of y if $u = 68°F$ and $w = 10$ days. What value of y will you predict if $u = 50°F$ and $w = 20$ days?

5. Continuation. The predictions for y using the equation in Problem 3 do not coincide exactly with the observed values; that is, an error term is associated with the equation. If you could use more variables than u and w to predict y, which one variable from the following list would you choose to add? Why?

a) average wind velocity (in miles per hour)
b) calendar days per month
c) days below 32°F

6. From a sample of 100 freshmen at a major university, administrators wish to predict first-year grade-point average (GPA, with values between 0 and 4) using verbal and mathematical scores on the Scholastic Aptitude Test (SAT). If

$$y = \text{first-year GPA},$$

$$u = \text{SAT math score},$$

$$w = \text{SAT verbal score},$$

what signs do you expect b_1 and b_2 to have in the following regression equation?

$$y = b_0 + b_1u + b_2w$$

7. Continuation. The least-squares equation computed from data for the 100 students is

$$y = 0.63 + 0.00019u + 0.00421w.$$

Do the coefficients agree in sign with your predictions in Problem 6?

8. Continuation. A student enters the university with SAT scores

$$u = 760,$$

$$w = 770.$$

Is the predicted GPA using the equation from Problem 7 a reasonable one? Explain your answer.

9. Continuation. What other variables might you wish to use in addition to SAT scores for predicting GPA?

13-2 LEAST-SQUARES COEFFICIENTS FOR TWO PREDICTORS

When we have just three variables, a dependent variable y and two independent variables u and w, our observations come in triples:

$$\text{observation 1:} \quad y_1, \; u_1, \; w_1,$$

$$\text{observation 2:} \quad y_2, \; u_2, \; w_2,$$

and so on. We regard a triple as a single observation. The general observation is

$$\text{observation } i\text{:} \quad y_i, \; u_i, \; w_i,$$

where the subscript i refers to the ith observation, and i runs from 1 through n, the total number of observations.

Just as in the case of one predictor, the least-squares method for two predictors minimizes the sum of squared residuals. We plan to fit an equation of the form

$$y = b_0 + b_1 u + b_2 w.$$

For specified values of b_0, b_1, and b_2, and a particular observation (y_i, u_i, w_i), the predicted value of y_i is

$$b_0 + b_1 u_i + b_2 w_i,$$

and the residual is

$$\begin{aligned} d_i &= \text{observed} - \text{predicted} \\ &= y_i - b_0 - b_1 u_i - b_2 w_i. \end{aligned} \tag{2}$$

(We use the notation d_i for distance instead of r_i for residuals because of possible confusion with the correlation coefficients.) We wish to choose the b's to minimize the sum of squares of the d's; that is, we minimize

$$\begin{aligned} S &= \Sigma d_i^2 \\ &= \Sigma (y_i - b_0 - b_1 u_i - b_2 w_i)^2. \end{aligned} \tag{3}$$

As we did in the one-predictor situation of Chapter 12, we use hats to label the b's that minimize S for the sample of n observations, as $\hat{b}_0$, $\hat{b}_1$, and $\hat{b}_2$. The minimum value of S is as follows:

$$\hat{S} = \Sigma (y_i - \hat{b}_0 - \hat{b}_1 u_i - \hat{b}_2 w_i)^2 \tag{4}$$

The values $\hat{b}_0$, $\hat{b}_1$, and $\hat{b}_2$ are called the least-squares estimates of the unknown regression coefficients, and the least-squares predicted value for the *i*th observation is

$$\hat{y}_i = \hat{b}_0 + \hat{b}_1 u_i + \hat{b}_2 w_i.$$

As we might expect, finding the values of $\hat{b}_0$, $\hat{b}_1$, and $\hat{b}_2$ to minimize expression (3) requires more effort than in the least-squares problem with a single predictor. Let us look at this minimization problem numerically in an example.

EXAMPLE 3 *Least squares for the ski trip expenses.* For the five observations dealing with the expenses for ski trips in Table 13-1, find the values of b_0, b_1, and b_2 that minimize S.

SOLUTION. Let us begin by assuming that we know the value of $b_0 = -5.2$. Then we need only find $\hat{b}_1$ and $\hat{b}_2$ to minimize

$$S = \sum_{i=1}^{5} (y_i + 5.2 - b_1 u_i - b_2 w_i)^2.$$

One way to find $\hat{b}_1$ and $\hat{b}_2$ is to look at several possible values of b_1 (say 71, 75, 79, and 83) and several possible values of b_2 (say 0.16, 0.18, 0.20, and 0.22), assuming for a moment these were in the neighborhood of the solution. Then we can compute S for each of the combinations of b_1 and b_2. These 16 values are given in Table 13-2. We see that the smallest value of S, 118.8, in this table is that for $b_1 = 79$ and $b_2 = 0.18$. As a next step toward getting the minimized value of $\hat{S} = 116.54$ we gave in Section 13.1, we need finer grids

TABLE 13-2

Values of $S = \Sigma(y_i - b_0 - b_1 u_i - b_2 w_i)^2$ for the data in Table 13-1, using $b_0 = -5.2$

	b_2			
b_1	*0.16*	*0.18*	*0.20*	*0.22*
71	6234.80	5613.20	5041.20	4518.80
75	1740.40	1426.00	1161.20	946.00
79	126.00	118.80	161.20	253.20
83	1391.60	1691.60	2041.20	2440.40

of values for b_1 near $b_1 = 79$ (say 78.8, 78.9, 79, 79.1, 79.2) and b_2 near $b_2 = 0.18$ (say 0.174, 0.176, 0.178, 0.180, 0.182, 0.184). Using these, we might arrive at the least-squares values $\hat{b}_1 = 78.9$ and $\hat{b}_2 = 0.176$.

For simplicity, the calculations in Table 13-2 for S are based on the minimizing value $\hat{b}_0 = -5.2$. If we did not know this in advance, we would need to do the same calculations for several different values of b_0 to find the minimizing value of b_0 as well.

FORMULAS FOR $\hat{b}_0$, $\hat{b}_1$, AND $\hat{b}_2$

To find the least-squares coefficients numerically by using a grid, as we did in Example 3, is both time-consuming and wasteful. As in the case of simple linear regression, the minimization of S can be done algebraically, and Appendix II in the back of the book contains formulas for the least-squares coefficients $\hat{b}_0$, $\hat{b}_1$, and $\hat{b}_2$. To use these formulas in practice, one will almost certainly want to use a calculator. Some pocket calculators have special buttons for computing these coefficients directly.

DOING LEAST SQUARES BY HIGH-SPEED COMPUTER

Fortunately, we have a better way to get the least-squares regression coefficients—we can use the high-speed computer. Because so many people want to compute least-squares regression coefficients in practical situations, almost every high-speed computer has a built-in computer program for doing regression calculations. To use these programs, usually we need only give the computer a list of our n observations (the triples: y_i, u_i, and w_i), tell it which is the dependent variable and which are the predictors, and then indicate that we want to compute the least-squares coefficients for the regression of y on u and w. A good computer program not only will compute $\hat{b}_0$, $\hat{b}_1$, and $\hat{b}_2$ but also will give us $\hat{S}$, the predicted values, $\hat{y}_i$, the residuals, $y_i - \hat{y}_i$, and other useful information.

PROBLEMS FOR SECTION 13-2

The data in Example 1 for the circumferences of picture frames are used for Problems 1, 2, and 3. For Problems 1 and 2, we take $b_0 = 0$.

1. Compute the value of S for the prediction formula $c = 2w + 2l$, which uses $b_1 = b_2 = 2$.

2. Find the four values of S corresponding to $b_1 = 2.0, 2.2$ and $b_2 = 1.8, 2.0$. Which values of b_1 and b_2 give the smallest value of S?
3. Compare the best predictor from Problem 2 with the performance of the least-squares estimator

$$y = 1.95 + 2.02w + 1.91l,$$

computed by using formulas in Appendix II in the back of the book. Note that the least-squares formula allows $\hat{b}_0$ to be nonzero.
4. For the data in Table 13-1, find the value of S for $b_0 = -5.2$, $b_2 = 0.176$, and (a) $b_1 = 78.8$, (b) $b_1 = 78.9$, and (c) $b_1 = 79.1$.
5. Continuation. Compare the values of S from (a) and (c) in Problem 4 with the least-squares value from (b).
6. Consider the following data for predicting first-year GPA from SAT scores:

Observation numbers:	1	2	3	4	5
(GPA) y:	2.4	2.5	1.9	3.2	3.8
(SAT math) u:	662	595	500	620	790
(SAT verbal) w:	565	600	450	675	740

Using a least-squares computer program or the formulas from Appendix II in the back of the book, find the least-squares equation for predicting GPA from SAT scores.
7. Continuation. Compute the value of $\hat{S}$, using equation (4), for your equation in Problem 6.
8. Continuation. Compute the value of S for the equation

$$y = 0.6 + 0.002u + 0.004v,$$

and compare it with the value of $\hat{S}$ from Problem 7.

13-3 ONE VERSUS TWO PREDICTORS

When we fit a two-predictor linear regression equation of the form

$$y = b_0 + b_1u + b_2w,$$

we often wish to compare our estimated coefficients with those that we would have computed using only one predictor, either u or w, and an estimating equation of the form

$$y = \hat{a}_0 + \hat{a}_1u$$

or one of the form

$$y = \hat{c}_0 + \hat{c}_2 w.$$

If we estimate the coefficients $b_0, b_1,$ and b_2 from actual data using least squares, the estimated coefficient $\hat{b}_1$ ordinarily will differ from $\hat{a}_1$, $\hat{b}_2$ will differ from $\hat{c}_2$, and b_0 will differ from both $\hat{a}_0$ and $\hat{c}_0$.

EXAMPLE 4 Using the data in Table 13-1, compute the least-squares regression coefficients for y versus u, and y versus w. Compare these with the coefficients in Example 2.

SOLUTION. For the regression of y on u, we have the following summary statistics:

$$\bar{y} = 328, \quad \bar{u} = 4, \quad SSu = 10, \quad \text{and} \quad Suy = 775.$$

Thus the least-squares regression coefficients are

$$\hat{a}_1 = \frac{775}{10} = 77.5 \quad \text{and} \quad \hat{a}_0 = 328 - (77.5)(4) = 18.$$

The fitted equation is

$$y = 18 + 77.5u.$$

Similarly, for the regression of y on w, we have

$$\bar{y} = 328, \quad \bar{w} = 100, \quad SSw = 12{,}000, \quad \text{and} \quad Swy = -4200.$$

Thus,

$$\hat{c}_2 = \frac{-4200}{12{,}000} = -0.35 \quad \text{and} \quad \hat{c}_0 = 328 - (-0.35)(100) = 363,$$

and the fitted equation is

$$y = 363 - 0.35w.$$

The coefficients using each predictor separately differ from those we computed earlier:

Equation	**Constant**	***u***	***w***
y versus u and w	−5.2	78.9	0.176
y versus u	18	77.5	—
y versus w	363	—	−0.35

Thus we see that the coefficients depend on which predictors are allowed to enter the equation. Even the signs change.

MINIMIZING *S*

Our aim in estimating the coefficients in the equation

$$y = b_0 + b_1 u + b_2 w$$

was to find a "good fit." The device we chose for achieving this aim was the method of least squares. This method requires minimizing

$$S = \Sigma(y_i - \hat{y}_i)^2, \tag{4}$$

where $\hat{y}_i$ is the predicted value. We can also compute the minimum value of S if $\hat{y}_i$ is a predicted value based on only one of the two variables u and w.

> **FACT:** The predicted values y_i based on $\hat{b}_0$, $\hat{b}_1$, and $\hat{b}_2$ always produce a value of S at least as small as that produced by the predicted values based on a linear equation using only one of the variables.

EXAMPLE 5 Compare the values of S for the one-predictor regression equations for the data in Table 13-1 with the value $\hat{S}$ based on the least-squares equation with two predictors.

SOLUTION. From Example 2, we have $\hat{S} = 116.54$. For the one-predictor equation with the predictor u,

$$y' = 18 + 77.5u,$$

we have $SSy = 60{,}530$ and $SSu = 10$; so

$$\begin{aligned} S' &= \Sigma(y_i - y_i')^2 \\ &= 60{,}530 - (77.5)^2 10 = 467.5. \end{aligned}$$

Similarly, for the other one-predictor equation with the predictor w,

$$y^* = 363 - 0.35w,$$

we have

$$\begin{aligned} S^* &= \Sigma(y_i - y_i^*)^2 \\ &= 60{,}530 - (-0.35)^2 12{,}000 = 59{,}060. \end{aligned}$$

If we had to choose only one predictor in this example, using *u* does a much better job than using *w*, because S' is much smaller than S^*. But $\hat{S}$ is still smaller (but not much smaller) than S'.

THE PROPORTION OF VARIABILITY EXPLAINED

We can easily extend the idea of Chapter 12 regarding the proportion of variability in the dependent variable y explained by the fitted regression line. For one predictor variable, we saw that the proportion of *SSy* that is removed by the fitted regression is as follows:

$$R^2 = \frac{SSy - \hat{S}}{SSy} \tag{5}$$

We use R^2 in place of the r^2 of Chapter 12 because we wish to use formula (5) with more than one predictor. Indeed, the same formula works for two predictors if $\hat{S}$ is the minimum sum of squared deviations:

$$\hat{S} = \Sigma(y_i - \hat{b}_0 - \hat{b}_1 u - \hat{b}_2 w)^2.$$

Just as we saw in Chapter 12, R^2 must be between 0 and 1; that is,

$$0 \leq R^2 \leq 1. \tag{6}$$

Let us use a special notation so that we can compare the values of R^2 for different equations:

Notation for R^2	Predictors in Equation
$R^2_{u,w}$	u, w
R^2_u	u
R^2_w	w

In this notation it is understood that we are predicting y. If there is any ambiguity about which variable is being predicted, we need a more complicated notation. From the FACT about $\hat{S}$ noted earlier in this section and formula (5), we can see the following:

FACT RESTATED:

$$R^2_{u,w} \geq R^2_u$$

$$R^2_{u,w} \geq R^2_w$$

This result means that by using a second predictor variable in our regression equation we never explain a smaller proportion of the variability in the dependent variable y.

EXAMPLE 6 Compare R^2 values for regression equations fitted to the data in Table 13–1.

SOLUTION. From Example 5 we have that

$$R^2_{u,w} = \frac{60{,}530 - 116.54}{60{,}530} = 0.998,$$

$$R^2_u = \frac{60{,}530 - 467.5}{60{,}530} = 0.992,$$

and

$$R^2_w = \frac{60{,}530 - 59{,}060}{60{,}530} = 0.024.$$

As we expected, $R^2_{u,w} > R^2_u$ and $R^2_{u,w} > R^2_w$.

Because R^2_u is almost the same as $R^2_{u,w}$ in value, and both are extremely close to the maximum value of 1, we can say that in terms of the variability of y explained, the addition of the predictor w to the regression of y on u has very little effect. Recall from Example 4 that the estimated regression coefficient for u also changed only a little when w was included in the equation, from 77.5 to 78.9. The addition of w did have a much larger impact on the estimate b_0 (going from 18 to -5.2), even though $\hat{b}_1$ and R^2 changed very little. Appendix II in the back of the book contains an alternative formula for the calculation of $R^2_{u,w}$.

MULTIPLE CORRELATION COEFFICIENT

We have already recognized that for one variable $R^2 = r^2$ and r is the correlation coefficient. What is new here is that when two variables predict y, R is called the **multiple correlation coefficient.** What is being correlated is y_i with its forecasted value $\hat{y}_i$, which is equal to $\hat{b}_0 + \hat{b}_1 u_i + \hat{b}_2 w_i$.

Whereas r can range between -1 and $+1$ and thus can be negative, R takes only values between 0 and 1. The formulas automatically make the sign positive when we fit the regression coefficients $\hat{b}_0$, $\hat{b}_1$, and $\hat{b}_2$.

PROBLEMS FOR SECTION 13-3

Table 13-3 contains data for 50 of the largest U.S. cities concerning their cancer mortality (y), the length of time their drinking water has been fluoridated (u), and the percentage of population age 65 or older (w). We are interested in examining the relationship between cancer mortality and the fluoridation of water. The data in Table 13-3 are used for the following 12 problems. Summary data are

$$n = 50, \quad \Sigma y = 9561, \quad \Sigma u = 330, \quad \Sigma w = 541.7,$$

$$SSy = 75{,}870.58, \quad SSu = 2{,}818, \quad SSw = 300.3322,$$

$$Suy = 5185.4, \quad Swy = 4103.826, \quad Suw = 278.08.$$

1. Using the first 10 and last 10 observations, make a plot of y against u.

2. Using the same observations, plot y against w.

3. Continuation. (a) In the plots of Problems 1 and 2, does y seem to be related to either variable? If so, describe how. (b) Are there any special features of the plots that you can note?

4. Compute the regression of y on u.

5. Compute the regression of y on w.

6. Continuation. Draw the estimated regression lines from Problems 4 and 5 on your plots from Problems 1 and 2. Do the relationships seem linear? Do there appear to be any outliers? If so, which cities are they?

★ **7.** Use a least-squares computer program to compute the values of R^2 for the two regressors in Problems 4 and 5.

★ **8.** Using the formulas from Appendix II in the back of the book, or a least-squares computer program, calculate the fit of the two-predictor regression equation, with predictors u and w.

★ **9.** Continuation. Use a least-squares computer program to calculate $\hat{S}$ and R^2 for your equation in Problem 8.

★ **10.** Continuation. Use a least-squares computer program to compare the R^2 values for the two-predictor regression equation and the regression of y on u.

TABLE 13-3

For the 50 largest cities in 1970: cancer mortality, number of years city water supply has been fluoridated, percent age 65 or more

City	y *Cancer mortality*	u *Number of years fluoridated*	w *Percent of population age 65 or more*
New York	215	5	12.1
Chicago	204	14	10.6
Los Angeles	174	0	10.1
Philadelphia	217	16	11.7
Detroit	213	3	11.5
Houston	131	0	6.5
Baltimore	223	17	10.6
Dallas	191	4	7.9
Washington, D.C.	200	18	9.4
Cleveland	219	14	10.6
Milwaukee	189	16	11.0
San Francisco	249	17	14.0
San Diego	132	0	8.8
San Antonio	137	0	8.4
Boston	227	0	12.8
Memphis	164	0	8.5
St. Louis	207	14	14.7
New Orleans	216	0	10.7
Phoenix	150	0	8.7
Columbus	174	0	8.5
Seattle	217	0	13.1
Pittsburgh	243	17	13.5
Denver	157	16	11.5
Kansas City, Mo.	187	0	11.9
Atlanta	179	0	9.2
Buffalo	248	15	13.3
Cincinnati	251	0	13.0
San Jose	118	0	5.6
Minneapolis	228	12	15.0
Fort Worth	169	5	9.6
Toledo	196	14	11.1
Newark, N.J.	166	0	8.0
Portland, Ore.	243	0	14.9
Oklahoma City	170	15	9.8
Louisville	230	18	12.4
Oakland	210	0	13.3
Long Beach, Calif.	205	0	14.1
Omaha	169	1	10.1

Table 13-3 (Cont.)

City	y *Cancer mortality*	u *Number of years fluoridated*	w *Percent of population age 65 or more*
Miami	266	18	14.5
Tulsa	159	16	9.1
Honolulu	115	0	6.9
El Paso	105	0	6.0
St. Paul	200	17	13.3
Norfolk	132	17	6.8
Birmingham	234	0	11.8
Rochester	215	10	13.7
Tampa	203	0	12.4
Wichita	145	0	8.8
Akron	197	1	11.5
Tucson	172	0	10.4

Note: Indianapolis, Nashville, and Jacksonville are not included because their boundaries changed markedly between 1960 and 1970; therefore the table lists 50 of the 53 largest cities.

Source: Data collected by W. Mason from various U.S. Government publications.

★ **11.** Continuation. Interpret the equation in Problem 8, explaining the effect of adding w to the regression of y on u.

12. Continuation. (a) Do your answers to Problems 1 through 11 shed any light on the possible causal links between the fluoridation of water supplies and cancer? (b) What other variables might you want to include in a further analysis of these data?

13-4 MULTIPLICATIVE FORMULAS WITH TWO PREDICTORS

Not all formulas with two predictor or independent variables have the nice additive form of our first examples. Some other examples of familiar formulas with two independent (predictor) variables are the following:

Dependent Variable	Formula	Independent Variables
(i) area of rectangle	$A = LW$	L, length; W, width
(ii) volume of a cylinder	$V = \pi r^2 H$	r, radius; H, height
(iii) pressure	$P = kT/V$	(k, constant); T, temperature; V, volume

The first two equations are examples of mathematical laws; the third is the ideal gas law from physics. By taking logarithms of both sides of these equations, we can write them in an additive form:

$$\text{(i)}\quad \log A = \log L + \log W,$$

$$\text{(ii)}\quad \log V = \log \pi + 2 \log r + \log H,$$

$$\text{(iii)}\quad \log P = \log k + \log T - \log V.$$

In all three of these equations we are predicting the value of a dependent variable, y, from the values of two independent variables, u and w, using a linear relation of the form

$$y = b_0 + b_1 u + b_2 w.$$

For the three examples, the variables are

	y	u	w
(i)	$\log A$	$\log L$	$\log W$
(ii)	$\log V$	$\log r$	$\log H$
(iii)	$\log P$	$\log T$	$\log V$

The coefficients are

	b_0	b_1	b_2
(i)	0	1	1
(ii)	$\log \pi$	2	1
(iii)	$\log k$	1	-1

In the social and biological sciences it is often natural to cast models with two or more predictors in a multiplicative form. Taking logarithms of both sides of the equation is thus a common practice.

EXAMPLE 7 *Computing the volume of trees.* In estimating the volume V of a cylinder from $\pi r^2 H$, we could have errors in measurement for r and H. But more generally, just as in linear regression with one predictor, we rarely expect, even without measurement error, perfect relationships; we expect only general trends. For example, foresters often want to predict the volume of tree trunks to be cut for lumber. To a first, and very rough, approximation, we might conjecture that tree trunks are roughly cylindrical, which suggests the possibility of using the equation

$$\text{cylinder formula:} \quad V' = \pi r^2 H. \tag{7}$$

Actually, large tree trunks are tapered, being wide at the bottom and narrow at the top. Thus, another possible estimation equation is that for a cone:

$$\text{cone formula:} \quad V^* = \pi r^2 H/3, \tag{8}$$

where r is the radius at the bottom of the cone.

In practice, foresters measure the circumference of a tree trunk at breast height and thus get a measure of the radius r. Table 13-4 contains measurements on 10 tree trunks; their volumes were determined by cutting the trees down, submerging them in a tank of water, and measuring the amount of water displaced. Note that the tree volume is much more like that of a cone than that of a cylinder. The magnitude of the difference between the cone formula prediction and the true volume increases as the volume increases, and the predictions are too small for the small trees and too large for the large ones. This can be seen in Table 13-4. Perhaps we can find a better relationship for predicting volume than either formula (7) or formula (8).

TABLE 13-4
Measurements on 10 tree trunks

Breast-height radius r *(ft)*	*Height* H *(ft)*	*Volume* V *(ft³)*	*Predicted volume* — *Cylinder formula* V′ *(ft³)*	*Predicted volume* — *Cone formula* V* *(ft³)*
7/12	16	6.5	17.1	5.7
7/12	20	7.1	21.4	7.1
7/12	14	5.6	15.0	5.0
7/12	16	5.9	17.1	5.7
8/12	32	14.2	44.7	14.9
8/12	24	10.3	33.5	11.2
8/12	20	9.6	27.9	9.3
9/12	36	17.9	63.6	21.2
9/12	40	22.8	70.7	23.6
10/12	48	25.0	104.7	34.9

Note: The radius was measured to the nearest whole inch, and the r value is the number of inches divided by 12.

We have just seen that taking logarithms in multiplicative equations like formulas (7) and (8) leads to linear equations of the form

$$y = b_0 + b_1 u + b_2 w.$$

By taking logarithms of both sides for formulas (7) and (8), we get two such equations, where

$$y = \log V, \quad u = \log r, \quad w = \log H, \quad b_1 = 2, \quad \text{and} \quad b_2 = 1.$$

The value of b_0 is $\log \pi$ for formula (7) and $\log(\pi/3)$ for formula (8).

Our prediction V^* from the formula for the cone, expression (8), underestimated when the volume was small and overestimated when it was large. Using a two-predictor linear regression equation, we can try to choose values of b_0, b_1, and b_2 to get better predictions for volume than the formula for a cone provides. For our calculations in this example we use natural logarithms, because in scientific problems they are more often used than common logarithms. The natural logarithms of the values of the three variables are listed in Table 13-5. We use the notation ln for logarithms to the base e and log for logarithms to the base 10. Table 13-5 also lists the values of

$$\text{formula for cone:} \quad y^* = \ln(\pi/3) + 2u + w$$
$$= 0.046 + 2u + w,$$

TABLE 13-5

Natural logarithms of radius, height, and volume for data on 10 trees in Table 13-5

$u = \ln r$	$w = \ln H$	$y = \ln V$	*Cone formula* $y^* = \ln V^*$	*Residual* $y - y^*$
−0.539	2.773	1.872	1.740	0.132
−0.539	2.996	1.960	1.960	0
−0.539	2.639	1.723	1.609	0.114
−0.539	2.773	1.775	1.740	0.035
−0.405	3.466	2.653	2.701	−0.048
−0.405	3.178	2.332	2.416	−0.084
−0.405	2.996	2.262	2.230	0.032
−0.288	3.584	2.885	3.054	−0.169
−0.288	3.689	3.127	3.161	−0.034
−0.182	3.871	3.219	3.552	−0.333

the predictions in the ln scale from our approximating formula for the volume of a cone. The sum of squared residuals using this equation is

$$S^* = 0.1826.$$

By using a computer program (or the formulas in Appendix II in the back of the book), we find that the least-squares fitted regression line for this problem is

$$\text{least squares:} \quad y = 0.152 + 1.44u + 0.883w,$$

and the minimizing value of sum of squared residuals is

$$\hat{S} = 0.0390.$$

As we expected, $\hat{S}$ is less than S^*. The proportion of variability in y explained by this equation is

$$R^2_{u,w} = \frac{2.786 - 0.0390}{2.786} = 0.986,$$

a value very close to 1. The numerical values we give here come from rounding values from the output of our computer program. If we use the formulas in Appendix II, our answers may differ slightly. The constant term b_0, which we compute using the formula

$$b_0 = \bar{y} - b_1\bar{u} - b_2\bar{w},$$

is sensitive to relatively small shifts in the values of the other two coefficients b_1 and b_2. When b_1 changes by 0.20 (from 1.44 to 1.64) and b_2 by 0.06 (from 0.883 to 0.823), the constant term b_0 goes from 0.152 to 0.427. The reason is that the three variables y, u, and w for the 10 observations in Table 13-5 are highly interrelated. This can be seen by looking at the R^2 values for the regressions of y on u and y on w,

$$R^2_u = 0.934,$$

$$R^2_w = 0.974,$$

and then looking at the square of the correlation between u and w: 0.886. Not only is y almost perfectly linearly related to u and to w, but also the relationship between u and w is almost linear. Although u and w are measuring different quantities (log-radius and log-height), the two are closely related for our range of observation. Thus we have trouble getting stable coefficients that are unaffected by minor changes in the data.

When two predictors are highly related, as in this example, we say that they are nearly collinear, and we must be careful how we interpret our estimated regression coefficients.

The improvement of the two-variable least-squares equation over the one-variable equations is measured by the differences in the R^2 values:
1. The improvement due to the addition of variable w, given that u is already in the equation

$$R^2_{u,w} - R^2_u$$
$$= 0.986 - 0.934 = 0.052.$$

2. The improvement due to the addition of variable u, given that w is already in the equation

$$R^2_{u,w} - R^2_w$$
$$= 0.986 - 0.974 = 0.012.$$

Looking at these differences in R^2's cannot tell us if we need to add the variable to the equation. To answer this question we need the estimated standard errors of the least-squares coefficients in the two-variable equation. For this problem, our computer program gives these:

Coefficient	Estimated Standard Error
$\hat{b}_1 = 1.44$	0.574
$\hat{b}_2 = 0.883$	0.171

We use the term *standard error* here in place of *standard deviation* because most computer programs that produce regression results do so. Thus, estimated standard error means estimated standard deviation.

To check on the plausibility of $b_1 = 0$ in the two-variable equation we compute the statistic

$$t = \frac{\hat{b}_1}{\text{estimated standard error of } \hat{b}_1}$$
$$= \frac{1.44}{0.574} = 2.51.$$

Because we have $n = 10$ observations in this example, and because we have estimated 3 parameters (b_0, b_1, and b_2), we compare the statistic t to values of the t distribution with $n - 3 = 10 - 3 = 7$ degrees of freedom. From Table A-2 in Appendix III in the back of the book for 7 degrees of freedom we find $t = 2.36$ gives a probability of 0.95 (two-sided), or the probability that a value exceeds 2.36 is 0.025. Because our observed value of 2.51 exceeds the tabled value 2.36, we have reasonable grounds for concluding that b_1 is different from zero. We conclude that we should not drop the predictor u from the equation.

Similarly, to check on the plausibility of $b_2 = 0$, we compute

$$t = \frac{\hat{b}_2}{\text{estimated standard error of } \hat{b}_2}$$

$$= \frac{0.883}{0.171} = 5.16.$$

We conclude that we should not drop variable w from the equation.

A Cautionary Note. In this example we first looked at the data in several forms, fitted three regressions and computed the corresponding R^2 values, and then computed the t statistics for the coefficients b_1 and b_2. The tail probabilities associated with the t tests do not have the clear-cut interpretation that we get for a single t test where the data have not been examined previously.

We are using the probabilities associated with the t values as approximate guides. The more ways that we look at the data, the less meaning such tail probabilities have in subsequent tests. This point is even more important to keep in mind when we have more than two predictors.

PROBLEMS FOR SECTION 13-4

Using a dozen hen's eggs from the grocery store, Dempster measured the long (L) and short (W) diameters and the volume (V). Dempster's goal was similar to the one we had in Example 7; he wished to see if the formula for an ellipsoid of revolution,

$$V = kLW^2,$$

where $k = \pi/6$, was a reasonable approximation for measuring the volume of eggs. By taking logarithms, the ellipsoid formula becomes

$$y = \ln(\pi/6) + u + 2w$$

$$= -0.647 + u + 2w, \qquad (9)$$

where

$$y = \ln V, \quad u = \ln L, \quad \text{and} \quad w = \ln W.$$

Dempster's data are given in Table 13-6, using natural logarithms.

TABLE 13-6

Measurements for a dozen hen's eggs

Egg number	u†	w	y
1	1.7635	1.4644	4.0296
2	1.6931	1.4271	3.9167
3	1.7076	1.4460	3.9467
4	1.7500	1.4460	4.0019
5	1.8101	1.4366	4.0296
6	1.7359	1.4175	3.8569
7	1.7838	1.4175	3.9743
8	1.7771	1.4366	3.9467
9	1.8165	1.4078	3.9467
10	1.7635	1.3981	3.9467
11	1.7704	1.4175	3.9467
12	1.7219	1.4366	3.9743

† $u = \ln L$, $w = \ln W$, and $y = \ln V$ (natural logarithms).

Source: Adapted from A. P. Dempster (1969). *Elements of Continuous Multivariate Analysis*, p. 151. Addison-Wesley, Reading, Mass.

1. Using the data in Table 13-6 and formula (9), calculate the predicted values of y.

2. Continuation. Plot the predicted values from Problem 1 against the observed. Do the points fall roughly along a straight line?

3. Continuation. Compute the sum of squared residuals using the prediction equation (9).

4. Continuation. What proportion of the variability in y is explained by equation (9)?

5. The actual least-squares fit for the multiple regression of y on u and w is

$$y = 0.0926 + 0.728u + 1.811w,$$

with $R^2 = 0.621$. How much of an improvement is the least-squares equation fit over equation (9), in terms of proportion of variability explained?

6. The estimated standard errors for the coefficients of u and w in the least-squares equations are 0.267 and 0.546. Compute the t ratios for $\hat{b}_1$ and $\hat{b}_2$, and check on the plausibility of (a) $b_1 = 0$, (b) $b_2 = 0$.

13-5 SUMMARY OF CHAPTER 13

Formula numbers correspond to those used earlier in the chapter.

1. With a dependent variable, y, and two predictors, u and w, the linear regression model is

$$y = b_0 + b_1 u + b_2 w + e, \tag{1}$$

where e is a random error with zero mean and unknown variance σ^2.

2. The least-squares estimates of the coefficients in (1), $\hat{b}_0$, $\hat{b}_1$, and $\hat{b}_2$, yield the minimum sum of squared residuals:

$$\hat{S} = \Sigma(y - \hat{b}_0 - \hat{b}_1 u - \hat{b}_2 w)^2. \tag{4}$$

3. The proportion of variability of the dependent variable, y, explained by the least-squares fit is

$$R^2 = \frac{SSy - \hat{S}}{SSy} = 1 - \frac{\hat{S}}{SSy}. \tag{5}$$

As with a single predictor, $0 \le R^2 \le 1$. If we use two predictors in our regression equation, we can always get a higher value of R^2 than by using only one of them.

4. R, the positive square root of R^2, is called the multiple correlation coefficient.

5. Multiplicative formulas with two or more predictors can be changed into formulas with linear equations by taking logarithms.

6. To check on the plausibility that a true regression coefficient, b_i, takes the value zero, we compute the t ratio

$$t_{\hat{b}_i} = \frac{\hat{b}_i}{\text{estimated standard error of } \hat{b}_i}$$

and compare the statistic with values from tables of the t distribution with $n - 3$ d.f.

SUMMARY PROBLEMS FOR CHAPTER 13

1. In an investigation of the deterrent effects of punishment on homicide rates, a sociologist planned to use state data to predict

$$y = \text{homicide rate}$$

from

u = number of homicide convictions divided by number of homicides reported

w = median time served.

If the certainty of punishment (variable u) and the severity (variable w) actually deter homicides, what signs do you expect to find for b_1 and b_2 in the following regression?

$$y = b_0 + b_1 u + b_2 w$$

2. Continuation. List three additional variables that might have effects on homicide rates in addition to certainty and severity of punishment.
3. Continuation. An analysis carried out using data from 48 states for 1960 and the model from Problem 1 yielded the following least-squares coefficients and estimated standard errors:

Coefficients	Standard Errors
$\hat{b}_1 = -0.0704$	0.032
$\hat{b}_2 = -0.062$	0.022

Check the plausibility of (a) $b_1 = 0$, (b) $b_2 = 0$.

4. Use the formulas from Appendix II in the back of the book or a least-squares computer program to fit an equation of the form $y = b_0 + b_1 u + b_2 w$ to the following data:

Observation:	1	2	3	4	5	6
y:	3	15	16	9	4	4
u:	3	15	13	8	3	2
w:	2	16	14	8	3	1

5. Continuation. Find the value of $\hat{S}$ and compute the proportion of variability explained, R^2.
6. A study predicting freshman GPA using SAT mathematics and verbal scores yields the following least-squares estimates for data from 20

college students:

Variable	Least-Squares Coefficient	Estimated Standard Error
SAT math (u)	0.0002	0.0001
SAT verbal (w)	0.0036	0.0007

Test the plausibility that (a) $b_1 = 0$, (b) $b_2 = 0$.

TABLE 13-7
Employment, price deflator, and GNP by years

Year	*Employment* y	*Price deflator* u	*GNP* w
1947	−4994	−18.7	−153,409
1948	−4195	−13.2	−128,272
1949	−5146	−13.5	−129,644
1950	−4130	−12.2	−103,099
1951	−2096	−5.5	−58,723
1952	−1678	−3.6	−40,699
1953	−328	−2.7	−22,313
1954	−1556	−1.7	−24,586
1955	702	−0.5	9,771
1956	2540	2.9	31,482
1957	2852	6.7	55,071
1958	1196	9.1	56,848
1959	3338	10.9	95,006
1960	4247	12.5	114,903
1961	4014	14.0	130,475
1962	5234	15.2	167,196

Note: The data have been centered on the sample means $\bar{y} = 65{,}317$, $\bar{u} = 101.7$, $\bar{w} = 387{,}698$.

Source: Adapted from Tables 1 and 2 in J. W. Longley (1967). An appraisal of least squares programs for the electronic computer from the point of view of the user, *Journal of the American Statistical Association* 62:819–841.

7. In a study predicting

$$y = \text{total derived employment}$$

from

$$u = \text{gross national product (GNP) price deflator}$$

$$w = \text{GNP}$$

for the years 1947 to 1962, the data in Table 13-7 on page 401 were gathered. Use a least-squares computer program to get an estimated equation for the linear regression of y on u and w.

8. Continuation. Test the plausibility that (a) $b_1 = 0$, (b) $b_2 = 0$.

9. Continuation. Calculate $\hat{S}$ and R^2 for your equation in Problem 7.

10. Continuation. Compare the R^2 value from Problem 9 with the value of R^2 for the regression of y on u alone.

REFERENCES

S. Chatterjee and B. Price (1977). *Regression Analysis by Example*, Chapter 3. Wiley, New York.

N. R. Draper and H. Smith (1980). *Applied Regression Analysis*, second edition, Chapter 4. Wiley, New York.

F. Mosteller and J. W. Tukey (1977). *Data Analysis and Regression*, Chapter 12. Addison-Wesley, Reading, Mass.

Multiple Regression

Learning Objectives

1. Extending the regression model and least squares to more than two predictors
2. Interpreting the output from computer programs for multiple regression analysis
3. Choosing subsets of predictors to be included in a regression model
4. Examining plots of residuals from a multiple regression analysis for patterns that relate to omitted predictors

14-1 MANY PREDICTORS

All of the basic ideas discussed so far carry over to the situation with more than two predictors. For example, the regression model for five predictors is

$$y = b_0 + b_1t + b_2u + b_3v + b_4w + b_5x + e, \tag{1}$$

where t, u, v, w, and x are the values of the predictors and e is a random error with mean 0 and unknown variance σ^2. This is the multiple regression model. Our observations now come in sextuples:

Observation						
	1	y_1	t_1	u_1	v_1	w_1	x_1
	2	y_2	t_2	u_2	v_2	w_2	x_2
	3	y_3	t_3	u_3	v_3	w_3	x_3
	⋮	⋮	⋮	⋮	⋮	⋮	⋮
	i	y_i	t_i	u_i	v_i	w_i	x_i
	⋮	⋮	⋮	⋮	⋮	⋮	⋮

In this chapter we discuss aspects of fitting multiple regression models to data and how to choose predictors for inclusion in the models being fitted. The calculations for multiple regression are complicated and extremely time-consuming unless one has a high-speed computer available. Thus we develop the material in this chapter as if we had available a computer program suitable for multiple regression—one that will do all of the calculations we require.

In Section 14-2, after describing some aspects of choosing multiple regression models, we develop a detailed practical example. Throughout this example we introduce new ideas and illustrate how they are used in the analysis of data. We conclude the chapter with a series of real-life examples that will guide us in discussing the interpretation and use of multiple regression.

COMPUTER OUTPUT AND ANALYSES

The least-squares estimators for the regression coefficients in equation (1) minimize

$$S = \Sigma(y_i - b_0 - b_1t_i - b_2u_i - b_3v_i - b_4w_i - b_5x_i)^2. \tag{2}$$

We label them $\hat{b}_0$, $\hat{b}_1$, $\hat{b}_2$, $\hat{b}_3$, $\hat{b}_4$, and $\hat{b}_5$. Our multiple regression computer program will compute these estimated coefficients as well as the following:

a) the minimum value of S,

$$\hat{S} = \Sigma(y - \hat{b}_0 - \hat{b}_1 t - \hat{b}_2 u - \hat{b}_3 v - \hat{b}_4 w - \hat{b}_5 x)^2 \quad (3)$$

b) the predicted values $\hat{y}_i$

c) the residuals $y_i - \hat{y}_i$

d) the estimated standard error for each of the regression coefficients

e) the proportion of variability explained by the estimated regression equation:

$$R^2 = 1 - \frac{\hat{S}}{SSy} \quad (4)$$

Some multiple regression programs produce even more information, such as plots of residuals similar to those we used when we had a single predictor. It takes considerable training and guidance in order to be able to interpret even those items listed here, such as residuals, and so we concentrate on these in this chapter.

USING *t* TESTS FOR REGRESSION COEFFICIENTS

It is possible to use a *t* test to see if a coefficient in the regression equation differs substantially from zero or if it is near enough to zero for us to consider the possibility of omitting that variable. For example, to check whether we wish to include the variable *u* in equation (1), we can compute the *t* ratio

$$t_{\hat{b}_2} = \frac{\hat{b}_2}{\text{estimated standard error of } \hat{b}_2}. \quad (5)$$

These *t* values often are included as part of standard computer output. The number of degrees of freedom depends on the total number of observations available, n, and the number of constants fitted, k, and equals $n - k$. We might want to keep a variable in a regression equation, however, even if the coefficient is not distinctly different from zero. For example, we may have theoretical reasons or experience showing that a variable should be kept. If a regression relating automobile fuel consumption to several variables did not give a significant coefficient to miles driven, we would still leave it in.

PROBLEMS FOR SECTION 14-1

1. This chapter does not give computational formulas for obtaining multiple regression equations. Why not?
2. To predict a person's weight from other variables, we might use a multiple regression based on

   y = weight in pounds

   t = height in inches

   u = chest circumference in inches

   v = waist circumference in inches

   w = hip circumference in inches

   x = sex (0 if male, 1 if female)

   Write out the sextuple of measurements for your own body.
3. How many constants are to be fitted in equation (1)?
4. What does e measure in equation (1)?
5. If you had a regression equation and could measure for many observations, what would the average value of e be, approximately?
6. Because e varies from one measurement to another, how can we estimate its variance?
7. If e_i is the error for the ith measurement, write equation (2) in terms of e's.
8. Explain the distinction between b_i and $\hat{b}_i$. Do we usually know or find b_i and $\hat{b}_i$?
9. Explain the distinction between y_i and $\hat{y}_i$. Do we usually know or find y_i and $\hat{y}_i$?
10. What measure of fit of the regression equation to the data is proposed?

14-2 CHOOSING A MULTIPLE REGRESSION EQUATION

LOOKING AT THE DATA

The data for multiple regression analyses are more extensive and more complex than those for the other statistical methods discussed in this book, but we cannot rely on automated computer programs to do all of our work for us. We still need to look at the raw data, as well as at various numbers produced as by-products of our analyses. After we fit a multiple regression model to our data, we should look at the residuals. They may have patterns. A typical way to look at residuals is to plot them against the estimated value of the y variable, $\hat{y}$, or against the predictors. By doing this we may find a pattern in the residuals. If there is a curvilinear trend, we might either introduce transformations into the variables or add a quadratic term to the prediction.

There may also be one or more very wild observations. If that happens, then at least one residual should be extremely large compared with the others. Then we may want to consider setting aside such points and redoing the multiple regression analysis on the remaining points.

FOCUSING ON A SUBSET OF PREDICTORS

How can we decide which variables actually belong in the regression equation? Expression (1) has five predictors, but could we actually settle for only two or three? We have several possible approaches for focusing on a subset of the possible predictors.

We might compute a *t* ratio for each of the estimated regression coefficients, as in equation (5). In each, we take the estimated regression coefficient and divide it by the estimate of its standard error. If the ratios are small, we are often inclined to omit the corresponding predictor from our equation. We must resist, however, the desire to conduct blindly significance tests at a prespecified level like 0.05 on the *t* ratio for every possible predictor. Especially with social science data, we may be unwise to delete these slight variables unless other variables are doing the job. These variables could have useful effects even though the effects are too small and our data are too few to prove firmly that they are useful variables. Remember that tail probabilities do not have the usual simple interpretation when we look at the data over and over again and then conduct several *t* tests.

As an alternative we might wish to run regression analyses on the same set of data to see how well we do when we eliminate some variables from the regression equation. Generally speaking, we prefer equations with fewer variables. Often we cannot readily tell the effect of removing a variable. To determine this we compare the sums of squared deviations, $\hat{S}$, for two different equations. We can do this directly by looking at $\hat{S}$ or R^2, remembering that as we delete variables from the equation, $\hat{S}$ increases in value, and thus R^2 decreases.

If we use 5 variables, we will have fitted 6 regression coefficients (including the constant term, b_0), and therefore we will have given up 6 degrees of freedom. This means that a more appropriate quantity to look at than $\hat{S}$ is the estimate of the error variance, σ^2, from the multiple regression model (1):

$$\hat{\sigma}^2 = \frac{\hat{S}}{\text{degrees of freedom}} \tag{6}$$

The degrees of freedom will be the sample size, n, minus the number of coefficients fitted. So if we have 20 points, $n = 20$, and if we fit 5 variables we will lose 6 degrees of freedom, leaving us with $20 - 6 = 14$ degrees of freedom. If we drop 2 variables from our equation, then we are fitting 4 instead of 6 coefficients. Thus the degrees of freedom increase from 14 to 16.

Remember, when we fit additional variables, the residual sum of squares, $\hat{S}$, must automatically be reduced, but the estimated error variance, $\hat{\sigma}^2$, need not be reduced because the denominator of $\hat{\sigma}^2$ will also be decreasing.

This is quite a bundle of ideas to absorb all at one time. But we can illustrate many of them in an example with actual data.

EXAMPLE 1 *Student achievement.* In the mid-1960s the U.S. Office of Education undertook a major research project involving some 570,000 school pupils and 60,000 teachers. It was intended to assess the availability of equal educational opportunities in public educational institutions in the United States. The report on the project became known as the Coleman report, after James S. Coleman, who headed the study. Table 14-1 contains data on a sample of 20 grade schools from the Northeast and Middle Atlantic states, drawn from the population of the Coleman report. Our aim with these data is to find reasonable explanations for why the average sixth grade verbal achievement scores, y, vary from school to school. Five predictors are available for our use here. Three of these pertain primarily to assessment of general background characteristics for each school:

u: sixth grade percentage of white collar (father)—(WHTC)

v: socioeconomic status—(SES)

x: sixth grade mothers' average education (1 unit = 2 years)—(MOM)

TABLE 14-1
Random sample of 20 schools from Northeast and Middle Atlantic states

School number	t† *SLRY*	u *WHTC*	v *SES*	w *TCHR*	x *MOM*	y *SCOR*
1	3.83	28.87	7.20	26.60	6.19	37.01
2	2.89	20.10	−11.71	24.40	5.17	26.51
3	2.86	69.05	12.32	25.70	7.04	36.51
4	2.92	65.40	14.28	25.70	7.10	40.70
5	3.06	29.59	6.31	25.40	6.15	37.10
6	2.07	44.82	6.16	21.60	6.41	33.90
7	2.52	77.37	12.70	24.90	6.86	41.80
8	2.45	24.67	−0.17	25.01	5.78	33.40
9	3.13	65.01	9.85	26.60	6.51	41.01
10	2.44	9.99	−0.05	28.01	5.57	37.20
11	2.09	12.20	−12.86	23.51	5.62	23.30
12	2.52	22.55	0.92	23.60	5.34	35.20
13	2.22	14.30	4.77	24.51	5.80	34.90
14	2.67	31.79	−0.96	25.80	6.19	33.10
15	2.71	11.60	−16.04	25.20	5.62	22.70
16	3.14	68.47	10.62	25.01	6.94	39.70
17	3.54	42.64	2.66	25.01	6.33	31.80
18	2.52	16.70	−10.99	24.80	6.01	31.70
19	2.68	86.27	15.03	25.51	7.51	43.10
20	2.37	76.73	12.77	24.51	6.96	41.01

† *y*, sixth grade verbal mean test score for all sixth graders (SCOR).
t, staff salaries per pupil (SLRY).
u, sixth grade percentage white-collar fathers (WHTC).
v, socioeconomic status composite: sixth grade means for family size, family intactness, father's education, mother's education, percentage white-collar fathers, and home items (SES).
w, mean teacher's verbal test score (TCHR).
x, sixth grade mean mother's educational level (1 unit = 2 school years) (MOM).

Source: Adapted from J. S. Coleman, E. Q. Campbell, C. J. Hobson, J. McPartland, A. M. Mood, F. D. Weinfeld, and R. L. York (1966). *Equality of Educational Opportunity*. U.S. Government Printing Office, Washington, D.C.

The other two relate to variables that are potentially subject to manipulation in order to produce equality of achievement:

t: staff salaries per pupil—(SLRY)

w: mean teacher's verbal test score—(TCHR)

This particular sample of 20 schools was chosen in part to illustrate the difficulties that arise when a variable like SES is used as a predictor along with other variables that are closely related to it, such as WHTC and MOM. These 20 schools are not typical of schools in the country as a whole, and we must be very cautious when we come to interpret the results of our analyses.

Our multiple regression analysis of these data involves several steps. We proceed by giving a series of exhibits consisting of pieces of actual computer output, and in between these exhibits we give brief explanations and interpretations of the reported analyses from the computer. One point we must recognize at the outset is that even our favorite statistical computer programs often include, as part of their output, information that we neither understand nor even wish to use in our analyses. Because the exhibits contain actual output, they will contain statistics we have not mentioned. Many of these we will simply ignore, because they do not play a central role in our discussion here. We do not delete them because they add a tang of reality to the computer printouts.

We begin by examining the correlations among the six variables—the five predictors and the dependent variable. Instead of using the variable labels *t*, *u*, *v*, *w*, and *x*, we adopt the practice of using three- or four-letter labels (known as mnemonics). These labels will save us the effort of continually flipping back to see if, for example, *v* stands for socioeconomic status or for staff salaries per pupil. Similarly, we describe the regression coefficients using the same labels; for example, b_1, the coefficient of *t*, is labeled b_{SLRY}.

The entries in Exhibit 14-A need some explanation.

1. To find the correlation coefficient between any two variables, we locate a row corresponding to one and a column corresponding to the other; if we hit a blank entry we simply reverse. For example, the correlation between SLRY and TCHR is found by entering the fourth row (for TCHR) and the first column (for SLRY), yielding $r = 0.5027$.
2. Some entries have strange labeling that tells us that the decimal place needs to be moved, and exactly where it should go. For example, the correlation between WHTC and TCHR is given as $r = .5106\text{E-}01$. "E-01" means "move the decimal point 1 space to the left." Thus $r = 0.05106$. This form of labeling is known as scientific notation. Similarly, "E+01" means "move the decimal point 1 space to the right."

EXHIBIT 14-A

Northeast achievement data—correlation coefficients of variables used

```
        NEXT? CORR
CORRELATION MATRIX
SLRY   1.000
WHTC   .1811     1.000
SES    .2296     [.8272]      1.000
TCHR   .5027     .5106E-01    .1833     1.000
MOM    .1968     [.9271]      [.8191]   .1238     1.000
SCOR   .1923     [.7534]      [.9272]   .3336     [.7330]    1.000
       SLRY      WHTC         SES       TCHR      MOM        SCOR
```

3. Note that $r = 1$ for the correlation of each variable with itself. This is as it should be, because SCOR, for example, is a perfect predictor of itself!

We can learn a considerable amount by examining the entries in Exhibit 14-A. First, if we look at the last row, we see that the predictor variable with the highest correlation with the dependent variables SCOR is SES, with $r_{\text{SCOR,SES}} = 0.9272$. This means that when SES is used all by itself in a regression equation to predict SCOR, it will account for

$$R^2 = (0.9272)^2 = 85.97 \text{ percent}$$

of the variability in SCOR. No other variable by itself accounts for more than 60 percent. Thus a regression equation with SES alone will give reasonably good predictions. The other two variables that are highly correlated with SCOR are WHTC and MOM.

Second, the correlations among the three background variables, WHTC, SES, and MOM, are all quite high:

$$r_{\text{WHTC,SES}} = 0.8272, \quad r_{\text{WHTC,MOM}} = 0.9271, \quad r_{\text{SES,MOM}} = 0.8191.$$

Each of these variables measures similar information. Perhaps we need to use only one of the three.

Third, the correlation between the other two variables, SLRY and TCHR, is also relatively large: $r_{\text{SLRY,TCHR}} = 0.5027$. Again, perhaps both variables are not needed as predictors.

Next we compute the multiple regression using all five predictors. From Exhibit 14-B we have the five-variable fitted equation (rounding to two decimal places), noting that BO means b_0, and using the labels:

$$\text{PREDICTED SCOR} = 19.95 - 1.79\text{SLRY} + 0.04\text{WHTC} + 0.56\text{SES} + 1.11\text{TCHR} - 1.81\text{MOM}.$$

(Note that we interpreted ".4360156E-01" as "0.04360156.") The display includes the estimated standard errors of the least-squares coefficients; for

EXHIBIT 14-B

Northeast achievement data—regression coefficients for predicting achievement score using all five predictors

```
REGS        SCOR ON    SLRY WHTC SES  TCHR MOM
VARIABLE       COEF'T              ST. ERROR              T VALUE
  B0        19.94857             13.62755                   1.46
  SLRY      -1.793333            1.233396                  -1.45
  WHTC       .4360156E-01         .5325887E-01               .82
  SES        .5557601             .9295638E-01              5.98
  TCHR      1.110168              .4337679                  2.56
  MOM       -1.810922            2.027389                   -.89
           DEGREES OF FREEDOM   =      14
           RESIDUAL MEAN SQUARE=  4.302704
           ROOT MEAN SQUARE     =  2.074296
           R-SQUARED            =      .9063
```

example,

$$\text{estimated standard error of } \hat{b}_{\text{MOM}} = 2.03.$$

The display also contains the t ratios. Only two t ratios (those for SES and TCHR) are larger than 2, and the variables with the smallest ratios are candidates to be dropped from the equation. Finally, we see that the estimated error variance is

$$\hat{\sigma}^2 = \text{RESIDUAL MEAN SQUARE}$$
$$= 4.30,$$

and

$$R^2 = 0.9063.$$

Thus, by using all 5 variables instead of only SES, we can explain only

$$R^2_{\text{SLRY,WHTC,SES,TCHR,MOM}} - R^2_{\text{SES}} = 0.9063 - (0.9272)^2 = 0.0466,$$

or 4.66 percent more of the variability of SCOR.

Before going on to omit some of the predictors from our equation, we look at the residuals from the full multiple regression equation in Exhibit 14-C. The first column in this exhibit, labeled CASE, contains the school numbers from Table 14-1, and the second column, labeled SCOR, contains the observed values of the dependent variable as given in the last column of Table 14-1. To get the residuals in the third column, the computer program calculates the PREDICTED SCOR for each case based on the least-squares regression line and then subtracts this from the observed value:

$$\text{RESIDUAL} = \text{SCOR} - \text{PREDICTED SCOR}.$$

EXHIBIT 14-C

Northeast achievement data—observed scores for the 20 schools, residuals, and studentized residuals of predicted scores using all five predictors; other quantities not treated here

NEXT? RESID

RESIDUALS SCOR ON SLRY WHTC SES TCHR MOM

CASE	SCOR	RESIDUAL	STUD. RES	V	DISTANCE	T
1	37.01	.3488	.2337	.4825	.0085	.23
2	26.51	-.3499	-.2353	.4862	.0087	-.23
3	36.51	-3.950	-2.0451	.1331	.1070	-2.35
4	40.70	-.4736	-.2508	.1712	.0022	-.24
5	37.10	.7809	.4153	.1781	.0062	.40
6	33.90	-.8570E-01	-.0584	.4999	.0006	-.06
7	41.80	.7188	.3972	.2391	.0083	.38
8	33.40	-.4343	-.2216	.1075	.0010	-.21
9	41.01	.6244	.3561	.2854	.0084	.34
10	37.20	.2104	.1641	.6180	.0073	.16
11	23.30	-2.208	-1.2646	.2914	.1096	-1.29
12	35.20	1.746	1.0894	.4027	.1334	1.10
13	34.90	-1.049	-.6366	.3693	.0396	-.62
14	33.10	-.3457	-.1765	.1088	.0006	-.17
15	22.70	-1.779	-1.0608	.3464	.0994	-1.07
16	39.70	1.297	.6812	.1569	.0144	.67
17	31.80	-1.440	-.8241	.2906	.0464	-.81
18	31.70	5.002	2.9362	.3256	.6936	4.56
19	43.10	1.123	.6398	.2846	.0271	.63
20	41.01	.2628	.1437	.2227	.0010	.14

DURBIN-WATSON= 2.2778

RESIDUAL SS = 60.23785266

PRESS = 117.7790268

In Section 12-7 we explained for simple regression how to estimate the standard error for any prediction, usually different for each prediction. By methods not given here, these formulas can be extended to multiple regression, and the computer calculates these standard errors. The column labeled STUD. RES contains the residuals, each divided by its estimated standard error (each standard error may differ from the rest), or **studentized residuals.** This division puts all the residuals on roughly the same footing, and the numbers should behave approximately like standard normal random variables. One of the things we look for in this column is large (either positive or negative) residuals—an indication of outliers. The two cases that stand out a little are case 3 (with a value −2.05) and case 18 (2.94). Given that we have 20 cases, it is not too shocking to find two with values larger than 2, but we shall keep our eyes on these observations in later analyses nonetheless. In figures not shown here, we plotted the studentized residuals against the predicted values, and against each of the predictors separately, to check for possible curvilinearity. We found no evidence to suggest that an equation adding squared terms for the predictors to take account of curvilinearity would be of much use.

In the discussion here we do not use the information in the final three columns of Exhibit 14-C, nor the quantities at the bottom of the display. They are included here simply because they were part of the output produced by the particular regression program that we used. If we use a different program, the information it provides on residuals may differ somewhat from that in Exhibit 14-C.

The two variables that are potentially subject to manipulation by a policymaker to produce equality of achievement are SLRY (staff salaries per pupil) and TCHR (mean teacher's verbal test score). Because the variable SLRY has the smaller t ratio in absolute size in Exhibit 14-B, we try omitting it from the equation, while retaining TCHR and the three background variables, WHTC, SES, and MOM. Exhibit 14-D gives the regression coefficient omitting SLRY. The new fitted equation is

$$\text{PREDICTED SCOR} = 22.37 + 0.04\text{WHTC} + 0.55\text{SES} + 0.80\text{TCHR} - 1.71\text{MOM},$$

and

$$R^2 = 0.8922,$$

only a slight change from the full equation (about 1 percent). The new estimate of the error variance is

$$\hat{\sigma}^2 = 4.62,$$

just a small increase, from 4.30 to 4.62.

From the t ratios we see that both WHTC and MOM have modest values, and they remain as candidates to be dropped from the equation. We

EXHIBIT 14-D

Northeast achievement data—regression coefficients for predicting achievement scores using the four predictors WHTC, SES, TCHR, and MOM

```
        NEXT? REGS SCOR ON WHTC SES TCHR MOM
REGS          SCOR ON    WHTC SES  TCHR MOM
VARIABLE        COEF'T             ST. ERROR           T VALUE
 B0           22.36708             14.01893               1.60
 WHTC         .3752068E-01         .5503077E-01            .68
 SES          .5519482             .9630820E-01           5.73
 TCHR         .8036499             .3929197               2.05
 MOM          -1.710551            2.100110               -.81
          DEGREES OF FREEDOM   =       15
          RESIDUAL MEAN SQUARE=   4.622268
          ROOT MEAN SQUARE     =  2.149946
          R-SQUARED            =      .8922
```

EXHIBIT 14-E
Northeast achievement data—regression coefficients for predicting achievement score using (a) three predictors, SES, TCHR, and MOM, (b) three predictors, WHTC, SES, and TCHR, (c) two predictors, SES and TCHR

(a)
```
          NEXT? REGS SCOR ON SES TCHR MOM
REGS        SCOR ON    SES  TCHR MOM
VARIABLE      COEF'T             ST. ERROR              T VALUE
  BO        18.39794             12.53800                 1.47
  SES        .5748537             .8873655E-01            6.48
  TCHR       .7418216             .3758644                1.97
  MOM       -.5943234            1.293191                 -.46
            DEGREES OF FREEDOM   =       16
            RESIDUAL MEAN SQUARE=  4.467673
            ROOT MEAN SQUARE     =  2.113687
            R-SQUARED            =      .8888
```

(b)
```
          NEXT? REGS SCOR ON WHTC SES TCHR
REGS        SCOR ON    WHTC SES  TCHR
VARIABLE      COEF'T             ST. ERROR              T VALUE
  BO        14.36298             9.892365                 1.45
  WHTC       .2578867E-02         .3410329E-01             .08
  SES        .5356886             .9322033E-01            5.75
  TCHR       .7551835             .3842810                1.97
            DEGREES OF FREEDOM   =       16
            RESIDUAL MEAN SQUARE=  4.525032
            ROOT MEAN SQUARE     =  2.127212
            R-SQUARED            =      .8874
```

(c)
```
          NEXT? REGS SCOR ON SES TCHR
REGS        SCOR ON    SES  TCHR
VARIABLE      COEF'T             ST. ERROR              T VALUE
  BO        14.58268             9.175407                 1.59
  SES        .5415607             .5004410E-01           10.82
  TCHR       .7498921             .3666401                2.05
            DEGREES OF FREEDOM   =       17
            RESIDUAL MEAN SQUARE=  4.260376
            ROOT MEAN SQUARE     =  2.064068
            R-SQUARED            =      .8873
```

next omit them one at a time, and then together. The results are given in Exhibit 14-E.

Omitting variables WHTC and MOM, either separately or together, leads to essentially the same value of R^2:

$$R^2_{\text{SES,TCHR,MOM}} = 0.8888,$$
$$R^2_{\text{WHTC,SES,TCHR}} = 0.8874,$$
$$R^2_{\text{SES,TCHR}} = 0.8873.$$

These differ by less than 0.2 of a percent, and they are about 0.5 of a percent smaller than 0.8922, the R^2 value with 4 variables from Exhibit 14-D. Little seems to be lost by removing both WHTC and MOM from the equation. Our new estimate of σ^2, the error variance, using only SES and TCHR as predictors, is now smaller than for any equation examined so far:

$$\hat{\sigma}^2 = 4.26.$$

Leaving out both WHTC and MOM as predictors makes sense from a different viewpoint. Socioeconomic status, variable SES, is a composite variable that combines several different social and economic background variables. Thus, variables WHTC and MOM measure background information that may already be largely taken into account in the variable SES, which we are keeping in the equation.

At this point we examine the residuals based on SES and TCHR in Exhibit 14-F(a). We note that cases 3 and 18 again have the largest absolute

EXHIBIT 14-F(a)

Northeast achievement data—observed scores for the 20 schools, residuals, and studentized residuals of predicted scores using two predictors, SES and TCHR; other quantities not treated here

(a)

```
          NEXT? RESID
RESIDUALS SCOR ON    SES   TCHR
CASE        SCOR      RESIDUAL   STUD. RES        V          DISTANCE       T
  1     37.01      -1.419        -.7345      .1238        .0254      -.72
  2     26.51      -.2838E-01   -.0152      .1781        .0000      -.01
  3     36.51      -4.017       -2.0548      .1029        .1615     -2.30
  4     40.70      -.8884        -.4600      .1244        .0100      -.45
  5     37.10       .5281E-01     .0264      .0577        .0000       .03
  6     33.90      -.2164        -.1416      .4516        .0055      -.14
  7     41.80       1.667         .8548      .1072        .0292       .85
  8     33.40       .1546         .0771      .0562        .0001       .07
  9     41.01       1.146         .5966      .1342        .0184       .58
 10     37.20       1.640         .9807      .3437        .1679       .98
 11     23.30      -1.948       -1.0811      .2378        .1215     -1.09
 12     35.20       2.422        1.2477      .1158        .0680      1.27
 13     34.90      -.6458        -.3232      .0629        .0023      -.31
 14     33.10      -.3100        -.1567      .0815        .0007      -.15
 15     22.70      -2.093       -1.1876      .2708        .1746     -1.20
 16     39.70       .6111         .3093      .0837        .0029       .30
 17     31.80      -2.978       -1.4804      .0502        .0386     -1.54
 18     31.70       4.472        2.3690      .1637        .3661      2.81
 19     43.10       1.248         .6485      .1310        .0211       .64
 20     41.01       1.132         .5854      .1229        .0160       .57
DURBIN-WATSON=     2.2978
RESIDUAL SS   =  72.42639387
PRESS         =  100.8276476
```

EXHIBIT 14-F(b)

Northeast achievement data—plot of studentized residuals versus predicted score using two predictors, SES and TCHR

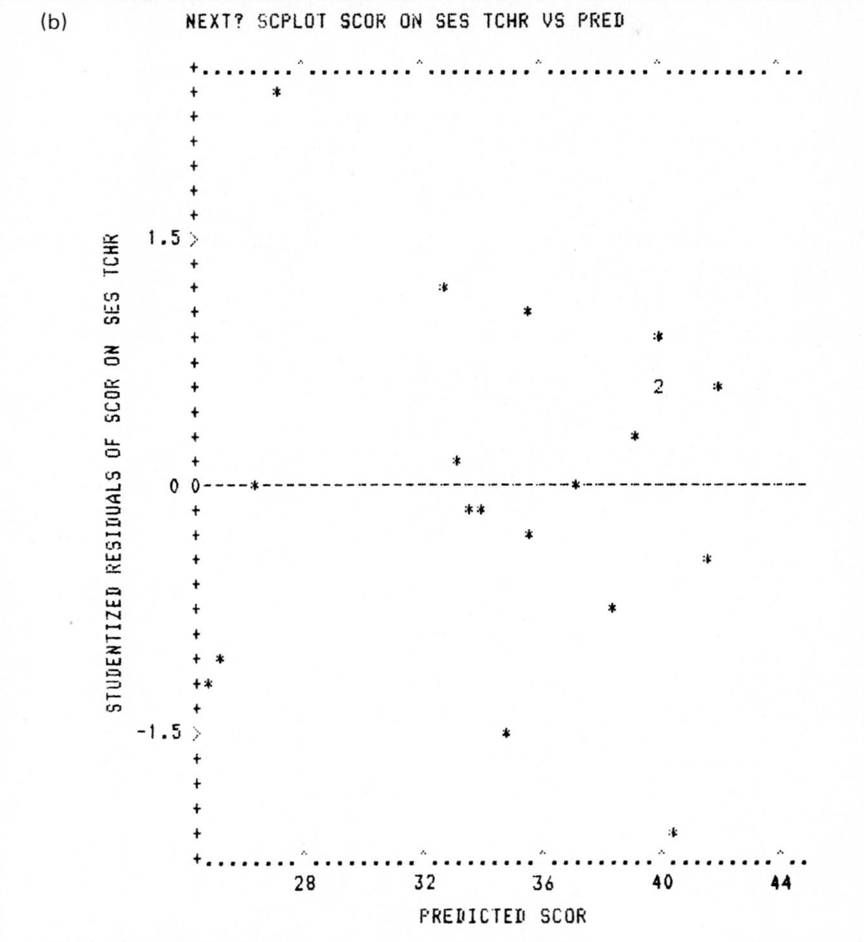

studentized residuals, but the one for case 18 is smaller than the corresponding one for case 18 in Exhibit 14-C. (This reduction in the size of the largest studentized residual is a good sign. Nonetheless, if we could, we would go back to the original observations for these cases to check on them further.)

At this point we also plot the studentized residuals against various variables. Exhibit 14-F shows two plots. Both show slight linear trends running downward from left to right. The linear trend in Exhibit 14-F(b)

EXHIBIT 14-F(c)

Northeast achievement data—plot of studentized residuals of SCOR on SES and TCHR versus SLRY

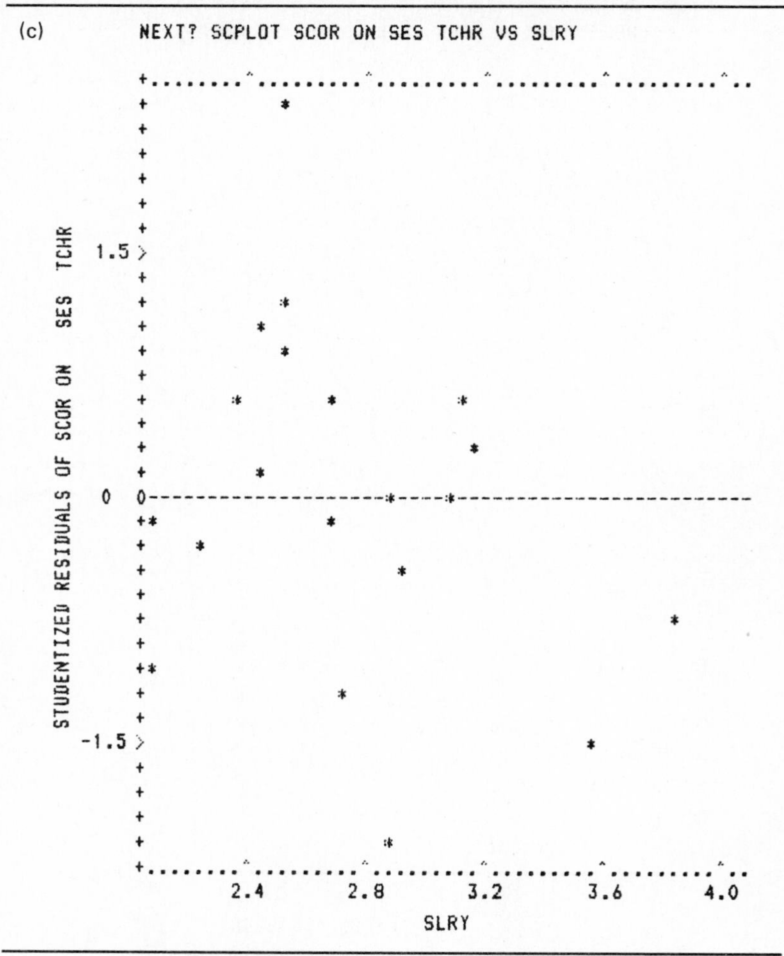

suggests that the fitted model may be inadequate, although it does not necessarily suggest why. The linear trend in Exhibit 14-F(c) suggests that the residuals contain some linear component due to the omitted variable plotted on the *x* axis—in this plot the variable SLRY. Thus, in Exhibit 14-G, we add SLRY back into our equation. The new regression equation is

$$\text{PREDICTED SCOR} = 12.12 - 1.74\text{SLRY} + 0.55\text{SES} + 1.04\text{TCHR},$$

EXHIBIT 14-G

**Northeast achievement data—
regression coefficients for predicting achievement
score using three predictors, SLRY, SES, and TCHR**

```
        NEXT? REGS SCOR ON SLRY SES TCHR
REGS        SCOR ON    SLRY SES   TCHR
VARIABLE       COEF'T              ST. ERROR                T VALUE
  B0          12.11951             9.036432                    1.34
  SLRY       -1.735809             1.182901                   -1.47
  SES          .5532101             .4907469E-01              11.27
  TCHR        1.035821              .4047864                   2.56
            DEGREES OF FREEDOM   =         16
            RESIDUAL MEAN SQUARE=  3.989707
            ROOT MEAN SQUARE     =  1.997425
            R-SQUARED            =      .9007
```

with $R^2_{\text{SLRY.SES.TCHR}} = 0.9007$. The t ratio for variable SLRY is -1.47, and we are left with the decision whether or not this value is large enough to keep the variable in the equation. At this point we have looked at the data so many times, in different ways, that it makes little sense to look up the t value in a table.

If we go back to our other criterion, making $\hat{\sigma}^2$ as small as possible, we may well decide to keep SLRY, because the value of $\hat{\sigma}^2$ in the three-variable equation,

$$\hat{\sigma}^2 = 3.99,$$

is smaller than the value 4.26 that we have for the equation using only SES and TCHR in Exhibit 14-E. A final check of the residuals for the regression of SCOR on SLRY, SES, and TCHR (not reproduced here) shows the same two large studentized residuals for cases 3 and 18, but no special patterns when the studentized residuals are plotted against anything else. This information mildly supports the use of the three-variable equation.

Conclusions. We began this example with five possible predictors: three background variables and two variables (SLRY and TCHR) that are subject to possible manipulation. Our final prediction equation,

$$\text{PREDICTED SCOR} = 12.12 - 1.74\text{SLRY} + 0.55\text{SES} + 1.04\text{TCHR},$$

accounts for about 90 percent of the variability in the dependent variable.

The sign of the coefficient for the variable SLRY is unexpected. It might be interpreted to say that if we control for background information (SES) and for the mean teacher's verbal test score (TCHR), an increase in staff salaries per pupil will produce a decrease in achievement. We have to be careful about such a causal interpretation, however, because we have not done an experiment that actually manipulated salaries for specific schools. Thus we should not conclude that we can increase achievement by reducing salaries. For example, teacher strikes could occur.

The other two variables do have the expected signs. When we introduce the other variables into the equation, differences in socioeconomic status (SES) are directly (and positively) related to achievement. Similarly for differences in mean teacher's verbal test score (TCHR). The latter has a potential policy interpretation, because it suggests that if we control for other relevant variables, increasing the teacher's ability to communicate may increase the pupil's achievement.

Before concluding this example, we recall that we are dealing with a small sample of schools from the Northeast and Middle Atlantic states. Although we have found a high R^2 for this sample, a much smaller correlation was found when similar analyses were done for the country as a whole. Thus, these data do not seem to be typical of the nation. We must always be cautious in generalizing the results of our analyses beyond the population from which we draw our samples.

In multiple correlation, although the size and sign of the coefficients may be suggestive, we cannot depend on them to be causally related to the output variable y. Thus deliberately changing the value of x in the real world may not have either the effect desired by the changers or that suggested by the equation. Nevertheless, the equation can still be useful for what it is supposed to do, namely, predict y from the values of the x's.

PROBLEMS FOR SECTION 14-2

In a study of geographic variation in the size of cricket frogs in the United States, data from several localities were gathered. One part of the study involved predicting mean body length from five environmental variables. Table 14-2 contains an excerpt of the data for 13 localities in the northeastern section of the country. The following problems are concerned with examining multiple regression output from an analysis of these data and drawing appropriate conclusions. We wish to predict body length of the frogs (LENG), using the predictors longitude (LONG), latitude (LAT), altitude (ALT), annual rainfall (RAIN), and annual temperature (TEMP).

1. Examine the correlations listed in Exhibit 14-H and identify (a) the two variables with the highest correlations with the dependent variable, LENG, and (b) those predictor variables that are highly intercorrelated (say $r > 0.6$ or $r < -0.6$).

TABLE 14-2

Data for predicting mean body length of male cricket frogs in 13 selected localities in the northeast United States

Location	*LAT*†	*LONG*	*ALT*	*RAIN*	*TEMP*	*LENG*
55	36.51	79.09	600	45	60	23.7
57	37.34	78.96	800	50	55	23.9
58	37.86	79.76	1450	45	50	23.4
59	38.41	79.03	1400	45	50	23.4
60	38.90	76.83	100	45	55	22.2
61	39.51	78.03	450	45	50	19.5
62	39.29	77.46	400	45	50	23.5
63	41.02	74.77	800	45	60	24.7
64	41.02	74.77	800	45	60	23.3
67	41.22	74.12	1100	45	50	21.7
68	41.28	73.98	500	40	50	21.3
69	41.87	74.08	400	40	50	20.2
70	41.87	74.08	400	40	50	22.1

† LAT, latitude
LONG, longitude
ALT, altitude
RAIN, annual rainfall
TEMP, annual temperature
LENG, body length of frog

Source: A. J. Izenman (1972). Reduced-rank regression for the multivariate linear model. Doctoral dissertation, Department of Statistics, University of California, Berkeley.

EXHIBIT 14-H

Cricket frog data—correlation coefficients of the variables used

```
      NEXT? CORR
CORRELATION MATRIX

LAT     1.000
LONG   -.9440       1.000
ALT    -.2598       .3572        1.000
RAIN   -.6869       .6492        .3382        1.000
TEMP   -.2254       .3211E-01  -.8519E-01   .3853        1.000
LENG   -.4378       .3400        .3993        .5263        .5793        1.000
          LAT       LONG          ALT         RAIN         TEMP          LENG
```

EXHIBIT 14-I

Cricket frog data—regression coefficients for predicting length (LENG) from all five predictors

```
         NEXT? REGS LENG ON LONG LAT ALT RAIN TEMP
REGS          LENG ON    LONG LAT   ALT  RAIN TEMP
VARIABLE         COEF'T               ST. ERROR            T VALUE
  BO          37.11349              82.49506                  .45
  LAT         -.3446890             .8003183                 -.43
  LONG        -.1753881             .6402401                 -.27
  ALT          .1499310E-02         .1048149E-02             1.43
  RAIN         .5026213E-01         .2078411                  .24
  TEMP         .1740991             .1161877                 1.50
            DEGREES OF FREEDOM  =        7
            RESIDUAL MEAN SQUARE=   1.674567
            ROOT MEAN SQUARE    =   1.294051
            R-SQUARED           =      .5830
```

2. Based on your examination of the correlations in Problem 1, would you expect both LONG and LAT to be useful as predictors in a multiple regression equation? Why or why not?

   Exhibit 14-I contains information on the regression using all five predictors.

3. Write out the multiple regression equation, rounding the coefficients to three decimal places.
4. What are the values of $\hat{\sigma}^2$ and R^2?
5. Find the degrees of freedom for the estimated error variance, and explain how the value was determined.
6. How much more of the variability of LENG is explained by using all five variables instead of the best single predictor identified in Problem 1(a)?
7. Which predictor variables appear to be reasonable candidates to be dropped from the equation in Problem 3?

In Exhibit 14-J we have a series of multiple regression equations, beginning with the one in Exhibit 14-I, and dropping variables one at a time (based on the t values), until only two remain.

8. What is an equation for the last regression in Exhibit 14-J?
9. How do the coefficients in the equation in Problem 8 compare in magnitude and sign with those for the same variables in your equation from Problem 3?
10. Make a table of the values of R^2, d.f., and $\hat{\sigma}^2$ for the four regression equations in Exhibits 14-I and 14-J.
11. Based on the information in the table from Problem 10, and on the other information in the Exhibits, choose a regression equation for prediction purposes.

EXHIBIT 14-J

Cricket frog data—regression coefficients for predicting length (LENG) of frogs using (a) four predictors, LONG, LAT, ALT, and TEMP, (b) three predictors, LAT, ALT, and TEMP, (c) two predictors, ALT and TEMP

(a)

```
            NEXT? REGS LENG ON LONG LAT ALT TEMP
REGS         LENG ON   LONG LAT  ALT  TEMP
VARIABLE       COEF'T              ST. ERROR              T VALUE
  BO          37.41682              77.47978                  .48
  LAT        -.3590558              .7496759                 -.48
  LONG       -.1510931              .5939364                 -.25
  ALT         .1559876E-02          .9560221E-03             1.63
  TEMP        .1851751              .1002996                 1.85
           DEGREES OF FREEDOM   =        8
           RESIDUAL MEAN SQUARE=  1.477487
           ROOT MEAN SQUARE     =  1.215519
           R-SQUARED            =     .5795
```

(b)

```
            NEXT? REGS LENG ON LAT ALT TEMP
REGS         LENG ON   LAT ALT  TEMP
VARIABLE       COEF'T              ST. ERROR              T VALUE
  BO          17.88967              9.975047                 1.79
  LAT        -.1758058              .1965678                 -.89
  ALT         .1480897E-02          .8559362E-03             1.73
  TEMP        .1991934              .7933204E-01             2.51
           DEGREES OF FREEDOM   =        9
           RESIDUAL MEAN SQUARE=  1.323946
           ROOT MEAN SQUARE     =  1.150629
           R-SQUARED            =     .5761
```

(c)

```
            NEXT? REGS LENG ON ALT TEMP
REGS         LENG ON   ALT  TEMP
VARIABLE       COEF'T              ST. ERROR              T VALUE
  BO          9.785579              4.129044                 2.37
  ALT         .1700912E-02          .8115809E-03             2.10
  TEMP        .2174479              .7589070E-01             2.87
           DEGREES OF FREEDOM   =       10
           RESIDUAL MEAN SQUARE=  1.297455
           ROOT MEAN SQUARE     =  1.139059
           R-SQUARED            =     .5384
```

12. Review your answers to Problems 1 and 2, and then note if you are surprised about your choice of equation in Problem 11. Explain your answer.

13. Examine the t values for the regression of LENG on ALT and TEMP. Do you think that either of these variables should be dropped from the equation? Explain your answer.

EXHIBIT 14-K

Cricket frog data—plot of studentized residuals of LENG using two predictors (ALT and TEMP) versus predicted LENG

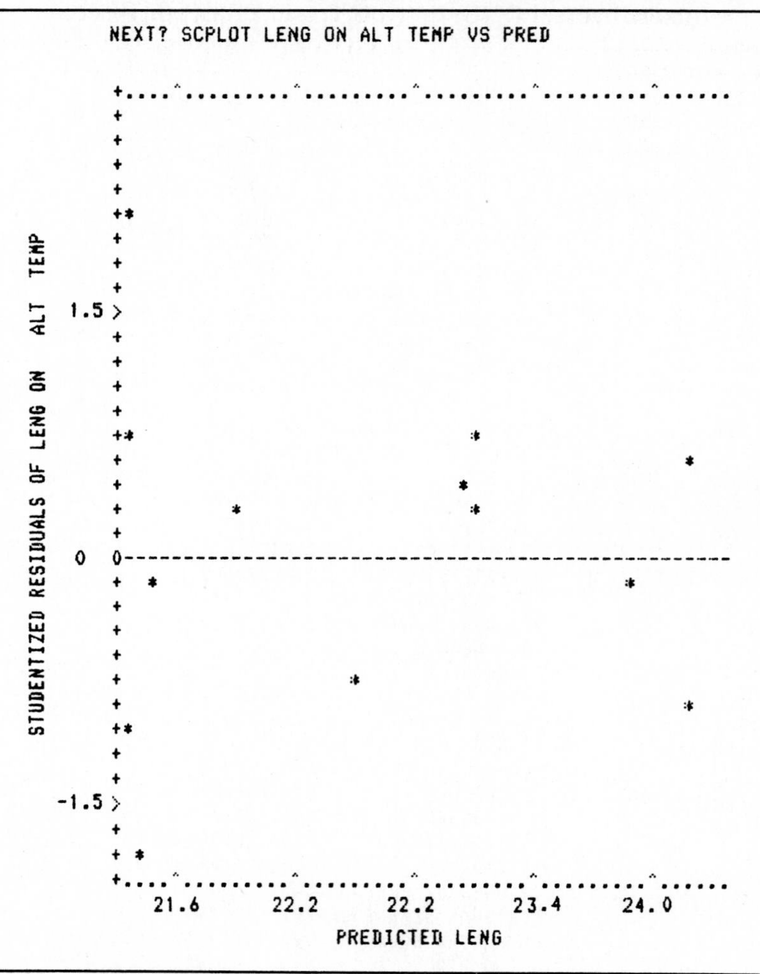

14. Exhibit 14-K contains a plot of the standardized residuals for the regression of LENG on ALT and TEMP. Is there any discernible pattern in this plot that should be cause for concern?

15. Exhibit 14-L contains the residuals whose values are plotted in Exhibit 14-K. Is there any evidence in this exhibit to suggest that one or more points may be outliers or wild observations?

EXHIBIT 14-L
Cricket frog data—observed LENG, residuals, and studentized residuals using two predictors, ALT and TEMP; other quantities not treated here

```
        NEXT? RESID
RESIDUALS LENG ON    ALT  TEMP
CASE        LENG      RESIDUAL  STUD. RES        V          DISTANCE      T
   1       23.70      -.1530      -.1594      .2895         .0034     -.15
   2       23.90       .7941       .7345      .0991         .0198      .72
   3       23.40       .2757       .3074      .3802         .0193      .29
   4       23.40       .3607       .3913      .3450         .0269      .37
   5       22.20       .2847       .2928      .2714         .0106      .28
   6       19.50      -1.923     -1.8414      .1591         .2138    -2.15
   7       23.50       2.162      2.0889      .1747         .3078     2.64
   8       24.70       .5068       .5315      .2992         .0402      .51
   9       23.30      -.8932      -.9367      .2992         .1248     -.93
  10       21.70      -.8290      -.8073      .1873         .0501     -.79
  11       21.30      -.2084      -.1980      .1460         .0022     -.19
  12       20.20      -1.138     -1.1000      .1747         .0854    -1.11
  13       22.10       .7617       .7360      .1747         .0382      .72
 DURBIN-WATSON=       2.4898
 RESIDUAL SS   =   12.97455429
 PRESS         =   19.56976845
```

16. Write a brief conclusion for the analysis in this data set that includes (a) the final prediction equation you would choose, (b) the values of R^2 and $\hat{\sigma}^2$ for your equation, and (c) an interpretation of the values of the estimated regression coefficients.

14-3 EXAMPLES OF ESTIMATED MULTIPLE REGRESSION EQUATIONS, THEIR INTERPRETATION, AND THEIR USE

Multiple regression analysis is used in many different fields, especially in situations in which the investigators are unable to collect true experimental data. In this section we review four additional examples of the use of multiple regression in four quite different areas, and we discuss problems with interpretation of the estimated coefficients and the use of the fitted equations for prediction purposes.

EXAMPLE 2 *Assessing the demand for fuel.* Using data for 1971 from the 48 contiguous U.S. states, Ferrar and Nelson (1975) used a multiple regression analysis to predict the demand for fuel. Their variables and fitted equation were as follows:

y = logarithm of total fuel consumption per capita

u = logarithm of mean personal income per capita

v = logarithm of price of heating fuel

w = logarithm of price of electricity

x = logarithm of average total heating degree-days (a measure of how cold it was)

$$y = -1.08 + \underset{(0.16)}{0.27u} - \underset{(0.09)}{0.28v} + \underset{(0.11)}{0.50w} + \underset{(0.04)}{0.50x} \quad (R^2 = 0.85).$$

The estimates of standard errors of the coefficients are reported in parentheses, and the proportion of variability of y explained by the regression equation is $R^2 = 0.85$.

This multiple linear regression equation is the linear version of what economists call a production function. They often model production Y with a product of several variables raised to powers, such as

$$Y = KU^aV^bW^c.$$

Then, by taking logarithms of both sides, they write

$$\log Y = \log K + a \log U + b \log V + c \log W.$$

Now, letting

$$y = \log Y, \quad k = \log K, \quad u = \log U, \quad v = \log V, \quad w = \log W,$$

we can write

$$y = k + au + bv + cw.$$

We have now a linear regression problem, where k, a, b, and c are the regression coefficients. Economists would interpret the fitted coefficients as indicating that the demand for heating fuel goes up with mean personal income, with the price of electricity, and with the duration and intensity of winter coldness, but goes down as the price of heating fuel itself goes up. The signs of the coefficients here are sensible, although it is difficult to interpret their sizes.

This equation is based on data gathered for 48 states for a single year. We would feel more confident of our interpretation of the equation for the demand for heating fuel if we had observed, in addition, several states over a period of many years and thus could see the actual impact of changes in the predictors on demand.

EXAMPLE 3 *Growth of Japanese larch trees.* Bliss (1970) used data from 26 Japanese larch trees to explore the relationship between the growth in height (y) and the percentage contents of nitrogen (u), phosphorus (v), and potassium (w) in their needles. His fitted linear equation is

$$y = -193.1 + \underset{(22.9)}{107.8u} + \underset{(165.1)}{304.2v} + \underset{(44.1)}{143.1w},$$

with $R^2 = 0.86$. Growth seems to be positively related to the percentage contents of all three minerals, although the t ratio of the phosphorus content (v), $t = 304.2/165.1 = 1.84$, is significant at the 0.10 level, but not the 0.05 level, when we compare it with values in the t table with $26 - 4 = 22$ d.f. These three components account for 86 percent of the variability in the heights of the trees. Because the three predictor variables are all measured on the same scale, we might wish to order the three minerals by the sizes of their regression coefficients and take this order to be an indicator of their relative importance. The difficulty with doing this is that the larger the estimated regression coefficient, the larger its estimated standard error.

In regression analyses such as this, we must avoid the temptation to interpret the size of a fitted coefficient as an indication of the importance of the corresponding predictor variable. Even though units are the same, the contributions may not be based on size. Furthermore, the units are not actually the same. Pounds of phosphorus are not pounds of nitrogen.

EXAMPLE 4 *Midterm congressional elections.* Tufte (1975) has tried to predict the nationwide vote in congressional contests in the United States for midterm elections (that is, elections held in years without a presidential contest). In every midterm election but one since the Civil War (until 1974) the political party of the incumbent president has lost seats in the House of Representatives. With

y = nationwide midterm congressional vote for party of the current president (percent)

as the dependent variable, Tufte considered the regression of y on

u = the average of the vote for the party of the president in the preceding eight elections

v = Gallup Poll rating of the president at the time of the election

w = the yearly change in real disposable income per capita (in 1958 dollars)

for the eight midterm elections from 1938 to 1970 (omitting 1942, because of special effects relating to wartime conditions). The fitted equation is

$$y = -12.71 + \underset{(0.10)}{1.03u} + \underset{(0.04)}{0.13v} + \underset{(0.007)}{0.04w},$$

and $R^2 = 0.93$. All three coefficients have the expected sign.

One difficulty with this use of regression analysis is that with only eight observations we have fitted four coefficients. Thus we have only 4 d.f. for the estimate of σ^2.

The data series here is simply too short for us to be very confident about predictions using the fitted equation.

EXAMPLE 5 *Property tax assessments.* In Ramsey County, Minnesota, the Department of Property Tax Assessment has begun to use multiple regression analysis to get estimates of the market values of single-family residences for tax purposes. They use an equation that incorporates two types of variables: (a) property characteristics, such as size of lot, number of square feet of living area, numbers of bedrooms, baths, etc., (b) neighborhood factors (two identical properties located in different parts of the county may be worth different amounts because of the relative desirabilities of the neighborhoods where they are located).

The tax assessors do not simply rely on the estimated values produced from the multiple regression equation for the tax bills they send out. They compare the predictions with estimated values that come from two other methods of assessment:

1. The cost approach, which is designed to calculate what it would cost to replace a home in today's building market,
2. Attribute matching (five houses are selected for which the assessor's office has recent sales information, houses that are most similar to the

house being valued; the median of the five sales prices, after some adjustments, is used as the estimated value).

When the three methods yield substantially different valuations for a property, appraisers are sent out to make a field inspection. In addition, homeowners have the right to challenge an estimated valuation and to ask for a field inspection.

When assessors apply this method to all of the homes in Ramsey County, inevitably some outliers occur—homes whose true values are not well described by the equation used. Predictions for these homes using the regression equation may be far different from the estimates using the other two assessment methods. Outliers can be found in both directions: overassessment and underassessment. When formulas such as these are used, taxpayers have ordinarily been given the right of appeal in the event that peculiar features distort the value. The same is true for other methods of assessment.

In the tax example, we see multiple regression in action directed toward its strength—estimation.

14-4 SUMMARY OF CHAPTER 14

Formula numbers correspond to those used earlier in this chapter.

1. With a dependent variable, y, and five predictors, t, u, v, w, and x, the multiple linear regression model is

$$y = b_0 + b_1 t + b_2 u + b_3 v + b_4 w + b_5 x + e, \tag{1}$$

where e is a random error with mean 0 and variance σ^2.

2. The least-squares estimates, $\hat{b}_i$, for the regression coefficients in (1) yield the minimum sum of squared residuals:

$$\hat{S} = \Sigma(y - \hat{b}_0 - \hat{b}_1 t - \hat{b}_2 u - \hat{b}_3 v - \hat{b}_4 w - \hat{b}_5 x)^2. \tag{3}$$

3. The proportion of variability of the dependent variable, y, explained by the least-squares fit is

$$R^2 = 1 - \frac{\hat{S}}{SSy}, \tag{4}$$

which is the square of the correlation r between the observed y and the least-squares estimate $\hat{y}$.

4. The estimate of the error variance, σ^2, for the multiple regression model is

$$\hat{\sigma}^2 = \frac{\hat{S}}{\text{degrees of freedom}} \tag{6}$$

where the degrees of freedom equal the sample size, n, minus the number of coefficients fitted.

5. We use the values of R^2, $\hat{\sigma}^2$, and the t ratios for the least-squares coefficients to help us decide when to omit variables from a multiple regression equation. We also use our understanding of any physical, social, and biological factors.

TABLE 14-3

1972 fuel consumption data for the 48 contiguous states

State	*TAX*†	*DLIC*	*INC*	*ROAD*	*FUEL*
1 ME	9.00	0.525	3751	1976	541
2 NH	9.00	0.572	4092	1250	524
3 VT	9.00	0.580	3865	1586	561
4 MA	7.50	0.529	4870	2351	414
5 RI	8.00	0.544	4399	431	410
6 CN	10.00	0.571	5342	1333	457
7 NY	8.00	0.451	5319	11868	344
8 NJ	8.00	0.553	5126	2138	467
9 PA	8.00	0.529	4447	8577	464
10 OH	7.00	0.552	4512	8507	498
11 IN	8.00	0.530	4391	5939	580
12 IL	7.50	0.525	5126	14186	471
13 MI	7.00	0.574	4817	6930	525
14 WI	7.00	0.545	4207	6580	508
15 MN	7.00	0.608	4332	8159	566
16 IA	7.00	0.586	4318	10340	635
17 MO	7.00	0.572	4206	8508	603
18 ND	7.00	0.540	3718	4725	714
19 SD	7.00	0.724	4716	5915	865
20 NE	8.50	0.677	4341	6010	640

TABLE 14-3 (Cont.)

State	*TAX*†	*DLIC*	*INC*	*ROAD*	*FUEL*
21 KS	7.00	0.663	4593	7834	649
22 DE	8.00	0.602	4983	602	540
23 MD	9.00	0.511	4897	2449	464
24 VA	9.00	0.517	4258	4686	547
25 WV	8.50	0.551	4574	2619	460
26 NC	9.00	0.544	3721	4746	566
27 SC	8.00	0.548	3448	5399	577
28 GA	7.50	0.579	3846	9061	631
29 FA	8.00	0.563	4188	5975	574
30 KY	9.00	0.493	3601	4650	534
31 TN	7.00	0.518	3640	6905	571
32 AL	7.00	0.513	3333	6594	554
33 MS	8.00	0.578	3063	6524	577
34 AR	7.50	0.547	3357	4121	628
35 LA	8.00	0.487	3528	3495	487
36 OK	6.58	0.629	3802	7834	644
37 TX	5.00	0.566	4045	17782	640
38 MT	7.00	0.586	3897	6385	704
39 ID	8.50	0.663	3635	3274	648
40 WY	7.00	0.672	4345	3905	968
41 CO	7.00	0.626	4449	4639	587
42 NM	7.00	0.563	3656	3985	699
43 AZ	7.00	0.603	4300	3635	632
44 UT	7.00	0.508	3745	2611	591
45 NV	6.00	0.672	5215	2302	782
46 WN	9.00	0.571	4476	3942	510
47 OR	7.00	0.623	4296	4083	610
48 CA	7.00	0.593	5002	9794	524

†TAX, 1972 motor fuel tax in cents per gallon.
DLIC, 1971 proportion of population with driver's license.
INC, 1972 per capita personal income in dollars.
ROAD, 1971 length of federal-aid primary highways in miles.
FUEL, 1972 fuel consumption in gallons per person.

Source: Data collected by C. Bingham from the *American Almanac for 1974* and the *1974 World Almanac and Book of Facts*, and reported in S. Weisberg (1980). *Applied Linear Regression*, Wiley: New York, p. 33.

SUMMARY PROBLEMS FOR CHAPTER 14

The following 10 problems are based on the data in Table 14-3 on pages 430 and 431.

For Problems 1 and 2, the correlation table for the five variables is as follows:

TAX	1.00				
DLIC	−0.29	1.00			
INC	0.01	0.16	1.00		
ROAD	−0.52	−0.06	0.05	1.00	
FUEL	−0.45	0.70	−0.24	0.02	1.00
	TAX	DLIC	INC	ROAD	FUEL

1. (a) Which variable has the highest correlation with the variable FUEL? (b) Which pairs of variables have correlations greater than 0.5 or less than −0.5?
2. Which variables do you expect to be of little use in predicting FUEL? Explain the reasons for your answer.

For Problems 3 through 7, the following information is produced by a computer program carrying out the regression of FUEL on TAX, DLIC, INC, and ROAD:

VARIABLE	COEF'T	ST. ERROR	T VALUE
BO	377.2911	185.541	2.03
TAX	−34.79015	12.9702	−2.68
DLIC	1336.449	192.298	6.95
INC	−0.6658875E-01	0.172218E-01	−3.87
ROAD	−0.2425889E-02	0.338918E-02	−0.72

DEGREES OF FREEDOM = 43
RESIDUAL MEAN SQUARE = 4396.511
ROOT MEAN SQUARE = 66.30619
R-SQUARED = 0.6787

3. Write out the multiple regression equation, rounding the coefficients to three decimal places.
4. What are the values of $\hat{\sigma}^2$ and R^2?
5. How much more variability of FUEL is explained by using all four predictors instead of the best single predictor?
6. Which predictor variables appear to be reasonable candidates to be dropped from the equation in Problem 3? Why?
7. Interpret the coefficients in the equation from Problem 3, and discuss the implications of the results of the analysis for the reduction of fuel consumption for the 48 contiguous states.

The following problems require a high-speed computer.

8. Use a least-squares multiple regression computer program to get results for the data of Table 14-3 for the following regressions:
   a) FUEL on TAX, DLIC, INC
   b) FUEL on TAX, DLIC
   c) FUEL on TAX, INC
   d) FUEL on DLIC, INC
9. Plot the residuals from Problem 8(b) versus ROAD and versus INC. Examine the plots for patterns.
10. Compute $\hat{\sigma}^2$ for each of the equations in Problem 8, and using your answer to Problem 4, decide which of the five equations has the smallest value of $\hat{\sigma}^2$.

REFERENCES

C. I. Bliss (1970). *Statistics in Biology, Vol. II*, pp. 308–310. McGraw-Hill, New York.

S. Chatterjee and B. Price (1977). *Regression Analysis by Example*, Chapters 3 and 9. Wiley, New York.

T. A. Ferrar and J. P. Nelson (1975). Energy conservation policies of the Federal Energy Office: economic demand analysis. *Science* 187:644–646.

F. Mosteller and J. W. Tukey (1977). *Data Analysis and Regression*, Chapter 13. Addison-Wesley, Reading, Mass.

E. R. Tufte (1975). Determinants of the outcomes of midterm congressional elections. *American Political Science Review* 69:812–826.

S. Weisberg (1980). *Applied Linear Regression*, Chapters 2, 3, and 8. Wiley: New York.

15

Analysis of Variance

Learning Objectives

1. Measuring variability and allocating it among sources
2. Comparing the locations of three or more groups
3. Breaking the total sum of squares for two-way tables of measurements into three components: rows, columns, and residuals
4. Interpreting computer output for the analysis of variance (ANOVA)

15-1 INTRODUCTORY EXAMPLES

In the analysis of variance, we measure variability and allocate it among its sources. We have already seen such analyses in tests of the differences between two observed means. For that problem, the sources of variation were (a) the difference between the population means and (b) the variation within the two samples. We used the variability within the samples to assess the size of the observed difference in means, that is, the variability between samples. We use these same ideas of variability within and variability between in this chapter, and we extend them to more groups. Then we consider problems in which the groups are organized in a two-way layout, and we allocate the variability between groups into pieces that go with the rows and columns. This analysis-of-variance approach gives us formal tools to accompany the exploratory methods for two-way tables in Chapter 4.

As we did in Chapter 14, we stress here the interpretation of analysis-of-variance methods applied to actual examples. Although the calculations can be done by hand, they are somewhat tedious. Fortunately there are computer programs that can spare us much of this drudgery, and we develop the material in this chapter as if we had such a program available. Our primary goal is to learn to understand what a printout from an analysis-of-variance computer program means.

ONE-WAY TABLE

In problems with more than two groups, organizing our approach to sources of variation is more complicated than in the two-group case.

EXAMPLE 1 *Geographic distribution of divorced men.* In Table 15-1 we give the percentages of divorced men by states, grouped by region. The regional averages run from a low of 2.20 for the Northeast to a high of 3.91 for the West. The question arises: Are the proportions of divorced men randomly distributed by region, or do the regions differ more than random sorting would suggest? It looks, at least superficially, as if the Southwest and West have substantially higher rates than the rest of the country. But perhaps a random sorting into collections of 11, 11, 12, 6, and 10 states would give us a similar amount of variation in means as that in Table 15-1.

In Table 15-2 we show a random rearrangement of the 50 numbers into five groups having the same numbers of states in the groups as before, but the 50 states are now sorted randomly rather than geographically. For example, in Table 15-1, Nevada, with 6.9, is in group 5; in Table 15-2 it is in group 3. The largest and smallest means are 3.04 and 2.35. In the real

TABLE 15-1

Percentages of divorced men by regions, state by state

Group 1 *Northeast*		*Group 2* *Southeast*		*Group 3* *Central*		*Group 4* *Southwest*		*Group 5* *West*	
Conn.	2.0	Ala.	2.5	Ill.	2.8	Ariz.	3.4	Alaska	4.0
Del.	2.4	Fla.	3.4	Ind.	2.1	Ark.	2.8	Calif.	4.4
Me.	3.0	Ga.	2.7	Iowa	2.2	Colo.	3.3	Hawaii	2.8
Mass.	2.1	Ky.	2.7	Kan.	2.8	N.Mex.	2.9	Idaho	3.3
N.H.	2.5	La.	2.1	Mich.	2.9	Okla.	3.8	Mon.	3.4
N.J.	1.6	Md.	2.3	Minn.	2.1	Texas	3.2	Nev.	6.9
N.Y.	1.7	Miss.	2.1	Mo.	3.0			Oreg.	4.1
Pa.	2.0	N.C.	1.8	Nebr.	2.2			Utah	2.6
R.I.	2.1	S.C.	1.7	N.D.	1.6			Wash.	4.0
Vt.	2.3	Tenn.	2.7	Ohio	3.0			Wyo.	3.6
W.Va.	2.5	Va.	2.2	S.D.	1.9				
				Wis.	2.2				
Average	2.20		2.38		2.40		3.23		3.91
n	11		11		12		6		10

regional grouping of Table 15-1, the largest and smallest means are 3.91 and 2.20. Thus, on the basis of one trial, the variability among the real regions looks larger than we would expect from random selection. We could repeat the random regrouping many times to see how the data behave. With high-speed computation this would be easy to do, but it would deflect us from developing the standard analysis of variance.

TABLE 15-2

Randomly chosen groups of states

	Group 1 *"Northeast"*	*Group 2* *"Southeast"*	*Group 3* *"Central"*	*Group 4* *"Southwest"*	*Group 5* *"West"*
	2.1	2.4	2.0	1.6	2.0
	2.3	3.0	2.5	1.8	2.1
	1.7	2.7	1.7	1.9	2.5
	2.9	2.1	2.5	3.3	2.2
	2.2	2.1	3.4	3.4	2.8
	2.2	2.8	2.7	4.1	2.2
	3.4	2.8	2.3		2.1
	2.9	4.0	2.1		1.6
	3.8	2.8	6.9		3.3
	3.2	2.6	3.0		2.7
	4.0	3.6	3.0		
			4.4		
Average	2.79	2.81	3.04	2.68	2.35

TABLE 15-3

Stem-and-leafs of percentages of divorced men by regions, from Table 15-1

	Group 1 *N*	*Group 2* *SE*	*Group 3* *C*	*Group 4* *SW*	*Group 5* *W*
6					9
6					
5					
5					
4					
4					0410
3				8	6
3	0	4	00	432	34
2	55	5777	889	89	86
2	041013	1312	12122		
1	67	87	69		

While we look at Table 15-1, we might examine the distributions of divorce rates in the several real regions using a stem-and-leaf diagram. Table 15-3 shows the results. In the West, the 6.9 for Nevada stands out as an outlier. Because many people go there to get divorced, we might wish to set it aside in any further analysis.

The layouts in Table 15-1, 15-2, and 15-3 are called one-way tables. Each column is composed of measurements that are not directly related to those in other columns. We want to know about variation between the column means and within the columns.

TABLE 15-4

Average 2-month temperatures for five large northeastern cities (row and column averages and effects)

City	*Jan./ Feb.*	*March/ April*	*May/ June*	*July/ Aug.*
Boston	28	40	62	71
Chicago	26	43	64	73
Detroit	25	40	62	72
New York	31	44	65	74
Philadelphia	36	49	68	77
Calendar averages	29.2	43.2	64.2	73.4
Calendar effects	−22.2	−8.2	+12.8	+22.0

TWO-WAY TABLE

In some situations we have a two-way layout. Then each entry in the table is connected with both a row and a column.

EXAMPLE 2 *City temperatures.* In Table 15-4 we give average temperatures for 2-month periods for five large cities. Table 15-4 shows two sources of variation: cities and calendar periods. A third source is residuals, which we discuss later. Looking at the row means, we see that the average temperatures vary from city to city; in particular, the more southern the city, the higher the mean temperature. We also get variation from calendar periods according to a pattern familiar for temperate regions in the Northern Hemisphere—colder November through April, warmer May through October. We can use these averages to develop residuals from an additive model, much as we did in Chapter 4.

Table 15-4 gives (a) city effects, which are the deviations of the cities' means from the grand mean, 51.4, and (b) calendar effects, which are the deviations of the calendar period means from the grand mean. We can estimate the value in the cell by using the fitted value obtained by adding

$$\text{grand mean} + \text{city effect} + \text{calendar effect} = \text{fitted},$$

and for Boston in January–February we get

$$51.4 - 1.7 - 22.2 = 27.5 = \text{fitted}.$$

To get the residual we compute

$$\text{residual} = \text{observed} - \text{fitted}$$

Sept./ Oct.	**Nov./ Dec.**	***City averages***	***City effects***
59	38	49.7	−1.7
60	34	50.0	−1.4
58	35	48.7	−2.7
61	39	52.3	+0.9
64	43	56.2	+4.8
60.4	37.8	51.4	
9.0	−13.6		

and get

$$\text{residual} = 28 - 27.5 = 0.5$$

for that cell, as we did in Chapter 4. The variation of these residuals from cell to cell is an additional source of variation.

We could have examples with additional sources of variation, but we shall not go further than the two-way table in this book.

In Chapters 9 and 10 we compared the means of two samples of data. In scientific inquiries we often ask whether several populations may reasonably have the same mean. For example, we may have air pollution data from several different time periods, and we want to know whether the pollution has been much the same across the periods. Usually, any difficulty in making this comparison arises because we have to take account of the variability within the groups of data.

The analytical tools developed in Chapter 10 will allow us to examine the difference between any pair of observed means, but those tools do not put us into position to estimate the variability among the set of means as a whole or to test for inequality among the whole set of population means. We turn next to the task of developing such methods.

As we progress with the analysis of variance, the reader may feel that we are analyzing means instead of variances. But it is the variation in means that draws our attention. We want to know how large that variation is and whether it is important or can be neglected.

PROBLEMS FOR SECTION 15-1

1. *One-way example.* From an almanac or other source, find and display a set of data appropriate to the one-way layout with more than two groups.
2. *Two-way example.* From an almanac or other source, find a set of data appropriate to a two-way layout.
3. Recompute the mean for the West, with Nevada removed, in Table 15-1 and the mean with Nevada removed from the "Central" in Table 15-2, and discuss how the ranges in means for the real data and the randomly arranged data now compare.
4. Continuation. For the recomputed data from Table 15-1, compute the sample variance for Northeast and West. Comment on whether or not the variance seems to be roughly constant for these groups.

5. Compute the residuals for the cells in Table 15-4 in the first row. Comment on the accuracy of prediction.
6. *Measuring variability.* In Table 15-4, is there more variation between cities or between calendar periods? How can you tell? How would you measure such variations?
7. *Residual trouble.* The sum of the row effects and the sum of the column effects should each add to zero in Table 15-4. Why don't they?

Olympic platform diving. Table 15-5 contains results from the finals of the women's platform diving event at the 1976 Montreal Olympics. The country of each judge and each diver is also listed. Each judge gives a score over 10 dives to each diver, incorporating the degree of difficulty. The entries in Table 15-5 are the total scores minus 1400, and these are the data for Problems 8 through 14. Problems 10 and 11 are for those with access to an analysis-of-variance computer program.

8. Use the mean scores for judges to calculate the judge effects.
9. Use the mean scores for divers to calculate the diver effects.
10. Compute the residuals for the cells in Table 15-5, using fitted value = grand mean + judge effect + diver effect.
11. Do any residuals stand out as being much larger than the rest?
12. For each judge, identify the highest and lowest scores assigned.
13. For each diver, identify the judges who gave the highest and lowest scores.
14. Is there any evidence of bias in the judging related to nationality?

15-2 SEVERAL SAMPLES

To compare the means of several populations, numbered for our convenience as 1, 2, . . . , k, we draw a sample from each. These samples will have sizes $n_1, n_2, \ldots, n_k$, and the total number of observations, n, for the k groups taken together is

$$\Sigma n_i = n_1 + n_2 + \ldots + n_k = n.$$

For each sample, we have a sample mean and a sample variance:

	Sample			
	1	**2**	**. . .**	**k**
mean:	$\bar{x}_1$	$\bar{x}_2$	. . .	$\bar{x}_k$
variance:	s_1^2	s_2^2	. . .	s_k^2

TABLE 15-5

Scores from finals of women's platform diving, Montreal Olympics, 1976 (each entry is the total score minus 1400)

	Diver								
Judge	*Canada* 1	*USA* 2	*Canada* 3	*USSR* 4	*Sweden* 5	*USSR* 6	*USA* 7	*E. Germany* 8	*Judge average*
USA 1	178	390	213	215	271	404	236	94	250
Sweden 2	160	184	93	179	327	280	35	82	168
Egypt 3	259	293	165	333	293	273	208	235	257
Australia 4	295	269	171	241	243	290	130	26	208
USSR 5	177	220	191	361	201	449	188	148	242
Colombia 6	238	223	201	255	280	309	189	158	232
Italy 7	218	321	143	273	243	272	195	109	222
Average	218	271	168	265	265	325	169	122	226

Source: Adapted from data reported by K. S. Brown (1979). Judging Judging. *Ontario Secondary School Mathematics Bulletin* 15 (3):16–18.

In the two-sample problem of Chapter 10 we compared the two sample means. In the several-sample situation we compare each mean with the grand mean, pooling the observations over all of the samples. We represent this grand mean by

$$\bar{x} = \frac{n_1 \bar{x}_1 + n_2 \bar{x}_2 + \ldots + n_k \bar{x}_k}{n_1 + n_2 + \ldots + n_k}. \tag{1}$$

It will be helpful to have a three-sample case with convenient numbers to be used for illustrative calculations throughout this section.

EXAMPLE 3 *A three-sample case.* Samples 1, 2, and 3 have observations as follows:

Sample 1: 1, 2, 6, Sample 2: 1, 1, 3, 3, 3, 7, Sample 3: 6, 7, 8

In this example, $k = 3$, and

$$n_1 = 3, \quad n_2 = 6, \quad n_3 = 3.$$

The three sample means and variances are as follows:

	Sample		
	1	**2**	**3**
mean:	3	3	7
variance:	7	4.8	1

If we pool the three samples, we get $n = 3 + 6 + 3 = 12$ observations, and the grand mean is

$$\bar{x} = \frac{3\bar{x}_1 + 6\bar{x}_2 + 3\bar{x}_3}{12} = 4.$$

We will refer various questions to this three-sample case.

THE ONE-WAY MODEL

In the one-way model for the analysis of variance, our data consist of samples from k populations, with unknown population means: $\mu_1, \mu_2, \ldots, \mu_k$. As we did in the regression model of Chapter 12, we can represent each observation as the sum of two parts, a true population value and a random error. Because the true population value is the population mean, we have the following:

$$\text{observation from sample } i = \mu_i + \text{error} \quad (2)$$

The errors are not directly observable unless we have the unusual situation of knowing the mean, μ_i.

To complete the one-way analysis-of-variance model, we make the following assumptions that resemble those of the regression situation:

1. The errors are independent random values from normally distributed populations.
2. Each population of errors has mean zero and the same variance σ^2 (unknown).

The role of the normality assumption is to assure that the percentage points of the tests of significance, described in Section 15-3, are correct. That assumption does not bear on the ideas of average values or of various sums of squares adding up correctly.

Because the errors have mean zero, our observations from sample i, on average, take the value μ_i. Thus we estimate μ_i by the ith sample mean, $\bar{x}_i$. We want to develop a way to assess differences among the μ_i's. We do this by comparing the sample means using information about the variability within the samples.

SUMS OF SQUARES

The sums of squares from various sources have useful relations, and so we emphasize them. To compare the sample means, we compute the squared deviation of each from the grand mean:

$$(\bar{x}_1 - \bar{x})^2, (\bar{x}_2 - \bar{x})^2, \ldots, (\bar{x}_k - \bar{x})^2.$$

Next, we calculate a summary measure of the difference among the means, the **between-means sum of squares:**

$$BSS = n_1(\bar{x}_1 - \bar{x})^2 + n_2(\bar{x}_2 - \bar{x})^2 + \ldots + n_k(\bar{x}_k - \bar{x})^2 \quad (3)$$

The sum of squares is a way of measuring how far the set of observed sample means is from the grand mean of all the measurements. Strictly, it is the square of the distance. Our plan is to compare BSS with a pooled estimate of error variance, σ^2, over all k samples. The variability in sample i is measured by s_i^2. To convert each s_i^2 back to a sum of squares, we multiply by the degrees of freedom (d.f.) in sample i, $n_i - 1$. Adding over the samples, we get the **within-groups sum of squares:**

$$WSS = (n_1 - 1)s_i^2 + (n_2 - 1)s_2^2 + \ldots + (n_k - 1)s_k^2 \tag{4}$$

or

$$WSS = \underset{\text{sample 1}}{\Sigma (x_1 - \bar{x}_1)^2} + \underset{\text{sample 2}}{\Sigma (x_2 - \bar{x}_2)^2} + \ldots + \underset{\text{sample } k}{\Sigma (x_k - \bar{x}_k)^2} \tag{5}$$

The total sum of squares is defined as the sum of squares of deviations of all measurements from the grand mean $\bar{x}$:

DEFINITION OF TOTAL SUM OF SQUARES

$$TSS = \underset{\text{sample 1}}{\Sigma(x_1 - \bar{x})^2} + \underset{\text{sample 2}}{\Sigma(x_2 - \bar{x})^2} + \ldots + \underset{\text{sample } k}{\Sigma(x_k - \bar{x})^2} \tag{6}$$

The total sum of squares can also be computed by adding the between-means sum of squares and the within-groups sum of squares:

$$TSS = BSS + WSS \tag{7}$$

Thus we have a numerical check that the definitional form of TSS, equation (6), gives the same result as the component form, equation (7).

EXAMPLE 4 *Verifying the sums of squares calculations.* For the three-sample problem of Example 3, the between-means sum of squares is

$$BSS = 3(3 - 4)^2 + 6(3 - 4)^2 + 3(7 - 4)^2 = 36.$$

Within each of the samples, the sums of squared deviations about the sample mean are

$$\underset{\text{sample 1}}{\Sigma (x_1 - \overline{x}_1)^2} = (1 - 3)^2 + (2 - 3)^2 + (6 - 3)^2 = 14,$$

$$\underset{\text{sample 2}}{\Sigma (x_2 - \overline{x}_2)^2} = 24, \quad \underset{\text{sample 3}}{\Sigma (x_3 - \overline{x}_3)^2} = 2.$$

Therefore the within-samples sum of squares is

$$WSS = 14 + 24 + 2 = 40.$$

Because $\overline{x} = 4$, using the definitional form of TSS in equation (6) we have

$$TSS = \underbrace{(1 - 4)^2 + (2 - 4)^2 + (6 - 4)^2}_{\text{sample 1}} + \underbrace{(1 - 4)^2 + \ldots + (7 - 4)^2}_{\text{sample 2}}$$
$$+ \underbrace{(6 - 4)^2 + (7 - 4)^2 + (8 - 4)^2}_{\text{sample 3}} = 76.$$

Finally, we use equation (7) to get a check on our answer:

$$\begin{aligned} BSS + WSS &= 36 + 40 \\ &= 76 \\ &= TSS. \end{aligned}$$

DEGREES OF FREEDOM

In Chapter 10 we learned that there are $n_i - 1$ d.f. associated with our estimate of variance in the ith sample, s_i^2. Thus, a pooled estimate of σ^2 using observations from all k samples should be based on the sum of the degrees of freedom for all of the samples:

$$\begin{aligned} \text{pooled d.f.} &= (n_1 - 1) + (n_2 - 1) + \ldots + (n_k - 1) \\ &= \Sigma n_i - k \\ &= n - k. \end{aligned} \tag{8}$$

To get our estimate of σ^2, we divide the WSS by the pooled degrees of freedom:

$$\text{estimate of } \sigma^2 = \frac{WSS}{n - k} \tag{9}$$

We refer to this estimate as the **within mean square** or WMS.

For the comparison among the means, we have k values of $\bar{x}_i$, but they are constrained to average to the overall mean, $\bar{x}$, as we saw in equation (1). Thus we lose 1 d.f. out of the k, yielding $k - 1$. We associate these $k - 1$ d.f. with our summary statistic comparing the $\bar{x}_i$, BSS. The **between mean square,** BMS, comes from dividing BSS by its degrees of freedom:

$$BMS = \frac{BSS}{k - 1} \tag{10}$$

The degrees of freedom associated with the TSS are $n - 1$, because we start with n observations and use up 1 d.f. to estimate μ by $\bar{x}$. Note that the degrees of freedom add up, just as do the sums of squares:

$$\underset{BSS}{(k - 1)} + \underset{WSS}{(n - k)} = \underset{TSS}{n - 1} \tag{11}$$

In Table 15-6 we summarize the formulas for degrees of freedom, the sums of squares, and the mean squares. Because TSS summarizes the overall variability in the data and the within mean square is our estimate of the error variance, σ^2, we call the summary in Table 15-6 an **analysis-of-variance table.** Table 15-6 contains a final column labeled F ratio, which we will discuss shortly.

For our three-sample case of Examples 3 and 4, Table 15-7 summarizes the analysis of variance. We see that $BSS = 36$, $WSS = 40$, and

$$\begin{aligned} BSS + WSS &= 36 + 40 \\ &= 76 = TSS, \end{aligned}$$

TABLE 15-6
Analysis-of-variance table (formulas)

Source	Degrees of freedom	Sum of squares	Mean squares	F ratio
Between means	$k-1$	BSS	$BMS = BSS/(k-1)$	BMS/WMS
Within samples	$n-k$	WSS	$WMS = WSS/(n-k)$	
Total	$n-1$	TSS		

TABLE 15-7
Analysis-of-variance table for our three-sample case

Source	d.f.	Sum of squares	Mean square	F ratio
Between	2	36	18	4.05
Within	9	40	4.44	
Total	11	76		

as we noted earlier. The value listed in the column for F ratio is

$$F = \frac{BMS}{WMS}$$

$$= \frac{18}{4.44} = 4.05.$$

We learn to interpret this value in the next section.

USING COMPUTER OUTPUT

The three-sample example used to illustrate the material in this section was designed for easy calculation by hand. For more complicated examples, the sums-of-squares calculations can still be done by hand or with a pocket calculator, but we usually need to take advantage of special computational formulas to keep the effort under control. Some of the references at the end of the chapter explain these computational formulas for the analysis of variance.

Most high-speed computers have programs readily available for carrying out the analysis of variance. Again, you should check these out first by giving them simple problems in which the work has already been carried out by hand or by hand calculator. Computer programs often provide valuable extra printout including the data, all shown in nice form. You likely will need instruction about what items printed out have special meaning for the problem at hand. For large jobs, such programs can be most helpful. Once you have invested the overhead to learn how to enter and submit problems, even *small* jobs go very quickly.

For much of the remainder of this chapter we act as if we have mastered the use of an analysis-of-variance computer program, and we focus on the output it produces.

EXAMPLE 5 *Analysis of variance for the geographic distribution of divorced men.* In Example 1 we examined the data in Table 15-1 on the percentages of divorced men by states. The states were organized into 5 regions. Construct an analysis-of-variance table for these data and discuss the results.

SOLUTION. We treat the data as $k = 5$ samples, one for each region. Exhibit 15-A contains the actual computer output from a one-way analysis of variance of these data.

The output in Exhibit 15-A contains two sets of information. In the bottom part we have the summary information for each of the 5 samples: the sample size n_j, the mean $\bar{x}_j$, and the standard deviation s_j. In the top part of the exhibit we have the analysis-of-variance table. The first column of this

EXHIBIT 15-A

Computer output for a one-way analysis-of-variance table of the data on percentages of divorced men (Table 15-1)

ANALYSIS OF VARIANCE

DUE TO	DF	SS	MS=SS/DF	F-RATIO
FACTOR	4	21.146	5.286	11.91
ERROR	45	19.979	0.444	
TOTAL	49	41.124		

LEVEL	N	MEAN	ST. DEV.
NE	11	2.200	0.397
SE	11	2.382	0.485
CENT	12	2.400	0.475
SW	6	3.233	0.361
WEST	10	3.910	1.198

POOLED ST. DEV. = 0.666

output differs somewhat from that of Table 15-6. To translate these terms, in place of DUE TO read SOURCE. The row labeled FACTOR corresponds to "between means," and the row ERROR corresponds to "within samples" (used to compute an estimate of the error variance). Thus we see from Exhibit 15-A that (rounding to one decimal place) $BSS = 21.1$, with 4 d.f., and $WSS = 20.0$, with 45 d.f.

The pooled estimate of variance for this example is

$$WMS = \frac{WSS}{n - k} = \frac{19.979}{45} = 0.444.$$

Taking the square root, we have $\sqrt{0.444} = 0.666$, which is listed at the bottom of the output as POOLED ST. DEV.

In the next section we compare the magnitude of the between mean square with that of the within mean square by taking their ratio, called F in honor of R. A. Fisher:

$$F = \frac{BMS}{WMS} = \frac{5.286}{0.444} = 11.91.$$

We use this value, found in the output in the column labeled F RATIO, to assess possible differences among the population means, the μ_i's.

To see how things might have turned out if we had used data where there are no differences among the μ_i's, let us recall the values in Table 15-2. This table was based on a random rearrangement of the 50 observations into 5 groups of size 11, 11, 12, 6, and 10. Exhibit 15-B contains the computer output for an analysis of variance of these data.

EXHIBIT 15-B

Computer output for a one-way analysis-of-variance table of the data in Table 15-2

ANALYSIS OF VARIANCE

DUE TO	DF	SS	MS=SS/DF	F-RATIO
FACTOR	4	2.704	0.676	0.79
ERROR	45	38.421	0.854	
TOTAL	49	41.124		

LEVEL	N	MEAN	ST. DEV.
NE	11	2.791	0.752
SE	11	2.809	0.575
CENT	12	3.042	1.411
SW	6	2.683	1.046
WEST	10	2.350	0.484

In Exhibit 15-B we see that $BSS = 2.704$, with 4 d.f., and $WSS = 38.4$, with 45 d.f. For the rearranged data the sample means all lie in an interval of length 0.7, a value that is less than the pooled standard deviation. Now the ratio of the between mean square to the within mean square is

$$F = \frac{BMS}{WMS} = \frac{0.676}{0.854} = 0.79.$$

Large values of F typically suggest differences among the samples and, by implication, differences among the populations from which they are drawn. Values of F near 1 suggest that the variation between observed means is primarily due to sampling variation.

PROBLEMS FOR SECTION 15-2

The following data for three samples form the basis for Problems 1 through 9. Save your results.

Sample 1: 2, 2, 8, Sample 2: 1, 3, Sample 3: 4, 4, 6, 6

1. Find the sample means and the grand mean.

2. Use s_i^2 to estimate separately the variances of the three populations from which the samples were drawn.

3. Continuation. *Estimation.* Assuming that the three populations have equal variances, estimate that common σ^2 by suitably weighting the s_i^2 in Problem 2.

4. Find the between means sum of squares, BSS.

5. Find the total sum of squares, TSS.

6. Find the within sample sum of squares, WSS.

7. Continuation of Problems 4, 5, and 6. Check that the relation $TSS = BSS + WSS$ applies to this example.

8. Lay out the analysis-of-variance table, and compute and interpret the F ratio.

9. If you have access to an analysis-of-variance computer program, input the data from this example and compare the output with the results you computed by hand (or calculator).

10. What happens to the BSS and WSS when the number of samples $k = 1$?

15-3 THE *F* RATIO

The F ratio gives us a way to make a statistical test for a difference among population means. The ideas follow. If the means for k different populations, $\mu_1, \mu_2, \ldots, \mu_k$, are all equal, then we expect the differences

$$\bar{x}_1 - \bar{x}, \bar{x}_2 - \bar{x}, \ldots, \bar{x}_k - \bar{x}$$

simply to reflect the variability from one sample to another. In fact, if

$$\mu_1 = \mu_2 \quad = \ldots = \mu_k,$$

then it can be shown that the average value of the BMS,

$$BMS = \frac{BSS}{k-1} = \frac{n_1(\overline{x}_1 - \overline{x})^2 + n_2(\overline{x}_2 - \overline{x})^2 + \ldots + n_k(\overline{x}_k - \overline{x})^2}{k-1},$$

averaging over all possible sets of samples, is σ^2. That is:

$$\text{ave}(BMS) = \sigma^2 \tag{12}$$

On the other hand, if the μ_i's differ, it can be shown that the average value of BMS is greater than σ^2.

It can also be shown that the average of the WMS over all samples is always σ^2. Consequently, the ratio of the mean squares

$$F = \frac{BSS/(k-1)}{WSS/(n-k)} \tag{13}$$

should give us a notion of whether or not the population means differ. When the means are all equal, the average over all samples of the numerator and that for the denominator are both σ^2. Thus a ratio of about 1 suggests equality of means. A larger ratio suggests that the population means are scattered.

The statistic F is usually described by relating it to the degrees of freedom in its numerator and in its denominator. For the k-sample situation, the numerator has $k - 1$ d.f. (one less than the number of samples), and the denominator has $n - k$ d.f., because that is the number of independent measurements available for estimating the common σ^2.

Because both the numerator and the denominator of F involve sums of squares, F can never become negative. Its possible values run from 0 to ∞. Further, just as t had a different distribution for each degree of freedom, F has a different distribution for each pair of degrees of freedom. Figures 15-1(a,b,c) show some illustrative distributions of F for various pairs of degrees of freedom. The F distributions are skewed. The probability is concentrated toward the left-hand tail instead of being symmetrically distributed.

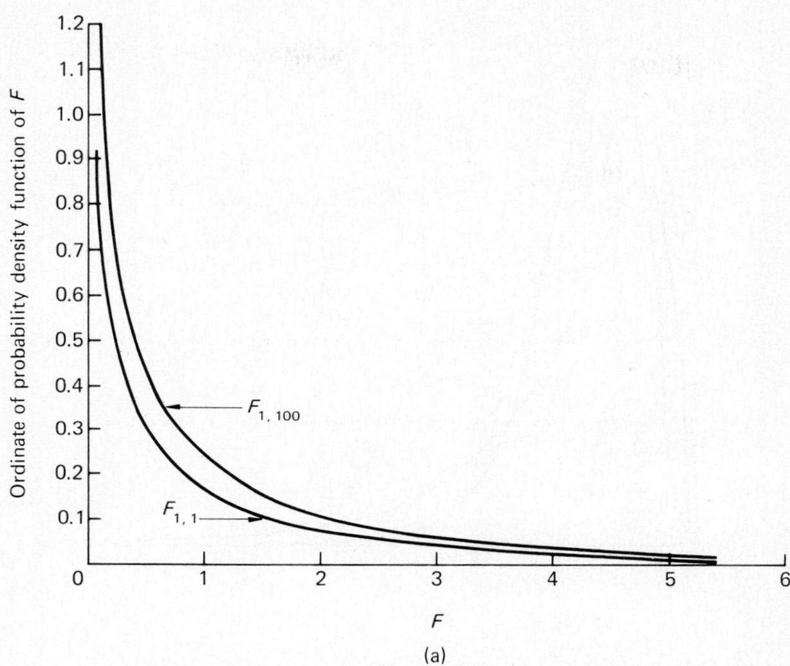

Figure 15-1(a) Curves of the probability density functions of $F_{1,1}$ and $F_{1,100}$.

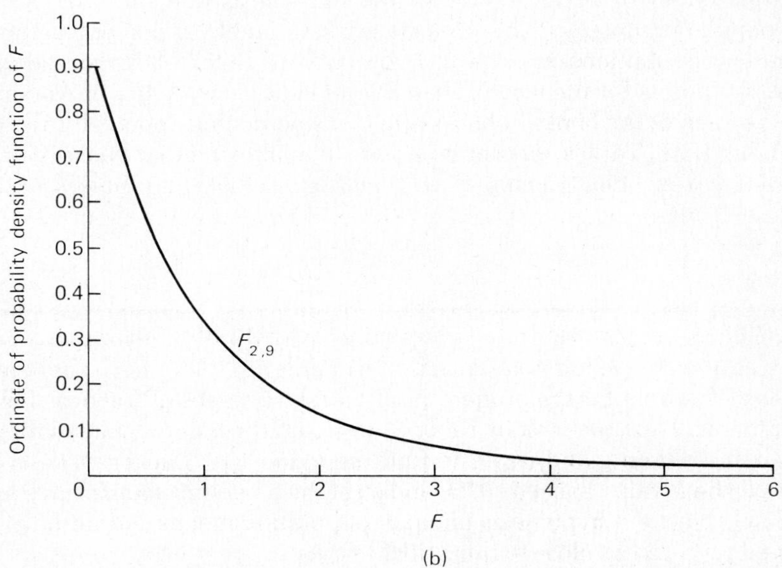

Figure 15-1(b) Curve of the probability density function of $F_{2,9}$.

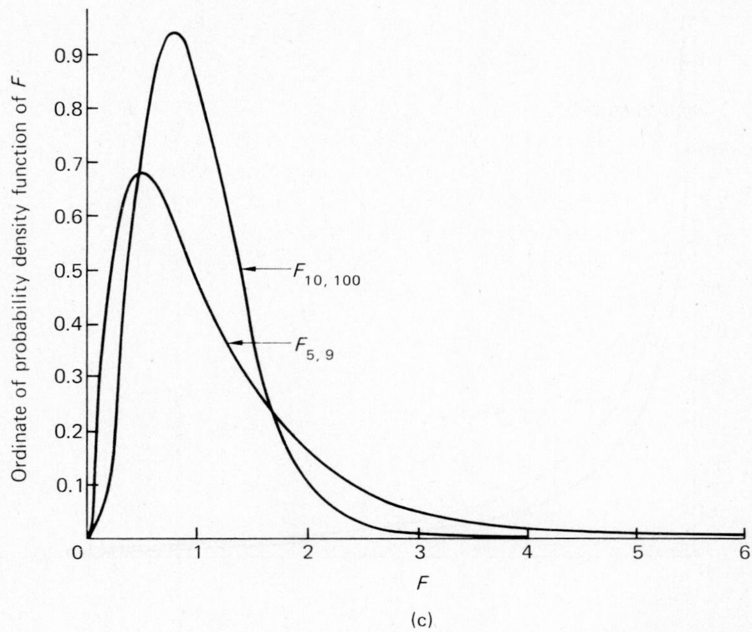

Figure 15-1(c) Curves of the probability density functions of $F_{5,9}$ and $F_{10,100}$.

Large values of F cast doubt on the hypothesis that the groups have equal population means. Table 15-8 gives a small table of percentage points for the F distributions for the 0.05 probability level. A larger table of percentage points for the 0.05 probability level is Table A-4(a) in Appendix III in the back of the book; Table A-4(b) gives percentage points for the 0.01 probability level. Tables of other levels are available in other books. See, for example, Owen (1962), Pearson and Hartley (1966), and Snedecor and Cochran (1980).

EXAMPLE 6 *F test for three-sample case.* In the three-sample case of Examples 3 and 4, which is summarized in Table 15-7, $F = 4.05$. There are 2 degrees of freedom for the numerator of F and 9 degrees of freedom for the denominator. Thus, we look in Table 15-8 to find the 5 percent level for $F_{2,9}$. We look in column 2 and row 9 to find the value 4.26. Because 4.05 is less than 4.26, the P value exceeds 0.05, and so at the 5 percent significance level we do not reject the hypothesis of equal population means. Admittedly, the observed ratio comes close to this critical value.

TABLE 15-8†

Values of a for which the probability is 0.05 that a value of $F_{u,v}$ is greater than a, where u is the number of degrees of freedom for the numerator and v is the number of degrees of freedom for the denominator in the F ratio (BMS/WMS)

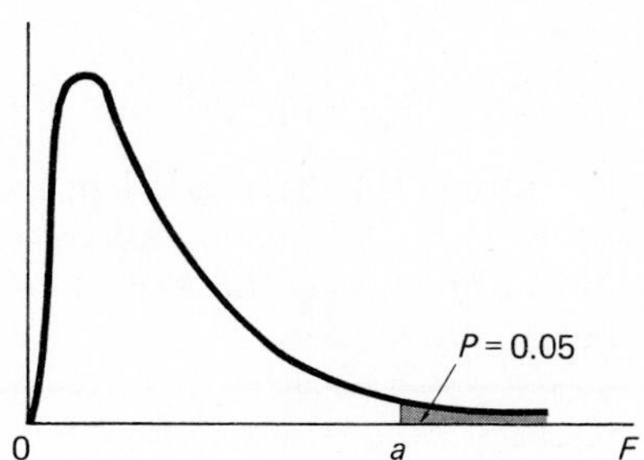

Degrees of freedom for denominator (v)	Degrees of freedom for numerator (u)				
	1	2	3	4	5
1	161	200	216	225	230
2	18.51	19.00	19.16	19.25	19.30
3	10.13	9.55	9.28	9.12	9.01
4	7.71	6.94	6.59	6.39	6.26
5	6.61	5.79	5.41	5.19	5.05
6	5.99	5.14	4.76	4.53	4.39
7	5.59	4.74	4.35	4.12	3.97
8	5.32	4.46	4.07	3.84	3.69
9	5.12	4.26	3.86	3.63	3.48
10	4.96	4.10	3.71	3.48	3.33
11	4.84	3.98	3.59	3.36	3.20
12	4.75	3.89	3.49	3.26	3.11
15	4.54	3.68	3.29	3.06	2.90
20	4.35	3.49	3.10	2.87	2.71
30	4.17	3.32	2.92	2.69	2.53
40	4.08	3.23	2.84	2.61	2.45
60	4.00	3.15	2.76	2.53	2.37
∞	3.84	3.00	2.60	2.37	2.21

† Larger tables are Table A-4(a) and Table A-4(b) in Appendix III in the back of the book.

EXAMPLE 7 *F test for geographic distribution of divorced men.* In Example 5 we computed the F ratio for the data of Example 1 on the percentages of divorced men by states. Carry out the F test for comparing the 5 regions.

SOLUTION. The F ratio from Exhibit 15-A is $F = 11.91$. When we look in Table 15-8 or Table A-4 in Appendix III in the back of the book to find the 5 percent level for $F_{4,45}$, we find the values

$$F_{4,40} = 2.61$$

$$F_{4,60} = 2.53.$$

Because the observed F ratio of 11.91 is well in excess of both of these values, the P value is much less than 0.05, and we conclude that the data provide sufficient evidence to support the existence of differences among regions in percentages of divorced men.

We recall that the assumptions needed to apply this F test are:

1. Random samples independently drawn from normally distributed populations.
2. Each population has the same variance σ^2.

When these assumptions are mildly violated, the results still may be approximately correct.

EXAMPLE 8 *Mathematics programs.* In a large school district, four different mathematics programs were used with sixth grade classes. The numbers of classes for each program were as follows:

	Program				**Total**
	I	**II**	**III**	**IV**	***n***
Number, n_i	4	5	8	6	23

The mean scores for the classes on a standardized mathematics test for sixth graders were determined at the end of the year, and these are given in Table 15-9. The classes were all about the same size. Carry out the analysis of variance and interpret the results.

TABLE 15-9
Data for comparing the mathematics programs

Program I	*Program II*	*Program III*	*Program IV*
69.5	61.0	72.6	51.8
90.4	57.3	70.4	78.4
80.9	60.8	64.5	62.3
67.5	60.6	69.8	74.0
	76.2	68.3	59.3
		72.5	81.7
		54.0	
		67.5	

EXHIBIT 15-C
Computer output for a one-way analysis-of-variance table of the mathematics test data in Table 15-9

```
ANALYSIS OF VARIANCE

DUE TO          DF          SS      MS=SS/DF        F-RATIO
FACTOR           3       445.7         148.6           1.86
ERROR           19      1521.4          80.1
TOTAL           22      1967.2

LEVEL            N        MEAN       ST. DEV.
C1               4       77.07          10.67
C2               5       63.18           7.44
C3               8       67.45           6.05
C4               6       67.92          11.85

POOLED ST. DEV. =          8.95
```

SOLUTION. Exhibit 15-C contains the computer output from a one-way analysis of variance of the data in Table 15-9.*

The F ratio from the ANOVA table in Exhibit 15-C is 1.86. We compare this with the 5 percent point for the F distribution with 3 and 19 degrees of freedom. In Table A-4 in Appendix III in the back of the book we find $F_{3,19}$ to be between 3.10 and 3.29. Because the observed ratio, 1.86, is much smaller than 3.10, we find that the P value is larger than 0.05 and conclude that the program results show no statistically significant difference at the 5 percent level.

* C1, C2, C3, and C4 in the computer printout for Exhibit 15-C correspond to Programs I, II, III, and IV, respectively.

Although Program I has a mean larger than the rest, it is based on a small sample. The estimated standard deviations (also from Exhibit 15-C) for the four programs are as follows:

I	II	III	IV
10.7	7.4	6.1	11.9

They do not seem to vary a great deal. Even though the program means differ by as much as 14 points, the inherent variability within the programs prevents us from drawing strong conclusions about differences. It would take much more data to establish these differences firmly than we have available in the present study.

SMALL VALUES OF *F*

Theory not given here shows that the "average mean square between" must be larger than or equal to the "average mean square within." We must remember that this is a long-run average under the assumptions. The F distributions displayed in Figs. 15-1(a,b,c) imply that an F ratio can have values less than 1. Even when we have true differences among the μ_i, the observed F ratio might be smaller than 1. Would we ever find it too small? That is, when might it give a value that would fall into the lower 5 percent of the distribution instead of the upper? Of course, this could happen in the 5 percent of the time that chance plays dirty tricks. But it might also be a signal that a mistake has been made.

EXAMPLE 9 *Mistakes in degrees of freedom.* An example of such a mistake might occur if we divided BSS by 100 instead of $k - 1$, or if we forgot to divide WSS by $n - k$. Suppose that we made both mistakes in our three-sample case. Then we would get

$$\text{mistaken } F \text{ ratio} = \frac{36/100}{40} = 0.009.$$

How can we look this up in Table 15-8? We cannot. What we need is an extra fact.

FACT ABOUT 1/*F*

Suppose the numerator and denominator of F have u and v degrees of freedom, respectively. The distribution of $1/F$ has the distribution $F_{v,u}$. And so the lower tail of $F_{u,v}$ corresponds to the upper tail of $F_{v,u}$.

Thus, to use our tables to study the lower tail, we need to invert F and interchange the degrees of freedom.

Because we do not give a table of lower limits, we need to flip the ratio over and make it an upper-tail problem. Then

$$1/(\text{mistaken } F \text{ ratio}) = 111.1.$$

The new degrees of freedom are (9, 2), in that order, and the 5 percent upper level for $F_{9,2}$, from Table A-4 in Appendix III in the back of the book is 19.38. The observed 111.1 exceeds 19.38, and so the original $F = 0.009$ is significantly small.

Systematic Effects. A systematic effect leading to a value of F near zero can occur when some hidden matching in the samples occurs that prevents the independence we assume between the samples. For example, suppose that two samples were made up of several sets of identical twins, one randomly chosen for the first sample and the other assigned to the second. Then the samples would be more alike than random samples from a population. Of course, this matching would be helpful for an experiment, but we should use the matching in the analysis as well. But then we are involved with a two-way table, rather than a one-way table. This type of matching is often called "blocking," and we discuss the link between blocking and analysis of variance in Chapter 17, Section 17-6.

EXAMPLE 10 *Omitting a variable.* In Table 15-4, for Example 2, suppose we forget that the columns match the cities for calendar months. Then we might run an analysis of variance as if we had 5 cities with 6 independent measurements each. We would expect the cities to show a small F ratio because of the strong matching. You will be asked to study this in Problem 4 of the summary problems for this chapter.

EXAMPLE 11 *Kansas City preventive patrol experiment.* With three different patrolling procedures, and several areas for each procedure, a study analyzed numbers of reported crimes for several types of crime. For bicycle larceny, the mean square between was smaller than the mean square within, 1.27 versus 6.22. This result presumably was due to geographic balancing in the design of the study. Similar results occurred for several other types of crime.

PROBLEMS FOR SECTION 15-3

1. *Weight loss among teenagers.* After physical examinations at the beginning of a term, 17 girls ages 16 to 18 were identified as being considerably overweight. All but two agreed to participate in an experiment designed to study the effects of two new diets. The physician in charge randomized them into 3 groups: 5 girls each to diets A and B and 5 girls to a control group (diet C). The control group received standard counseling on weight loss but no specific diet. After 3 months their weight losses to the nearest pound were as shown in Table 15-10. Carry out the analysis of variance. Explain the results and interpret them.

TABLE 15-10
Weight losses for teenagers

	Subjects				
Diets	***1***	***2***	***3***	***4***	***5***
A	9	8	4	3	6
B	8	11	6	14	11
C	1	2	6	0	1

2. *Sugar.* Among 21 samples of size 5 each from different stores, the Within Sum of Squares for the net weight of a "pound" of sugar was 25 ounces^2. The Between Sum of Squares was 5.50 ounces^2. Estimate σ^2, the common variance, and σ, the common standard deviation.

3. The Between Mean Square is 4, and the Within Mean Square is 6, with 3 and 150 d.f. If you are testing for differences among the μ_i's, is it worth looking in the F table? Why or why not?

4. The Between Mean Square is 6, and the Within Mean Square is 4, with 3 and 150 d.f. Compute the F ratio, and interpret it using the F table, Table A-4(a) in Appendix III in the back of the book.

5. *Small* F. With an $F_{3,9}$ an investigator observed $F = 0.06$. He wonders if this is unusually small. Use Table A-4(a) in the back of the book to find out, and explain your calculation.

One-way analysis of variance for $k = 2$ *samples.* Problems 6 through 12 demonstrate the link between the F ratio for the two-sample case and the two-sample t test from Chapter 10, using the following data. An x_1 sample is drawn from one population and an x_2 sample from another, with the following values:

x_1 sample		x_2 sample
2, 3, 4	and	4, 4, 6, 10, 6, 6

6. Find the sample means and the grand mean.

7. Compute BSS and TSS.

8. Lay out the analysis-of-variance table, and compute and interpret the F ratio.

9. Compute the pooled estimate of σ^2, based on WSS.

10. Carry out a t test for the equality of the means of the x_1 and x_2 populations, using the pooled estimate of σ^2.

11. Verify for this example that the square of the t value from Problem 10 equals the F ratio from Problem 8; that is, $F = t^2$.

12. Check that the 5 percent point for $F_{1,7}$ in Table 15-8 is identical with the 5 percent point obtained by squaring the two-sided 5 percent point of t_7.

Project based on Problems 13 through 16: Explore the analysis of variance in a situation in which you secretly know the μ_i's. If you sum three random digits, you get numbers that are approximately normally distributed for many purposes.

13. Use the random number table, Table A-7 in Appendix III in the back of the book, to construct 30 numbers by summing successive sets of three digits (no overlap). Break these into three samples of 10 each using the first 10, the second 10, and the third 10.

14. Carry out the analysis of variance on them, estimating σ^2 (whose true value is 24.75, or about 25) and computing the F ratio and interpreting it.

15. You have now carried out an analysis of variance in a situation in which you know there is no difference among the μ_i's. To assess σ^2, compute

$$F = \frac{WSS/(n - k)}{\sigma^2}$$

and compare it to values in the F table with $n - k$ and ∞ d.f. (you know σ^2 exactly, and so you have ∞ d.f.).

16. Continuation. Add -5 to the ten numbers in your first set and $+5$ to the ten numbers in your third set. Leave the second set unchanged. Now carry out the same steps as in Problem 14. How has the estimate of σ^2 changed?

15-4 TWO-WAY ANALYSIS OF VARIANCE WITH ONE OBSERVATION PER CELL

The temperature data for northeastern cities in Table 15-4 introduces an extra "way" in the table. Thus, we want to break the total sum of squares into three parts:

1. between calendar periods (columns)
2. between cities (rows)
3. residual (interaction plus error)

To give formulas, we need a little more notation than we have given up to now. Let

$$\begin{aligned} r &= \text{number of rows} \\ c &= \text{number of columns} \\ x_{ij} &= \text{entry in row } i\text{, column } j \text{ (rows are stated first)} \\ x_{i+} &= \text{total for row } i \\ x_{+j} &= \text{total for column } j \\ x_{++} &= \text{grand total.} \end{aligned}$$

The grand mean is simply

$$\bar{x} = \frac{x_{++}}{rc}.$$

The total sum of squares is defined as before as the sum of squares of deviations of all measurements from the grand mean $\bar{x}$. In our current setting, with the new notation, this is the following:

Total sum of squares:

$$TSS = \sum_{\text{all values}} (x_{ij} - \bar{x})^2 \qquad (14)$$

The calculation of the between-rows sum of squares is just like the *BSS* calculation from Section 15-2. We only need to keep in mind that the number of observations in each row is the same, that is, c:

$$c\left(\frac{x_{1+}}{c} - \bar{x}\right)^2 + c\left(\frac{x_{2+}}{c} - \bar{x}\right)^2 + \ldots + c\left(\frac{x_{r+}}{c} - \bar{x}\right)^2$$

or we can write more briefly

Between-rows sum of squares:

$$BRSS = c \sum_i \left(\frac{x_{i+}}{c} - \bar{x} \right)^2 \tag{15}$$

We use a similar calculation to get

Between-columns sum of squares:

$$BCSS = r \sum_j \left(\frac{x_{+j}}{r} - \bar{x} \right)^2 \tag{16}$$

Because sums of squares for the two-way table add up to the total as they did for the one-way case, we have

Residual sum of squares:

$$RSS = TSS - BRSS - BCSS \tag{17}$$

Special computational formulas are available to simplify the work in computing these sums of squares, but as before we rely on the computer to do our calculations. The analysis-of-variance table based on these sums of squares can be laid out as in Table 15-11. Note that there the lines for *BRSS* and *BCSS* replace the single line for *BSS* that we had in the one-way case.

The degrees of freedom for each sum of squares are also listed in Table 15-11. The only new value is the residual sum of squares. The $(r - 1)(c - 1)$ d.f. for the *RSS* can be derived most directly by recalling that the degrees of freedom must also add up:

$$\underset{BRSS}{(r - 1)} + \underset{BCSS}{(c - 1)} + \underset{RSS}{(r - 1)(c - 1)} = \underset{TSS}{rc - 1} \tag{18}$$

TABLE 15-11

Two-way analysis-of-variance table with one observation per cell

Source	d.f.	Sum of squares	Mean square	F ratio
Between rows	$r - 1$	$BRSS$	$BRSS/(r - 1)$	$\dfrac{BRSS/(r-1)}{RSS/(r-1)(c-1)}$
Between columns	$c - 1$	$BCSS$	$BCSS/(c - 1)$	$\dfrac{BCSS/(c-1)}{RSS/(r-1)(c-1)}$
Residual	$(r - 1)(c - 1)$	RSS	$RSS/(r - 1)(c - 1)$	
Total	$rc - 1$			

By dividing each sum of squares by its degrees of freedom, we get the mean squares. If the cells were all drawn from the same population, then the between-rows mean square, the between-columns mean square, and the residual mean square would be identical, on the average, though not in specific examples. The ratio of the between-rows mean square to the residual mean square, when the cells all come from a population with the same mean, has an F distribution with degrees of freedom $r - 1$ and $(r - 1)(c - 1)$. Similarly, the distribution of the ratio of the between-columns mean square to the residual mean square is F, with $c - 1$ and $(r - 1)(c - 1)$ d.f. A large value of the F for rows suggests that the row means differ. A similar interpretation applies to the F for columns.

In the two-way analysis, we think of a measurement as being made of four parts:

μ = grand mean

α_i = effect of row i $(\Sigma\alpha_i = 0)$

β_j = effect of column j $(\Sigma\beta_j = 0)$

e_{ij} = error with mean zero and variance σ^2.

To have accuracy with F ratios, we need the e_{ij} to be approximately normally distributed. To sum up, the relation between the observation x_{ij} and the four components given earlier is the following:

Analysis-of-variance model:

$$x_{ij} = \mu + \alpha_i + \beta_j + e_{ij} \tag{19}$$

EXAMPLE 12 *City temperatures.* Carry out the analysis of variance for the temperature data of Example 2, Table 15-4.

SOLUTION. In Exhibit 15-D we give the computer results of the analysis-of-variance calculation. The huge effect of calendar periods is expected and obvious. Northeasterners do not need analysis of variance to know that winter comes and is substantially different from summer. But by taking out the effects of period, we get a better measure of the basic variability in the residuals to help us appraise the variation between cities. From Table A-4(a) in Appendix III in the back of the book, the 5 percent level of F with 4 and 20 d.f. is 2.87, and our observed F, which is equal to $54.03/1.77 = 30.5$, is very much higher than this. We therefore have firm evidence of differences in temperatures among these cities averaged over the year. Because all the cities are situated near water and are not very much removed north to south, we might not have been able to detect the differences reliably without controlling for the effects of period.

The estimate of the residual variance is 1.77, so that the residual standard deviation is 1.33 degrees Fahrenheit. This is the standard deviation of the residuals after we have estimated the cell values on the basis of the grand mean and the row and column effects. What is it composed of?

EXHIBIT 15-D

Computer output of the analysis-of-variance calculations for the temperature data (Table 15-4)

ANALYSIS OF VARIANCE

DUE TO	DF	SS	MS=SS/DF
CITY	4	216.13	54.03
CALEN	5	7369.37	1473.87
ERROR	20	35.47	1.77
TOTAL	29	7620.97	

OBSERVATIONS
ROWS ARE LEVELS OF CITY   COLS ARE LEVELS OF CALEN

	1	2	3	4	5	6	ROW MEANS
1	28.00	40.00	62.00	71.00	59.00	38.00	49.67
2	26.00	43.00	64.00	73.00	60.00	34.00	50.00
3	25.00	40.00	62.00	72.00	58.00	35.00	48.67
4	31.00	44 00	65.00	74.00	61.00	39.00	52.33
5	36.00	49.00	68.00	77.00	64.00	43.00	56.17
COL. MEANS	29.20	43.20	64.20	73.40	60.40	37.80	51.37

POOLED ST. DEV. = 1.33

Several things:

1. Rounding error: These cells were averages rounded to the nearest degree.
2. Variability of the cells: The cell values are no doubt based on several years' experience, but they are subject to sampling fluctuations.
3. Interaction: No law says that even when measured perfectly the row and column effects will predict the cells additively. The failure to produce that additivity is called interaction.

These and perhaps other effects combine to produce the residual mean square that we use to gauge the row and column effects.

PROBLEMS FOR SECTION 15-4

1. An experiment is done to study the effects of seat height (24, 26, 30 inches) and tire pressure (40, 45, 50, 55 pounds per square inch) on the time required for a college student to ride a bicycle over a two-mile course. She completes the course a total of 12 times, once on each of 12 successive days for each of the 12 combinations of seat height and tire pressure. Set up an analysis-of-variance table to display the results of such an experiment.
2. *Air pollution extremes.* The following table gives extreme value indices of air conditions (that is, the worst values) for each of 3 years for 3 sizes of communities:

Cities	**1968**	**1969**	**1970**
small	1035	768	641
medium	661	410	334
large	1156	799	666

   Carry out the analysis of variance and discuss the results.
3. *Boiling water.* A laboratory experiment in home economics compares household appliances. One experiment measures the effects of the brand of pot and brand of stove on the time required to boil two quarts of water. The experiment involves 3 pots of identical size from different manufacturers and 2 makes of stoves. The observations are time in seconds to boiling.

Brand of Stove	Brand of Pot A	B	C
X	199	194	203
Y	182	185	193

Analyze the data and discuss the results.

4. *Scholastic test performance.* In a national high school study in 1955, investigators gathered data on high schools, their communities, and the career and college plans of 35,000 seniors. Background information on the seniors' families made it possible to classify them into 5 fifths according to an index of socioeconomic status. The investigators classified the school itself according to the percentage of seniors' families falling into the top two-fifths of the socioeconomic index.

		High School Climate
low	I:	0 to 20% in top two-fifths
	II:	20–29% in top two-fifths
	III;	30–39% in top two-fifths
	IV:	40–49% in top two-fifths
high	V:	50% and over in top two-fifths

The two-way data in Table 15-12 show the percentage of seniors performing above the median on their scholastic aptitude tests. Make and interpret a two-way analysis of variance for this table.

TABLE 15-12

Percentage of seniors scoring in the top half of the scholastic aptitude test broken down by socioeconomic status of the family and the high school climate

Family status	*High school climate* (high) V	*IV*	*III*	*II*	*I* (low)
5 (high)	78	74	66	57	44
4	67	59	51	47	37
3	62	51	47	46	32
2	58	48	46	38	31
1 (low)	47	39	37	32	22

5. Continuation. Compute residuals for the cells, and see if you observe any pattern to them.
6. Continuation. What interpretation can you make of the residuals?

15-5 SUMMARY OF CHAPTER 15

1. The purpose of analysis of variance is to measure variability and assign it to its sources.
2. One-way analysis of variance breaks the total sum of squares, TSS, into the between-means sum of squares, BSS, and the within-groups sum of squares, WSS, with degrees of freedom $n - 1$, $k - 1$, and $n - k$, respectively. In one-way analysis of variance, the grand mean is

$$\bar{x} = \frac{\Sigma n_i \bar{x}_i}{n}. \tag{1}$$

The total sum of squares, TSS, is

$$TSS = \underset{\text{sample 1}}{\Sigma (x_1 - \bar{x})^2} + \underset{\text{sample 2}}{\Sigma (x_2 - \bar{x})^2} + \ldots + \underset{\text{sample } k}{\Sigma (x_k - \bar{x})^2}. \tag{6}$$

The between-means sum of squares, BSS, is

$$BSS = n_1(\bar{x}_1 - \bar{x})^2 + n_2(\bar{x}_2 - \bar{x})^2 + \ldots + n_k(\bar{x}_k - \bar{x})^2, \tag{3}$$

and the within-groups sum of squares is

$$WSS = \underset{\text{sample 1}}{\Sigma (x_1 - \bar{x}_1)^2} + \underset{\text{sample 2}}{\Sigma (x_2 - \bar{x}_2)^2} + \ldots + \underset{\text{sample } k}{\Sigma (x_k - \bar{x}_k)^2}. \tag{5}$$

3. For calculations, we use

$$TSS = BSS + WSS. \tag{7}$$

4. An estimate of σ^2 is

$$\frac{WSS}{n_1 + n_2 + \cdots + n_k - k}.$$

5. To test for differences among group means, we compute

$$F = \frac{BSS/(k-1)}{WSS/(n-k)} \tag{13}$$

and refer it to an F table with $k - 1$ and $n - k$ d.f. We reject equality of means when F is too large.

6. To test whether F is too small for an $F_{u,v}$ variable, we take the reciprocal of F and see if it is too large for an $F_{v,u}$ variable (note interchange of subscripts).
7. Programmable hand calculators and high-speed computers may conveniently use the definitional formulas for the sums of squares of deviations required in the analysis of variance.
8. The two-way analysis of variance deals with measurements classified two ways, and so it can allocate variability to row effects, column effects, and residual effects:

$$TSS = BRSS + BCSS + RSS.$$

The residual effects are owing to at least sampling and rounding errors and nonadditivity (interaction). If we had more than one observation per cell, we could assess sampling error directly in the two-way analysis.

SUMMARY PROBLEMS FOR CHAPTER 15

1. An experiment was set up to compare 5 groups of size 4. Some of the results are displayed in the following ANOVA table:

DUE TO	DF	SS	MS=SS/DF	F RATIO
Factor	4	89.037	22.259	—
Error	—	10.102	0.673	
Total	—	99.139		

Complete the table by filling in the missing entries, and compare the F ratio to the appropriate 5 percent tail value from the F table.

Data for Problems 2 and 3: In an experiment to compare the weight gains of baby chicks reared on four different forms of tropical feed, 5 chicks were assigned to each form of feed. The resulting weight gains for the 20 chicks were the following:

Feed Type			
1	**2**	**3**	**4**
55	61	42	169
49	112	97	137
42	30	81	169
21	89	95	85
52	63	92	154

(Source: Query 70, *Biometrics* 5:250, 1949.)

2. Compute the sample mean and variance for each of the four feed types.

3. Analyze the results of this experiment using an analysis-of-variance table. Compute and interpret the F ratio.

4. *Omitting a matching variable.* Using the results of Exhibit 15-D in Section 15-4, carry out a one-way analysis of variance for the city temperature data of Table 15-4, dropping the distinction between the calendar periods, and thus behaving as if each city has six independent measurements. Does F turn out to be small?

5. Using the data in Table 15-5 and your calculations in Problems 8 through 10 following Section 15-1, construct an analysis-of-variance table for the results of the 1976 Montreal Olympics women's platform diving competition. Interpret the results.

6. A study of automobile performance involved 3 different models of cars and 5 blends of gasoline. Each car was driven over a 50-mile course 5 times, once using each gasoline. The results in miles per gallon were as follows:

	Blend of Gasoline				
Car	***A***	***B***	***C***	***D***	***E***
I	18.9	20.2	19.0	17.1	20.3
II	18.5	21.2	20.1	18.0	20.3
III	19.5	19.1	19.3	19.8	21.8

The data are summarized in the following two-way analysis-of-variance table:

DUE TO	DF	SS	MS=SS/DF
Cars	2	1.65	0.82
Blends	4	11.56	2.89
Error	8	6.99	0.87
Total	14	20.20	

Compute the F ratios for cars and blends, and discuss the results of this study.

7. An experiment is designed to produce information on the yield of three different types of popcorn kernels. Two equal-size samples of each type were tested, one using corn oil and a traditional popper, the second using a new hot-air popper that requires no oil. Lay out an analysis-of-variance table for this experiment, filling in all entries for which information is available.

8. Continuation. What are the degrees of freedom for the F ratio for popcorn types in Problem 7?

9. Continuation. The experiment in Problem 7 is extended to include a third type of popper that uses a small amount of peanut oil. All three types of kernels are tested with this new popper as well, and the results are combined with those from the earlier experiment. What are the changes in the degrees of freedom for the analysis-of-variance table?

REFERENCES

G. E. P. Box, W. G. Hunter, and J. S. Hunter (1978). *Statistics for Experimenters*, Chapters 6 and 7. Wiley, New York.

D. B. Owen (1962). *Handbook of Statistical Tables*, pp. 64–87. Addison-Wesley, Reading, Mass.

E. S. Pearson and H. O. Hartley (editors) (1966). *Biometrika Tables for Statisticians*, vol. I, third edition, pp. 169–175. Cambridge University Press, Cambridge, England.

T. A. Ryan, Jr., B. L. Joiner, and B. F. Ryan (1976). *MINITAB Student Handbook*, Chapter 10. Duxbury, North Scituate, Mass.

G. W. Snedecor and W. G. Cochran (1980). *Statistical Methods*, seventh edition, Chapters 12 and 16, and pp. 480–487. Iowa State University Press, Ames.

16

Nonparametric Methods

Learning Objectives

1. Applying methods that defend themselves against wild observations
2. Replacing measurements or comparisons by their signs (+ or −)
3. Using ranks in place of measurements to get alternatives to the t test for two groups, and analysis of variance for more than two groups

16-1 WHAT ARE NONPARAMETRIC METHODS?

The best-known methods of statistical inference, methods that yield probability levels, such as regression analysis and analysis of variance, assume that we know the shape of the probability distribution that the measurements take. For example, if we assume that the measurements are drawn from a normal distribution, then we can proceed to construct confidence limits for its parameters, such as the mean μ or the variance σ^2, or check on the plausibility of specific values. Frequently, however, we cannot confidently make an assumption about the shape of the distribution of measurements. The evidence may not support a given shape or, still worse, may sharply deny it.

In such circumstances we may prefer to use methods whose strengths do not depend much on the precise shape of the distribution. For example, we may want to compare properties of distributions even when we know little about their shapes. So we turn to methods based on **signs** of differences, **ranks** of measurements, and **counts** of objects or events falling into categories. The behavior of such methods may not rest heavily on the shape of the distribution, and for this reason they are called **nonparametric methods.**

The term *nonparametric* is somewhat misleading, because nonparametric statistics do deal with parameters such as the median of a distribution or the probability of success p in a binomial. Indeed, the word *nonparametric* as commonly used does not lend itself to a precise definition.

The main point is that many of the methods that we now describe defend themselves against wild observations and stand up well against various shapes of distributions and failures of assumptions. Statisticians use such words as *robust* and *resistant* for methods that have these properties. To illustrate resistance, consider the medians and means of two samples of 5 measurements:

10, 22, 35, 38, 45 median 35, mean 30

10, 22, 35, 38, 945 median 35, mean 210.

The huge change from 45 to 945 in one of the measurements changed the median not at all. The mean, though, changes from 30 to 210. Similarly, if we ranked these measurements from 1 to 5, both 45 and 945 would be given the same rank in their respective samples. If we picked some cutoff number and assigned plus signs to larger numbers and minus signs to smaller numbers, then for any cutoff less than 45, both 45 and 945 would get the same sign. This illustrates the idea of resistance of medians, ranks, and signs.

The word *robust* means that if the populations do have shapes appropriate for parametric methods, then we lose only a little information by using the robust method. If the data come from normal distributions, the sample mean has a variance about 64 percent as large as that of the sample median. So when we use the sample median, we say that we lose 36 percent of the information if we are sampling from a normal distribution. Some statistics for estimating location lose less information than the median, and they are more robust. An example is the average of the observations remaining after deleting the largest 20 percent and the smallest 20 percent of the data. In our first sample of 5 measurements given earlier, this would leave 22, 35, and 38, whose average is $95/3 = 31.7$, as would the second sample. Thus resistance to wild measurements means that a statistic does not change much when a small proportion of the measurements change, and robustness means that the statistic preserves much of the information present whether or not ideal assumptions are true.

Sometimes data come only in dichotomies ("plus or minus signs," "better or worse," "dead or alive," "success or failure") or in grades ("high, middle, or low," "excellent, good, fair, or poor"). Then nonparametric methods are essential for making significance tests or constructing confidence limits. We used some nonparametric methods in Chapter 9 on contingency tables and counts: the sign test based on the binomial distribution and the chi-squared test for independence.

Although ease of application and ease of calculation seem to be attractive reasons for using nonparametric methods, it is more important that good, rather than easy, methods be used.

The methods we present in this chapter correspond to further applications of the binomial distribution of Chapter 7, to a two-sample nonparametric test comparable to the *t* test in Chapter 10, and to an analysis of variance by ranks comparable to the method of Chapter 15.

16-2 THE SIGN TEST

When we want to know whether one treatment more often gives better results than another, the sign test offers a method for deciding. We have already used the method repeatedly in connection with counts in Chapter 9. The data may arise in various ways, as the following examples suggest.

EXAMPLE 1 *Matched pairs.* On the basis of family background, age, and severity of delinquency, 100 delinquent boys were paired. One randomly chosen member of each pair attended summer camp; the other participated in a police athletic league. After a period, 25 of the athletic league members were rated as performing less delinquently than their paired "mates," 15 of the summer camp boys were performing better than their "mates," and 10 pairs were tied.

EXAMPLE 2 *Time periods.* Are body excretions of certain chemicals higher during the day or during the night (12 hours each)? In 3 of 8 individuals they were higher at night.

EXAMPLE 3 *Gains.* Black plastic mulch for tomatoes costs more than dirt. An extra yield of 5 percent or more would pay for the mulch. Among 18 matched plots, the extra yield exceeded 5 percent in 12 plots and was less in 6 plots.

The Name. The sign test gets its name from our replacing the attributes or measurements or ratings or differences or comparisons by plus (+) or minus (−) signs. When the comparisons are independent, we use the binomial distribution with

$$\begin{aligned} p &= \text{probability of a plus sign} \\ &= \frac{1}{2} \end{aligned}$$

as a basis for judgment. If the two treatments are about the same in performance, the number of plus signs will be about half the total number of signs; if one treatment is better, the number tends to differ from half. When it differs enough, we prefer to conclude that one of the treatments is more often successful.

We illustrate the method by applying it to Example 1. The big idea is, of course, that the athletic league boys did better than the summer camp boys, and so if we have to choose one of these treatments now, if costs and politics are similar, we choose the athletic league. Nevertheless, we may also ask if the 25–15 outcome is compelling.

What about the ties? Although they tell us a lot about how often the performances are close together, they do not tell us which treatment more often wins, and for making such a test, we set them aside. We proceed as follows:

We arbitrarily assign a plus sign to pairs when one group wins or a difference is positive and a minus sign when the other group wins or the difference is negative, and we regard the sample size, *n*, for comparison purposes as the number of nonzero differences (here 40).

We use the cumulative binomial table, Table A-5 in Appendix III in the back of the book, for $p = \frac{1}{2}$ to compute the probability of a split as extreme as or more extreme than the one observed. Or we could use the normal approximation, as we did earlier for the binomial. To use Table A-5, we find the column corresponding to the sample size *n*, here 40. It shows the following:

Number in the Less Frequent Class

n	10	11	12	13	14	15	16	17	18	19
40	001	003	008	019	040	077	134	215	318	437

The numbers in our classes are 25 and 15, with 15 the less frequent, and so we read the number 077. Each number in the table is to be understood as preceded by a decimal point. Therefore the probability of 15 or fewer observations in the class when $p = \frac{1}{2}$ is 0.077. To get the probability that one or the other class has a count less than or equal to 15, we must multiply by 2, and we get $2 \times 0.077 = 0.154$. This two-sided calculation seems sensible, because we were not given either treatment as a standard. What we have found is that although a 25–15 split sounds impressive, splits at least this wide occur more than 15 percent of the time when $p = \frac{1}{2}$.

PROBLEMS FOR SECTION 16-2

1. In Example 1, if $n = 40$, use Table A-5 in Appendix III in the back of the book to find the probability of a split of 30–10 or wider.
2. Apply the method to Example 2, and find the probability of a split at least as wide as that observed.
3. Apply the method to Example 3, and find the probability of a split at least as wide as that observed.
4. When $n = 50$, what split gives a two-sided probability level of just less than 5 percent?
5. Because we know that in real life *p* is not likely to be exactly $\frac{1}{2}$, what are we testing when we look at the binomial probabilities? That is, what are we trying to decide?
6. If you did not have Table A-5 in Appendix III in the back of the book, how could you get an approximate answer in Problem 1?

7. *Methods of memorizing*. Two methods of memorizing difficult material are tried to see which gives better retention. Pairs of students are matched for both IQ and academic performance. They then receive instruction on the same material, but one member of a pair uses method A to learn it, and the other member uses method B. The students are then tested for recall, and the following scores are obtained:

Pair:	1	2	3	4	5	6	7	8	9	10	11	12
A score:	71	82	59	78	92	85	81	79	77	54	61	83
B score:	64	72	61	75	88	84	88	82	70	49	64	81

Use the t test of Chapter 10 to analyze these data.

8. Continuation. Reanalyze the data from Problem 7 using the sign test, and compare the observed significance level with that from the t test.
9. Continuation. Suppose we were to add two additional pairs of scores to the data in Problem 7:

Pair:	13	14
A score:	90	79
B score:	90	79

How do these new observations affect the sign test of Problem 8?

16-3 THE MANN-WHITNEY-WILCOXON TWO-SAMPLE TEST

When we have two independent samples, we may want to know if the populations have much the same location or if they are separated.

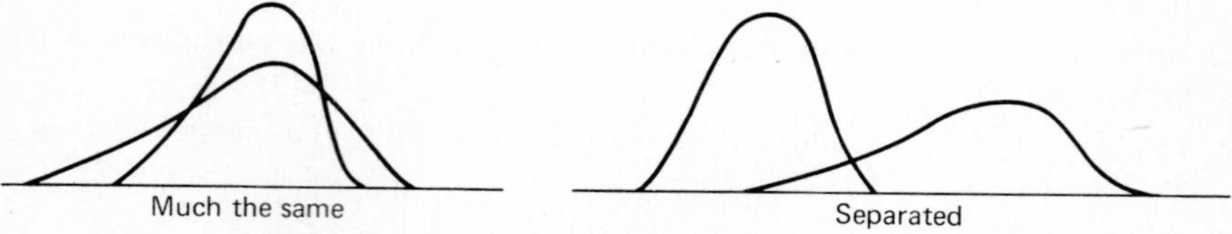

When we are willing to suppose that our measurements are approximately normally distributed without wild observations, the two-sample t test suits us well for this purpose. But when we have no such comfortable views about the samples and their populations, we may prefer an approach in which a few wild observations will cause only limited damage. The Mann-Whitney-Wilcoxon two-sample rank test offers such an approach.

EXAMPLE 4 *Samples of sizes 2 and 4.* Sample *A* contains two measurements, 6 and 24; sample *B* has four measurements, 14, 33, 74, and 105. Compare the samples for evidence that sample *B* comes from a population slipped to the right of that of sample *A*.

SOLUTION. The ranking approach considers all six measurements as a population. Ranks are assigned from least to greatest, here rank 1 to the measurement 6, rank 2 to 14, and so on up to rank 6 for 105. Then we form all possible situations that divide the six into two samples of sizes 2 and 4. Finally, we compute the distribution of their summed ranks for the samples of size 2.

This program has been carried out in Table 16-1. In all, we have 15 possible samples of size 2, and the ranks associated with the samples of size 2 have been summed. Equivalently, we could have summed those for the sample of size 4, but it is more trouble. The extremeness of any sample is judged by how far it is in the tails of the frequency distribution.

The rank sum of our sample *A* is 4. The chance of a rank sum being this small or smaller if the sample is a random choice from these six measure-

TABLE 16-1

Possible samples of two from six measurements together with the sums of ranks

Measurements:	***6***	***14***	***24***	***33***	***74***	***105***	
Rank:	***1***	***2***	***3***	***4***	***5***	***6***	
Samples of size 2							***Sum of two ranks***
1	1	2					3
2	1		3				4
3	1			4			5
4	1				5		6
5	1					6	7
6		2	3				5
7		2		4			6
8		2			5		7
9		2				6	8
10			3	4			7
11			3		5		8
12			3			6	9
13				4	5		9
14				4		6	10
15					5	6	11

Distribution

Sum of two ranks:	3	4	5	6	7	8	9	10	11	Total
Frequency:	1	1	2	2	3	2	2	1	1	15

ments is 2/15. If we also include the possibility of rank sums being as large as 10 or larger, we have 4/15 as the probability of a rank sum at least this far from the middle of the distribution in either direction. Even the most extreme rank sum 3 (or 11) would have given us a two-sided probability of 2/15. Thus our samples are not large enough to indicate a very rare event. But they have served to illustrate the idea of the test.

The general procedure for sample A with n measurements and sample B with m measurements, $n \leq m$, is as follows:

1. Rank the $n + m$ measurements from least to greatest: rank 1, 2, . . . , $n + m$.
2. Total the ranks for sample A to obtain the total of its ranks, t.
3. Look in Table A-6 in Appendix III in the back of the book to find the probability of a total at least as extreme as t, $P(T \leq t)$ or $P(T \geq t)$ if t is in the upper tail. For a two-sided problem, double the probability.

EXAMPLE 5 *Sample sizes 4 and 8.* Two samples of sizes $n = 4$ and $m = 8$ have a rank sum $t = 15$ for the n measurements. Find from Table A-6 in Appendix III in the back of the book the probability of a value of T this small or smaller when sampling is done at random.

SOLUTION. We enter Table A-6 in Appendix III in the back of the book with $n = 4$ and $m = 8$ on the line reading 15 036 37. The probability of a rank sum less than or equal to 15 is 0.036, which solves Example 5. Furthermore, the probability of a rank sum greater than or equal to 37 is 0.036; that of a rank sum at least this far from the middle of the distribution is $2 \times 0.036 = 0.072$.

The general idea is that if the selected population has slipped to the right, its observations will tend to have larger ranks on the average; if it has

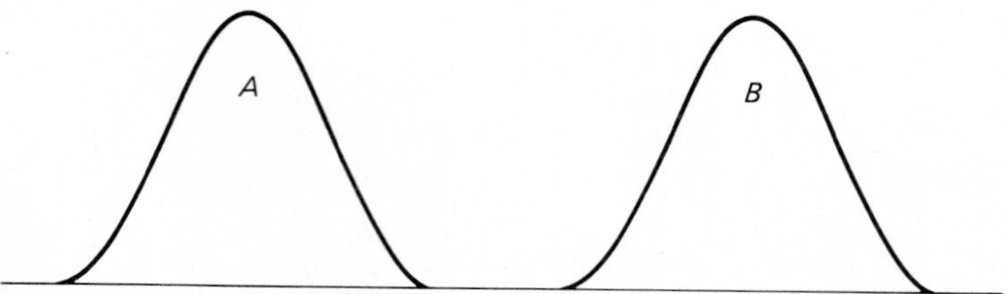

Figure 16-1 Population A has slipped to the left of population B.

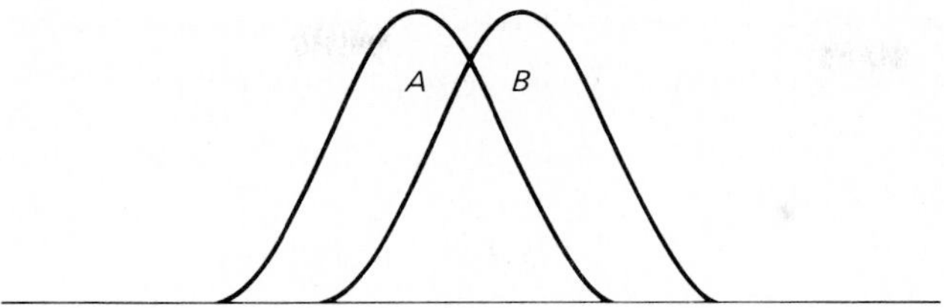

Figure 16-2 The populations heavily overlap.

slipped to the left, its observations will tend to have smaller ranks, as suggested in Fig. 16-1. When the populations are positioned in much the same place, the ranks of the two samples will have much the same average, as suggested in Fig. 16-2.

EXAMPLE 6 *Rosier futures. Tabular approach.* How much better will the future be? A social psychologist surveyed samples of people in many countries, asking what concerned them, how satisfied they were, and how satisfied they expected to be 5 years later. He rated their degrees of satisfaction on a scale from 0 to 10. People from all countries thought the future would be better than the present. Table 16-2 compares the improvements in satisfaction ratings expected in three industralized countries with those expected in a number of less-industrialized countries. The two samples are of sizes $n = 3$ and $m = 8$. The sum, t, of the ranks in the smaller sample is $t = 4 + 1 + 2 = 7$. Table A-6 in Appendix III in the back of the book tells us that for $n = 3$ and $m = 8$, when the populations are identical, the chance of a sum of 7 or less in the sample of size 3 is 0.012. This is a very small probability, and we are inclined to think that the industrialized countries are not expecting as much improvement as the others and that this difference is not accounted for by sampling variation.

NORMAL APPROXIMATION

We can use a normal approximation for values of n and m at least as large as 3, using the approach shown in the following box. Here μ_T and σ_T are the mean and standard deviation of the distribution of the rank sum T for the smaller sample when the two samples are drawn from identical populations and thus have identical locations. We do not give their derivations.

NORMAL APPROXIMATION FOR TWO-SAMPLE RANK TEST

If T is the rank sum for the sample of n randomly distributed among $m + n$ measurements, then the mean of T is

$$\mu_T = \frac{n(m + n + 1)}{2}, \tag{1}$$

and the variance of T is

$$\sigma_T^2 = \frac{mn(m + n + 1)}{12}. \tag{2}$$

We compute

$$z = \begin{cases} (t + \tfrac{1}{2} - \mu_T)/\sigma_T & \text{for } P(T \le t) \\ (t - \tfrac{1}{2} - \mu_T)/\sigma_T & \text{for } P(T \ge t) \end{cases} \tag{3}$$

and refer z to the standard normal table to get a probability.

EXAMPLE 7 *Normal approximation for rosier futures.* For the data given in Table 16-2, $n = 3$, $m = 8$,

$$\mu_T = \frac{n(m + n + 1)}{2} = \frac{3(12)}{2} = 18$$

$$\sigma_T^2 = \frac{mn(m + n + 1)}{12} = \frac{3(8)(12)}{12} = 24.$$

TABLE 16-2

Expected improvements in industrialized and less-industrialized countries together with their ranks

Industrialized countries		*Rank*	*Less-industrialized countries*		*Rank*
United States	1.2	4	Yugoslavia	1.7	6
West Germany	0.9	1	Philippines	1.8	7
Japan	1.0	2	Panama	2.2	8
			Nigeria	2.6	9
			Brazil	2.7	10
			Poland	1.1	3
			India	1.4	5
			Dominican Republic	4.2	11

To get the approximate probability of $T \le 7$, we first compute

$$Z = \frac{t + \frac{1}{2} - \mu_T}{\sigma_T} = \frac{7 + \frac{1}{2} - 18}{\sqrt{24}}$$

$$= \frac{-10.5}{4.90} = -2.14.$$

The normal table, Table A-1 in Appendix III in the back of the book, gives $P(Z \le -2.14) = 0.0162$. This 0.0162 approximates the 0.012 we got from Table A-6 in Appendix III in the back of the book. On the one hand, it is close, within 0.006, but on the other hand, it is 35 percent larger. This is one reason we like exact tables for the smaller n's and m's.

Ties. For the sign test, we dealt with ties in a simple way—we ignored them, setting the corresponding observations aside. For the Mann-Whitney-Wilcoxon test, we do not ignore them:

1. When two or more measurements are tied, we assign to them the average rank available to the tied measurements.
2. We use the normal approximation for T, the rank sum for the sample of n, but we replace formula (2) for the variance by

$$\sigma^2_{T,\text{ties}} = \frac{mn(m + n + 1)}{12} - \text{correction}. \tag{4}$$

For the correction term, we let $W = w^3 - w$, where w is the number of tied observations at a given level. Table 16-3 gives W for various values of w. Then, if ΣW is the sum of W over all tied sets,

$$\text{correction} = \frac{mn}{12(m + n)(m + n - 1)} \Sigma W. \tag{5}$$

TABLE 16-3

Helping table for ties, $W = w^3 - w$†

w:	1	2	3	4	5	6	7	8	9	10
W:	0	6	24	60	120	210	336	504	720	990

†w = number of tied measurements in a set of ties.

EXAMPLE 8 *Sample sizes $n = 3$, $m = 8$, with ties.* Two samples, of sizes $n = 3$ and $m = 8$, have one set of 5 tied measurements consisting of all 3 from the n sample and 2 from the m sample, and these 5 are the lowest measurements. Find the variance.

SOLUTION. The 5 lowest ranks are 1, 2, 3, 4, and 5, and their average is 3. This average is used for the 3 tied ranks from the n sample in computing t. To correct for the ties, we need

$$W = w^3 - w = 5^3 - 5 = 120.$$

From formula (5),

$$\text{correction} = \frac{3(8)}{12(11)10} \times 120 = \frac{24}{11}.$$

Therefore,

$$\sigma^2_{T,\text{ties}} = \frac{3(8)(12)}{12} - \frac{24}{11}$$
$$= 21.8.$$

PROBLEMS FOR SECTION 16-3

For Problems 1 and 2, use the normal approximation to find the probability indicated, and compare the result with that given in Table A-6 in Appendix III in the back of the book.

1. $P(T \geq 20)$, when $m = 7$, $n = 3$.
2. $P(T \geq 29)$, when $m = 8$, $n = 4$.
3. *Masses of planets* (World Almanac, *1970*). The following table gives the masses of the planets in terms of the mass of Earth, which is taken as 1 unit. The B planets are farther from the sun than is Earth, and the A planets are nearer.

***A* planets**			***B* planets**					
Mercury	**Venus**	**Earth**	**Mars**	**Jupiter**	**Saturn**	**Uranus**	**Neptune**	**Pluto**
0.05	0.81	1	0.11	318	95	15	17	0.18

What probability does the Mann-Whitney-Wilcoxon test give for a rank sum as small as or smaller than that for sample A? Does this support or weaken the hypothesis of identical locations for the populations leading to groups A and B? Does the test accept, at the 10 percent level, the hypothesis of identical distributions for groups A and B?

4. Are famous men more likely to die before or after reaching their birthmonths? One might want very much to live until one's next birthday. Is it possible that some people can live a little longer if they have a goal? The following data were obtained by a researcher (Phillips, 1978), who investigated the relationship between the months in which 1251 famous men died and their birthmonths (i = number of months after the birthmonth).

i:	−6	−5	−4	−3	−2	−1	0	+1	+2	+3	+4	+5
Number of deaths:	90	100	87	96	101	86	119	118	121	114	113	106

Use these data and the Mann-Whitney-Wilcoxon test to decide whether or not famous men usually die after rather than before their birthmonths.

SOLUTION. Note that the given data offer several possible interpretations. It seems natural to let $i > 0$ indicate "after birthmonth" and $i < 0$ indicate "before birthmonth." But how shall we deal with the birthmonth itself, where $i = 0$? We might interpret "before birthmonth" and "after birthmonth," respectively, by (1) $i \leq 0$, $i > 0$, or by (2) $i < 0$, $i \geq 0$, or by (3) $i < 0$, $i > 0$, in which we omit the $i = 0$ record.

For each of these three interpretations, the data suggest that famous men die after their birthmonths. In order to get precise information about the strength of the evidence, we shall apply the Mann-Whitney-Wilcoxon text, using interpretation (1).

Let the A sample include the numbers of deaths in the months at or before the birthmonth ($i \leq 0$), and let the B sample include the numbers of deaths in months after the birthmonth ($i > 0$). Rank the numbers, and indicate the ranks belonging to sample B:

86	87	90	96	100	101	106	113	114	118	119	121
1	2	3	4	5	6	7	8	9	10	11	12
						B	B	B	B		B

Let n be the number of B's, and test whether their ranks are too high for a random sample. Here $n = 5$, $m = 7$, $t = 7 + 8 + 9 + 10 + 12 = 46$,

$$\mu_T = \frac{5(13)}{2} = 32.5, \quad \sigma_T = \sqrt{\frac{7(5)(13)}{12}} = 6.16, \quad z = \frac{46 - \frac{1}{2} - 32.5}{6.16} = 2.11.$$

From Table A-1 in Appendix III in the back of the book, we find $P(Z \geq 2.11) = 0.017$. This casts doubt on the notion of a random sample and gives strong evidence that famous men die after their birthmonths.

5. Continuation. Use interpretation (2) in Problem 4 to make your decision.

6. Continuation. Use interpretation (3) in Problem 4 to make your decision.

7. Carry out the Mann-Whitney-Wilcoxon approach to analyze the rates of word usage in Table 10-6, and compare the results with those obtained by the approximate t test there.

8. If $m = 5$ and $n = 4$ and the pooled 9 measurements have the "ranks" 2, 2, 2, 5.5, 5.5, 5.5, 5.5, 8.5, and 8.5, find the variance of T.

9. A physics teacher had two small classes taking the same course. On the final examination, the following marks actually occurred:

Class A:	92	88	88	82	75	70	66	62
Class B:	95	88	52	52	48			

Use the normal approximation to the Mann-Whitney-Wilcoxon test to see if the results from the two classes could be regarded as samples from the same distribution.

10. *Enzyme activity.* The activity of an enzyme (beta glucuronidase) was measured for the sweat glands of a group of diseased patients and a group of control patients. The results were as follows:

Diseased patients: 1.2, 2.9, 1.9, 1.5, 0.5, 1.5, 1.4, 2.1, 3.3, 1.4, 2.2, 1.6, 1.8

Control patients: 2.7, 3.1, 2.2, 1.3, 4.2, 2.2, 2.3, 2.9, 2.2, 2.2, 2.7, 2.8, 1.7, 4.1, 1.9

Use the normal approximation for the Mann-Whitney-Wilcoxon test to see if these measurements might reasonably have come from identical distributions.

16-4 ANALYSIS OF VARIANCE BY RANKS: THE KRUSKAL-WALLIS TEST

In addition to the two-sample Mann-Whitney-Wilcoxon test, which is the nonparametric version of the two-sample t test, we would like a nonparametric version of the one-way analysis of variance for more than two groups. What we need is something corresponding to the between-means sum of squares. We rank all the observations in the pool of all the samples just as we did for the Mann-Whitney-Wilcoxon. Let

n_i = number of observations in sample i

R_i = sum of ranks in sample i

$N = \Sigma n_i$.

If we take the average rank in each sample and measure its departure from the mean of all the ranks, we get for sample i

$$\frac{R_i}{n_i} - \frac{N+1}{2}.$$

These quantities are analogous to $\bar{x}_i - \bar{x}$ in the analysis of variance. For the between-means sum of squares, we squared such quantities and multiplied by n_i to get $n_i(\bar{x}_i - \bar{x})^2$, and then we summed them. Here we get

$$D = n_1\left(\frac{R_1}{n_1} - \frac{N+1}{2}\right)^2 + n_2\left(\frac{R_2}{n_2} - \frac{N+1}{2}\right)^2 + \ldots + n_C\left(\frac{R_C}{n_C} - \frac{N+1}{2}\right)^2,$$

where C is the number of groups. We use $H = 12D/[N(N+1)]$ as a measure of departure from equality when we have no tied ranks.

When we deal with random samples and the n_i are not too small, the distribution of H is approximately chi-square with $C - 1$ d.f. For purposes of calculation, we can use the following form:

$$H = \frac{12}{N(N+1)}\left(\sum_{i=1}^{C} \frac{R_i^2}{n_i}\right) - 3(N+1) \quad \text{(no ties)} \qquad (6)$$

Large values of H cast doubt on the hypothesis that the C distributions have much the same positions. Very small values mean that the samples agree too well. This latter situation often arises from mistakes in arithmetic.

EXAMPLE 9 *Graduate admission rates.* Table 16-4 gives percentages of admissions to the Graduate School of Arts and Sciences at Harvard University in 1966–1967 for the 11 departments receiving the most applications, grouped by division. Use the Kruskal-Wallis test to decide if there is

TABLE 16-4

Graduate admission rates to Graduate School of Arts and Sciences, Harvard University, 1966–1967

Natural sciences		*Social sciences*		*Humanities*	
% admitted	*Rank*	*% admitted*	*Rank*	*% admitted*	*Rank*
34.5	9	23.0	3	25.4	6
36.0	10	26.6	7	24.6	4
25.0	5	20.3	2	18.8	1
28.2	8			58.3	11
Total	32		12		22

substantial variation among the three divisions of the graduate school in admission practices.

SOLUTION. The sample sizes and ranks are

i:	1	2	3	
n_i:	4	3	4	$N = 11$
R_i:	32	12	22	

$$H = \frac{12}{11(12)}\left(\frac{32^2}{4} + \frac{12^2}{3} + \frac{22^2}{4}\right) - 3(12)$$

$$= \frac{425}{11} - 36 = 2.6.$$

Because $C = 3$, we are dealing with chi-square with 2 degrees of freedom. Interpolating in Table A-3 in Appendix III in the back of the book, we find for $\chi^2 = 2.6$ that $P = 0.3$, and so we have little evidence for different locations among the three divisions. Observe by eye that one department has a rather higher rate than the rest.

Ties. The data for multiple samples frequently are afflicted with ties. We assign to each observation the mean of the ranks of the observations tied for that value, as we did for the Mann-Whitney-Wilcoxon test in Section 16-3. The mean of the ranks of those tied values is either an integer or a half-integer. In addition, to get H we need a simple correction. Let $W = w^3 - w$, where w is the number of tied observations at a given level, as for the Mann-Whitney-Wilcoxon test with ties. Then we correct H by dividing it by the divisor for ties:

$$\text{divisor for ties} = 1 - \frac{\Sigma W}{N^3 - N} \qquad (7)$$

where ΣW is the sum of W over all the tied sets. Thus we get the following:

$$H^* = H\bigg/\left(1 - \frac{\Sigma W}{N^3 - N}\right) \qquad (8)$$

EXAMPLE 10 *Ties.* In Table 16-5 we analyze some artificial data constructed to display the problem of ties. The table contains five 0's, which have ranks allotted 1, 2, 3, 4, 5. The average rank is 3. It has three 1's, allotted ranks 6, 7, 8, or average rank 7. Three 2's at 9, 10, 11 have average rank 10. Now, for $w^3 - w$, we have, from Table 16-3, for five ties 120 and for three ties 24, and so the sum $\Sigma W = 120 + 24 + 24 = 168$. We have three groups, and so $C = 3$. The probability of a larger value of H^* is about 0.66, and so we have no evidence of substantial difference among the locations of the populations. Note that the huge observation 88 in treatment 1 ultimately contributed little more than the observation 15. The ranks are resistant. Note also that the correction for ties changed matters very little (about 3 percent).

TABLE 16-5

Example of the H statistic for ranked analysis of variance

Treatment 1		*Treatment 2*		*Treatment 3*	
Observation	*Rank*	*Observation*	*Rank*	*Observation*	*Rank*
0	3	6	14	2	10
3	12	5	13	8	15
0	3	2	10	0	3
1	7	1	7	2	10
12	16	0	3	0	3
15	17	1	7		
88	18		—		—
	$R_1 = 76$		$R_2 = 54$		$R_3 = 41$
	$n_1 = 7$		$n_2 = 6$		$n_3 = 5$

Ties: 5 at 0, 3 at 1, 3 at 2.
$\Sigma W = 120 + 24 + 24 = 168.$
$N^3 - N = 18^3 - 18 = 5{,}814.$

$$H^* = \frac{\dfrac{12}{18\,(19)}\left[\dfrac{76^2}{7} + \dfrac{54^2}{6} + \dfrac{41^2}{5}\right] - 3\,(19)}{1 - \dfrac{168}{5814}}$$

$$= \frac{0.8015}{0.9711} = 0.825.$$

$P(\chi^2 > 0.825) = 0.66$ (2 d.f.).

Source: F. Mosteller and R. R. Bush (1954). Selected quantitative techniques. In *Handbook of Social Psychology*, edited by G. Lindzey. Addison-Wesley, Reading, Mass., p. 320. © 1954. Reprinted with permission.

In a large problem with many ties, it can be a nagging backache to get the rankings all correct. We have found that a stem-and-leaf diagram with the leaves ordered and widely spaced helps a great deal. We can mark on it both the correct ranks to assign and the cumulative rank. Table 16-6 shows the layout applied to the data of Table 15-1 for the state-by-state data on divorced males broken down by region of the country. After the layout is

TABLE 16-6

Stem-and-leaf diagram to aid the ranking and keep track of ties applied to data of Table 15-1 on divorced males. The middle row for each stem gives the ordered leaves of the stem and leaf; the top row gives the rank, counting from the smallest number left to right; the bottom row for each stem indicates the ties and the average rank to be assigned these ties.

Stem	Row														
	Rank	50													
6	Leaf	9													
	Assigned	50													
	Rank	46	47	48	49										
4	Leaf	0	0	1	4										
	Assigned	46½		48	49										
	Rank	44	45												
3	Leaf	6	8												
	Assigned	44	45												
	Rank	35	36	37	38	39	40	41	42	43					
3	Leaf	0	0	0	2	3	3	4	4	4					
	Assigned	36			38	39½		42							
	Rank	22	23	24	25	26	27	28	29	30	31	32	33	34	
2	Leaf	5	5	5	6	7	7	7	8	8	8	8	9	9	
	Assigned	23			25	27			30½				33½		
	Rank	7	8	9	10	11	12	13	14	15	16	17	18	19	20	21
2	Leaf	0	0	1	1	1	1	1	1	2	2	2	2	3	3	4
	Assigned	7½		11½						16½				19½		21
	Rank	1	2	3	4	5	6								
1	Leaf	6	6	7	7	8	9								
	Assigned	1½		3½		5	6								

TABLE 16-7
States with their rankings of percentages of divorced men sorted by region of the country

				Northeast				
7½	Conn.	2.0	23	N.H.	2.5	11½	R.I.	2.1
21	Del.	2.4	1½	N.J.	1.6	19½	Vt.	2.3
36	Maine	3.0	3½	N.Y.	1.7	23	W.Va.	2.5
11½	Mass.	2.1	7½	Pa.	2.0			
				Southeast				
23	Ala.	2.5	11½	La.	2.1	3½	S.C.	1.7
42	Fla.	3.4	19½	Md.	2.3	27	Tenn.	2.7
27	Ga.	2.7	11½	Miss.	2.1	16½	Va.	2.2
27	Ky.	2.7	5	N.C.	1.8			
				Central				
30½	Ill.	2.8	33½	Mich.	2.9	1½	N.D.	1.6
11½	Ind.	2.1	11½	Minn.	2.1	36	Ohio	3.0
16½	Iowa	2.2	36	Mo.	3.0	6	S.D.	1.9
30½	Kans.	2.8	16½	Neb.	2.2	16½	Wis.	2.2
				Southwest				
42	Ariz.	3.4	39½	Colo.	3.3	45	Okla.	3.8
30½	Ark.	2.8	33½	N.M.	2.9	38	Tex.	3.2
				West				
46½	Alaska	4.0	42	Mont.	3.4	46½	Wash.	4.0
49	Calif.	4.4	50	Nev.	6.9	44	Wyo.	3.6
30½	Hawaii	2.8	48	Oreg.	4.1			
39½	Idaho	3.3	25	Utah	2.6			

made, the ranks are transferred back to be associated with the appropriate states and regions. Table 16-7 shows the final layout. A high-speed computer may have a program for ordering the measurements.

PROBLEMS FOR SECTION 16-4

1. Given six independent samples and no ties, with $H = 9.2$, find approximately $P(H \geq 9.2)$ if the samples are randomly drawn from identically distributed populations.

2. Three chemical sprays for killing flies are tested, and the percentages of kills are recorded as follows:

brand A:	72	65	67	75	62	73
brand B:	55	59	68	70	53	50
brand C:	64	74	61	58	51	69

Compute H. What do you conclude about the differences among the brands?

3. Suppose that we have Scholastic Aptitude Test scores for three groups of students:

group 1:	772	764	600	564	
group 2:	792	612	592		
group 3:	752	680	624	580	572

Assuming that these groups represent three independent samples, test these samples to see if they could reasonably have come from distributions with the same location.

4. To test four different diets, 24 young turkeys were randomly divided into 4 groups of 6 each, and each group was fed a different diet. At the end of the experiment, the gain in weight for each turkey was recorded. The results were as follows:

Weight Gained in Pounds

diet A:	20.2	18.5	17.7	17.2	17.0	15.7
diet B:	18.7	17.3	16.1	14.6	13.9	12.2
diet C:	21.7	19.9	19.6	18.8	18.3	17.5
diet D:	19.4	18.0	17.8	16.5	15.0	12.5

Compute H and the probability of a larger value of H for these data if all four diets produced the same distribution of gains.

5. Davies and Sears (1955) gave data on fuel-oil consumption for 5 buses during a 6-month period. Disregarding possible differences between months, and thereby treating the measurements as independent, compute the value of the Kruskal-Wallis statistic H and its statistical significance. What conclusion do you draw about differences between buses in fuel consumption? Fuel consumption, gallons per 1000 miles:

	Bus				
Month	**1**	**2**	**3**	**4**	**5**
May		114	111		
June		108	109	107	
July		110	112	106	105
August		110	110	116	98
September			114	111	102
October	118	110	115	102	

6. *Reading scores.* The average reading test scores reported for grade 7 by schools in six Manhattan districts for 1969 are tabulated below. (The national norm for grade 7 is 7.7 for this test. The low scores are said to be partly due to a long teachers' strike.)

District					
1	**2**	**3**	**4**	**5**	**6**
5.9	5.6	6.4	6.3	5.9	7.1
5.6	5.6	6.4	6.0	6.5	5.6
5.6	7.5	7.3	5.6		5.4
8.0					5.4

Use the Kruskal-Wallis test to assess differences among districts 1 through 6. What conclusion do you draw?

7. In an analysis of the behavior of stocks of various companies, an economist estimated the percentage of total variance of stock prices contributed by the behavior of the market. Use the Kruskal-Wallis statistic to test whether there is industry-to-industry variation.

Percentages of variances		
Oil	**Light Manufacturing**	**Heavy Manufacturing**
20.0	8.6	11.2
23.7	12.1	14.1
	7.1	31.9
	46.9	
	7.6	
	11.9	

8. Why is the mean rank for a group taken as $(N + 1)/2$?

★ 9. Prove that $\Sigma n_i[R_i/n_i - (N + 1)/2] = 0$, when n_i is the number of measurements in sample i, $N = \Sigma n_i$, and R_i is the sum of the ranks of the n_i measurements in sample i.

10. For Table 16-7 on rates of divorced men by regions of the country, the Kruskal-Wallis approach is especially attractive for two reasons. First, we can include the outlier state without having it ruin the analysis. Second, the idea of random rearrangements on which the Kruskal-Wallis test is based is exactly the kind of null hypothesis appropriate to the discussion. Carry out the analysis and interpret the result. The sums, R_i, and region sizes and ties are summarized as follows:

	Northeast	Southeast	Central	Southwest	West	Total
R_i:	165.5	213.5	246.5	228.5	421.0	
n_i:	11	11	12	6	10	$N = 50$

Ties

Number of tied observations, w:	2	3	4	5	6
Frequency:	7	4	2	0	1

16-5 SUMMARY OF CHAPTER 16

Formula numbers cited in these items correspond to those used earlier in this chapter.

1. Nonparametric methods based on signs, counts, and ranks offer alternative analyses to those based on assumptions about the shapes of distributions.

2. Nonparametric methods are resistant to outliers and robust in that they lose only a modest amount of information if used when ideal assumptions hold.

3. The sign test can be used to test differences to see if pluses and minuses split about 50–50 or more extremely. Binomial tables or the normal approximation can provide the probabilities.

4. For robust two-sample tests of location, we use the Mann-Whitney-Wilcoxon approach:

a) Take two samples of sizes n and m $(n \leq m)$.
b) Pool the measurements and rank them from least to greatest.
c) Compute

$$t = \text{sum of the ranks}$$

for the sample of size n.
d) When n and m are small, we use Table A-6 in Appendix III in the back of the book.

5. The mean and variance for the sum, T, when the samples come from identical distributions are

$$\mu_T = \frac{n(m + n + 1)}{2} \tag{1}$$

$$\sigma_T^2 = \frac{mn(m + n + 1)}{12}. \tag{2}$$

6. For large samples, we use a normal approximation for the distribution of T. We compute

$$z = \begin{cases} (t + \frac{1}{2} - \mu_T)/\sigma_T & \text{for } P(T \leq t) \\ (t - \frac{1}{2} - \mu_T)/\sigma_T & \text{for } P(T \geq t) \end{cases} \tag{3}$$

and refer z to the standard normal table. See Section 16-3 of the text for ties.

7. To handle the one-way analysis of variance for C groups robustly, we use the Kruskal-Wallis statistic:

$$H = \frac{12}{N(N + 1)} \sum_{i=1}^{C} \frac{R_i^2}{n_i} - 3(N + 1) \quad \text{(no ties)}, \tag{6}$$

where R_i is the sum of the ranks for the ith group, whose sample is of size n_i, and $N = \Sigma n_i$.

8. When the n_i are not too small, the distribution of H is approximately chi-square, with $C - 1$ d.f.

9. When ties occur in the ranks for the Kruskal-Wallis test we compute ΣW, where $W = w^3 - w$, and w is the number of tied observations at a given level. Then we use the corrected statistic

$$H^* = H/\left(1 - \frac{\Sigma W}{N^3 - N}\right). \tag{8}$$

SUMMARY PROBLEMS FOR CHAPTER 16

1. Why do we sometimes prefer nonparametric to parametric methods?
2. What can we use in place of the original measurements when we employ nonparametric methods?
3. The two-sample t test and its extension to the sample analysis of variance have what counterparts in this chapter?
4. In using the sign test, we are assessing the value of a parameter even though the test is called nonparametric. What is the parameter we study?
5. Although in the sign test we have usually taken as the null hypothesis $p = \frac{1}{2}$, other values are sometimes useful. In a medical investigation, suppose that in the past the probability of improvement has been 0.3. For a new treatment, 7 out of 10 improve. Is this result a significant improvement at the 5 percent level?
6. In the Kruskal-Wallis test, suppose there are two groups of observations, and $n_1 \le n_2$, with no ties. Then for each of n_1, n_2, N, R_1, R_2, give the corresponding notations from the Mann-Whitney-Wilcoxon two-sample test. (You may use the fact that the sum of all the ranks is $N(N+1)/2$.)
7. When the Kruskal-Wallis test is applied to two groups without ties, the value of H is exactly the square of

$$z = \frac{T - \mu_T}{\sigma_T}$$

for the Mann-Whitney-Wilcoxon two-sample test. Check this numerically for the special case where the samples are of sizes $n = 2$ and $m = 4$, and the corresponding ranks are for the n sample 1, 3, and for the m sample 2, 4, 5, 6.

REFERENCES

A. P. Davies and A. W. Sears (1955). Some makeshift methods of analysis applied to complex experimental results. *Applied Statistics* 4:48.

F. Mosteller and R. E. K. Rourke (1973). *Sturdy Statistics.* Addison-Wesley, Reading, Mass.

D. P. Phillips (1978). Deathday and birthday: an unexpected connection. In *Statistics: A Guide to the Unknown*, second edition, edited by J. M. Tanur, F. Mosteller, W. H. Kruskal, R. F. Link, R. S. Pieters, G. R. Rising, and E. L. Lehmann (special editor). Holden-Day, Inc., San Francisco, pp. 71–85.

17

Ideas of Experimentation

Learning Objectives

1. Distinguishing an experiment from an observational study
2. Devices for strengthening experiments
3. Troubles that weaken experiments
4. The four principal one- and two-group designs
5. Linking blocking in an experiment to two-way analysis of variance

17-1 ILLUSTRATION OF EXPERIMENTS

In many simple situations we can predict accurately the effect of an action or a change. For example, hit a glass jar with a hammer and it will shatter, or stir a little salt into water and it will dissolve. We are confident of these and many other outcomes because of our previous experiences, or because the events are simple enough that a well-established theory tells us what to expect.

In more complicated situations we may require careful definitions of the events and elaborate experiments to discover the direction and magnitude of the effects. To illustrate such questions: Do people who smile at others get more smiles in return than those who don't? If we increase the money supply, will we create increased jobs and inflation? Does washing wounds with alcohol instead of water reduce the frequency of infections? These are not questions that we plan to answer here, but rather questions whose answers might require investigations of the form we call experiments. We may each have our opinions about the outcomes of such studies, but it is something else again to define the problems carefully and gather convincing data about them. The questions that data gathering helps us answer are these: What do we believe? What is the evidence for the belief?

In this chapter we give the simplest forms of controlled experiments, and we describe some of the devices and precautions that have been developed through the years to add to their strength or to defend them from threats to their validity. Let us begin with two dietary examples.

EXAMPLE 1 *Daniel and the diets.* Although the Daniel of the Bible is famous for returning from the lions' den, he also designed a very early dietary experiment. Daniel and his young noble friends were hostages in Nebuchadnezzar's palace and were being treated well with the Babylonian king's diet of wine and rich food. Daniel complained that he wanted to be fed the kind of food he ate at home, primarily vegetables. The man in charge of the hostages feared for his own life if, after he gave Daniel and his friends the kind of food they asked for, they wound up in poorer condition than the others. In one translation of the Bible, Daniel said, "Test your servants [Daniel and his three friends] for ten days; let us be given vegetables to eat and water to drink. Then let our appearance and the appearance of the youths [the other hostages] who eat the king's rich food be observed by you, and according to what you see, deal with your servants [Daniel and his three friends]." It turned out that at the end of 10 days they looked better to the man in charge than those who had the rich foods, and so Daniel and his friends were allowed to eat their own diet.

Discussion. A number of issues will have occurred to the reader:

1. Maybe Daniel and his friends were in better shape than the other young men to start with. More generally, we ask: Were the conditions comparable initially?

We would call Daniel's group the experimental group, because from the point of view of the caretaker, the diet Daniel wanted was new and untested. The other group was the control group; in this instance that group had the same treatment as before, sometimes called "the standard treatment," meaning "what is ordinarily done." We usually want the control and experimental groups to be as alike as possible initially.

2. Maybe Daniel's group exercised more. So we ask: Was the treatment really the vegetable and water diet, or were other things at work? This matters if we base future actions on the effect of the diet.

Following an experiment, often someone claims that the treatment said to cause the effect is not the real cause, because some other variable has changed as well as the treatment. Then new experiments are required. Redoing experiments in psychology, biology, chemistry, and the other quantitative sciences is commonplace, because new variables are always being discovered. The idea that experiments are done once and that the matter is then settled is an oversimplification.

3. Is 10 days a reasonable length of time for such a diet to show its effect? More generally, have we chosen the experimental conditions properly?

4. Was the experiment of sufficient size? Daniel's friends, called Shadrach, Meshach, and Abednego by the Babylonians, made only four in the Judean sample. We are not told how large the other group was.

5. Do we have a good measure of health? More generally, have we outcome measures for the variables we want measured? This is a grave stumbling block in many investigations.

EXAMPLE 2 *James Lind on scurvy.* When sailors were long at sea before 1800 A.D., the disease called scurvy slew them by the thousands. In the British Navy, more sailors died from scurvy than from all other causes, including battle. When Anson sailed around the world, 1740–1744, he lost about four-fifths of his 961 sailors to death from scurvy.

The experiment. In 1747, James Lind, a physician in the British Navy on board H.M.S. *Salisbury*, chose 12 men with scurvy and assigned the following six treatments to pairs of sailors:

1 quart of cider per day

25 drops of elixir vitriol three times per day

2 teaspoons of vinegar three times per day

1/2 pint of seawater per day

2 oranges and 1 lemon per day

mixture of selected herbs

Results. The outcome was that in 6 days the two sailors treated with oranges and lemons were back on duty. The cider seemed to help the patients somewhat, but the other treatments did nothing. The conclusion was that oranges and lemons provide dramatic relief, because before treatment the sailors could hardly move.

How can the result for two sailors be so compelling? When we have a lot of experience with a situation and know how it will continue if left alone, we are impressed when it changes following a treatment. In Lind's case, physicians were convinced from long experience and from the reports of others that unless some help arrived, the scurvy patient would continue to deteriorate.

Design. Lind's experiment had six treatments, one applied to each of two sailors. The large effect of the citrus fruit would be called a "slam-bang effect." The body responds remarkably well to a resupply of vitamin C. This strong experimental result, plus extensive historical reading, led Lind to recommend that the British Navy carry citrus juices on warships. It took the navy 50 years to implement Lind's recommendation, and in the meantime thousands of men were lost each year. Once the regulation of issuing citrus juice was established, scurvy disappeared from the Royal Navy.

A previous recommendation. Lest it be supposed that 50 years is a long time for public policy to change when firm information is available, we should note that Sir Richard Hawkins had recommended the use of citrus fruits in 1593, about 150 years before Lind.

A previous controlled study. In 1601, Captain James Lancaster of the East India Company performed an experiment, giving 3 teaspoons of lemon juice daily to each of the 202 men on his ship, but none to the 108, 88, and 82 men on the other three ships of his fleet bound for India. Essentially, nobody on his ship got scurvy, and practically all the sailors on the other three ships did, and 105 of 278 of them died from it on the trip.

Comments. Lind had six experimental groups, and we are not told of any treatment being regarded by him as not being expected to have any effect (placebo). Thus, any five of these groups can act as controls for the sixth. In addition, all his patients were on the same ship.

Lancaster's experiment, although controlled, did not have the same ship for all the men. Thus, if someone claims there was something special about the flagship that prevented the disease, we have trouble making a convincing case for the orange and lemon juice. The experiment would be a bit stronger (a) if some sailors on each ship had the treatment and some did not or (b) if we had more ships, some employing lemon juice and some not.

In either case (Lind or Lancaster), we wish that the choice of individuals for the treatment group could have been assured to be independent of the characteristics of the sailors. We would not want the brawny ones getting the juice and the sickly ones nothing, or vice versa.

PROBLEMS FOR SECTION 17-1

1. Regarding the Biblical example: (a) Who were the members of the control group, and who were the members of the experimental group? (b) What was the experimental treatment? What was the control treatment? (c) What bad consequence of the new diet did the investigator fear?

2. Suggest two types of measurements that might have been made on the Biblical dietary groups to see whether or not the experimental group improved. (Remember, they had no watches, thermometers, or blood pressure measurement devices in those days.)

3. Consider Lind's experiment on 12 sailors from H.M.S. *Salisbury* and Lancaster's experiment for the East Indian fleet. Which do you find provides more compelling evidence in favor of citrus juice as a preventer of, or cure for, scurvy? Why?

4. Imagine two theories of shooting foul shots in basketball. How could you use the members of your class to see which method is preferable?

5. Continuation. In Problem 4, how does the design equate for previous experience or athletic excellence among the class members?

6. *Typewriting.* Two kinds of typewriters are to be compared for their speeds when used by secretaries. Should they be given some experience with the typewriters first? Would it be better to use the same secretaries on both machines or two different sets of secretaries?

7. Continuation. In Problem 6, it is decided to use the same secretaries for both machines. Do you see any objections to testing the secretaries all first on machine A and then all on machine B? What objections do you have?

17-2 BASIC IDEA OF AN EXPERIMENT

An experiment is a carefully controlled study designed to discover what happens when the values of one or more variables are changed. Large uncontrolled variations are quite common in business situations and in many settings in the biological and social sciences. In an experiment, we try to control the application of treatments in order to distinguish their effects from the uncontrolled sources of variation. For example, to determine the effects of a new drug on patients afflicted with cancer, we need an experiment.

In the simplest investigations, a level or rate is determined, sometimes for comparison with another number already known. Some laboratory chemistry and physics experiments do this. Perhaps we know at what temperature distilled water boils when it is heated at a given altitude, and we would like to know the change in boiling point when certain amounts of a salt are added. In such circumstances, it may be adequate to add the salt, boil the water, and get the boiling temperature.

If we are concerned that the boiling temperature at a given place might change a bit from day to day, we might try to be even more careful. For example, we might not only boil the salted water but also boil some distilled water side by side, just to make sure today is like other days, and if it is not, we can correct for that. Such an extra investigation is often called a "control" on the experimental treatment, which here consists of the addition of salt.

More generally, in what we call experiments, the investigator makes changes in some process and observes or measures the effects of these changes. Here we call these changes "treatments." This method contrasts with observational studies or sample surveys, where ordinarily the investigator measures things as they stand and does not make changes in the process.

The reason controlled experimentation has achieved high status in science, engineering, and medicine is that other, easier methods have so often failed to deliver the goods. Nature has left such wide and well-disguised traps that even bright, well-trained people sometimes fall into them. Observational studies are especially likely to contain such pitfalls.

In many experiments, we have a number of **experimental units** and a group of **treatments.** For a study of the effects of a new drug on cancer, the experimental units might be a random sample of individuals afflicted by a particular form of cancer. There would be at least two treatments in the experiment: one using the new drug, another using the old standard drug. In testing new medications, if there is no standard treatment, experimenters frequently give one group (known as the control group) no treatment or an ineffective treatment called a placebo. The placebo often is designed to resemble the new treatment.

The experiment consists of our applying each treatment to several experimental units and then making one or more observations on each unit. For example, does the cancer get worse or improve? Two key features of the experiment are the way that we assign the treatments to the experimental units and the number of units we decide to use. In an observational study we have no control over which treatments are administered to which units or subjects.

The concept of "no treatment" may be difficult to imagine in a social investigation. In a spelling class at school, some method of teaching will be used, and so "no treatment" would mean "the usual teaching." Often the idea of "no treatment" is troublesome. In medicine, some diseases have no remedy, and some patients may have no disease and yet insist on treatment. What is the physician to do? To handle this problem, physicians invented the placebo—classically a harmless bread or sugar pill with no direct medicinal effect except to provide a treatment and thereby comfort the patient. Some people disapprove of the use of such devices. Others see no other way to deal with the patient whose disease has no remedy or is imaginary.

Our interest here in the idea of a placebo is to provide a treatment that looks like another treatment. If we give a pill to one patient and none to another, pill taking can be regarded as part of the treatment, just as visiting the doctor is itself part of the treatment. When some patients get pills and others do not, the groups may have different expectations that interfere with measuring the effect of the new treatment. Indeed, just knowing that one is in an experiment and being paid attention can affect one's behavior. This effect is called "the Hawthorne effect," named for a Western Electric plant where it was discovered by industrial experimenters early in this century. Workers responded productively to attention. Perhaps the reader has noticed that most people do.

In experiments, then, we often want all groups to appreciate that they are in an experiment and, where possible, not to be able to distinguish between the treatments given. Thus, everyone may get an injection (some with saline solution) or evil-tasting medicine or may go to the clinic for examination to keep the basic effects of the outward symptoms of the treatment constant. If there is a standard treatment that works, then we want to compare the new treatment with the standard, rather than with a placebo. Sometimes several new treatments are compared with one another.

We distinguish "control groups" from "control." When a new drug is tried and a placebo given, those getting the placebo are in the control group. But the general idea of control is keeping other things constant. Except for the treatments, the groups are treated alike.

To summarize, we review the key ingredients of an experiment. An experiment is a carefully controlled study designed to discover the effects of changing the values of one or more variables. The changes are called treatments. In many experiments we have a number of experimental units and a group of treatments. Usually, one of the treatments is "no treatment" or a standard treatment or a placebo. The experiment consists of our applying each treatment to several experimental units and then making one or more observations on each unit. The key to the experiment is that we control the application of treatments to distinguish their effects from the uncontrolled sources of variation.

PROBLEMS FOR SECTION 17-2

1. Suppose that Boy Scouts make traffic counts at the same corner twice at the same time of year, once on a clear Thursday and once on a rainy Thursday. (a) What features are controlled in this investigation? (b) What is the treatment? (c) In the language of this section, is this investigation called an experiment? Why or why not?

2. A lady says that she can usually tell by tasting whether the cream is poured into the tea or the tea into the cream. She is presented 10 pairs of cups, with one cup in each pair prepared one way, one the other way. She tastes the pairs and decides for each cup which way the cup was prepared.

a) What features are controlled in this investigation?
b) What are the treatments?
c) Is this an experiment, in the language of this section? Why or why not?
d) How many guesses would she have to get correct before you thought that she had some ability, not necessarily perfect, to distinguish the preparations?

3. Sometimes an experiment is designed to compare a treatment with an abstract standard. For example, we speak of a fair coin as one that comes up heads half the time when it is tossed. Coins can also be spun by poising them perpendicular to a smooth table top and holding them lightly in place with a finger on top. Then with the other hand the side of the coin is flicked with a finger so that the coin spins like a top. After a while it comes to rest, and "head" or "tail" is recorded, depending on which side is up.

a) What are the control features in this investigation?
b) What is the treatment?
c) Try this experiment 20 times with a Kennedy half dollar, a Washington quarter, and a 1964 Denver penny, if you can find one. (Note: Tossing these coins is of no interest. The table top must be very smooth—glass or plastic is recommended. If the coin falls on the floor or is stopped by striking something, the spin does not count. It should spin like a top for a while before coming to rest.)

4. A researcher tossed a die many times and recorded for each throw whether the result was odd or even. He would throw a die of one brand 20,000 times and then take a new die of that brand and throw 20,000 times until he had several blocks of 20,000. Brands A, B, and C were expensive dice; brand X were inexpensive, with holes drilled for the pips. The average numbers of even throws in blocks of 20,000 were as follows:

	Brand				Simulation with Random Numbers
	A	B	C	X	
Average:	10,009	9,999	10,004	10,145	9,993
Number of blocks:	100	30	31	58	100

a) What are the control features in this investigation?
b) What are the treatments?
c) Is this an experiment, in the language of this section?
d) Discuss the outcome in terms of bias.

17-3 DEVICES FOR STRENGTHENING EXPERIMENTS

To strengthen experiments, investigators have developed a number of devices. We review some important ones:

A. Control by Stratification. If experimental units differ, we may break the units into subgroups, called strata, and experiment with each stratum separately. This method, called stratification, is widely used. When the strata are few, this approach can be very effective. For example, a study of hormones might treat men and women separately. When many variables can be influential (stage of disease, sex, age, occupation, race, region of country), the number of possible strata quickly becomes enormous. If we had 10 such influential variables, at only 2 levels each, this would produce $2^{10} = 1024$ categories. When the number of strata becomes unmanageably large, we need another technique.

B. Control by Randomization. Sometimes investigators use random numbers to assign treatments to the experimental units. This approach is called randomization. It seems strange at first to introduce randomness into a situation in which we are striving for control and comparability, but the randomness provides just the help we need.

After we have chosen whatever strata we plan to use, we then pick items or individuals randomly from within a stratum to control for the variables that still may differentiate the individual experimental units. Suppose we wish to study the effects of hormones on men only and plan not to stratify further. Then we allow the assignment based on random numbers to balance the treatment groups. This approximately equates for the variables we have not dealt with by stratification, such as occupation and age, as well as other variables that could affect the outcome of the treatments but that we may not know about, such as other medications the men have used recently. An important feature of this approach is that randomization approximately balances all the variables simultaneously.

What are the purposes of randomization?

1. It offers an active way of controlling for variables, known or unknown, not otherwise controlled.
2. It offers objectivity to insulate our actions from our prejudices, known or unknown.
3. By methods described in Chapters 9 and 10, it gives a way of assigning confidence limits or statistical significance levels based on the results of the investigation.

C. Control by Matching. An additional way to achieve control is by matching experimental units. The idea of matching is to improve comparability. In dealing with humans, we often try to use identical twins, because heredity can make substantial differences. In some instances we even try to study variations within individuals.

EXAMPLE 3 *Hand lotion.* We may put a hand lotion on one hand and not on the other and compare hands after washing dishes for 2 weeks. If we always put the lotion on the same hand for all our experimental subjects, say the right one, we might get into trouble. What if all the subjects are right-handed? Then the hands that get the lotion may be the busy underwater ones attacked by soap and garbage. And so some subjects should have lotion on the left hand, others on the right. In doing this, we combine matching with stratification. We look at the difference in performance for the two hands.

Let us consider a set of hypothetical data under both unmatched and matched circumstances. Consider the scores of 10 subjects in the lotion experiment when their hands are not matched. In Table 17-1 the left panel shows roughness scores for dishpan hands, comparing untreated with treated. In the first two columns the treated and untreated hands have been paired randomly. The third column shows the differences. They vary from −4 to 4.

In the second panel the pairings are for hands of the same persons. Here, by matching, we take advantage of the fact that some people, for various reasons, have rougher hands than other people. The left columns of both panels are identical. The middle columns are the same except for rearrangement to match the hands to the same person. We see by eye the

TABLE 17-1
Comparison of roughness scores for treated and untreated hands, unmatched and matched

Unmatched			*Matched*		
Treated	*Untreated*	*Difference*	*Treated*	*Untreated*	*Difference*
5	2	−3	5	6	1
0	4	4	0	2	2
3	3	0	3	3	0
4	5	1	4	5	1
4	0	−4	4	4	0
5	6	1	5	6	1
1	4	3	1	0	−1
0	2	2	0	2	2
2	6	4	2	4	2
1	3	2	1	3	2
25	35	10	25	35	10

tendency in matched pairs for both hands to score high or both low. The result is that the differences are smaller. They run now from −1 to 2, instead of −4 to 4. This reduction in variation increases the sensitivity of the investigation. This means that smaller differences can be more reliably detected with the same sample sizes than if matching were not used. How the detection is carried out was treated in Chapter 9.

To make matching work, we need one or more extra variables that produce subgroups that are more homogeneous. The extra variable in our example is the name of the individual, and it matches many important dimensions simultaneously. In a way, matching is an extreme form of stratification.

D. Control by Blocking. In agriculture, the matching idea is often used by breaking a field into plots and breaking the plots into subplots so that all treatments can be tried in a plot. The idea is that fertility may differ from one part of the field to another. By using small plots, with all the treatments represented, we gain additional control.

The use of blocking effectively introduces a new variable into the analysis of our experiment. This new variable is blocks. In Section 17-6 we discuss how to carry out the analysis of variance in experiments where we control by blocking.

In both matching and blocking, the treatments may be randomly assigned within the block.

E. Covariates. In addition to stratification, matching, and blocking, we can use an extra variable (such as age of patient) to help us gain precision. The method is an analytical device called covariance analysis that reduces variation from extraneous causes in a manner comparable to blocking.

F. Blindness. To control for biases and expectations, we often keep the information about which units receive the different treatments secret from those involved in the conduct of the experiment.

In a chemistry experiment, the investigator may think he sees detectable patterns on a TV screen when certain samples are run through a testing system, and no special patterns for other samples. If the patterns are somewhat random and hard to describe, he will get an assistant to keep secret from him which samples are being tested, and he will classify the samples from the patterns on the screen. When doing the classification, he is "blind" to all information about the samples being tested. Thus, investigators can protect themselves from being fooled by random variation and personal biases.

It may be necessary to keep the assistant blind to the identity of the sample also, lest by some special behavior the assistant inadvertently tips off the investigator to the identity of the sample. When neither knows the identity of the samples, we speak of double-blind experiments. Of course, a third party can decode the labels.

In medicine, sometimes the administering physician, the patient, and the evaluating physician are all ignorant of the treatment. The fear is that the patient, knowing the treatment, may react in some way to this knowledge rather than to the actual treatment. Suggestion is often effective in patients, and physicians also must protect themselves from their own prejudices about treatments when they can.

G. Sample Size. Once we have well-controlled investigations, then the size of the sample becomes important. As we learned in Chapter 6, multiplying the sample size by 4 decreases the standard deviation by a factor of ½. Thus, from information about the size of the standard deviation of the differences we are studying, we can get an idea of the reliability of our study. Sometimes we can see from preliminary information that we cannot afford the study.

EXAMPLE 4 To reliably detect a departure of 0.01 from $p = 0.5$, the standard deviation of the estimate, $\sqrt{p(1 - p)/n}$, needs to be smaller than 0.01. If we had samples of 100, the standard deviation would be 0.05, because $\sqrt{.51(.49)/100} \approx 0.05$. This is too large if we are to detect differences of size 0.01, because we want (difference/standard deviation) to be sizable, say 2 or more. Samples of 10,000 would give us, by a similar calculation, a standard deviation of 0.005, which would give a ratio of 2. We are now in the correct ballpark. That is, we have a fair chance of detecting a difference of 0.01 in the presence of a standard deviation of 0.005. We now know that with the approach under consideration we cannot hope to assess the differences sufficiently accurately with samples near 100, but 10,000 could do a good job. In many situations, this sort of information will tell us that we cannot afford to make the study because 10,000 is too costly.

On the other hand, if we were trying to assess differences of the order of 0.05, then samples which are modest multiples of 100 would do us some good.

PROBLEMS FOR SECTION 17-3

1 through **7**. List seven devices for strengthening experiments, and indicate in a sentence how each might be applied in an experiment to see whether summer camp is more effective than police athletic league participation in reducing juvenile delinquency. The juveniles may be of different ages, from different parts of the city, and of different socioeconomic status.

8. In Table 17-1, check that the same numbers appear in the two treated columns and the two untreated columns. How do the two panels differ? Because the treated and untreated columns have the same sums in both panels, what was the advantage of the matching?

9. How could you use the method of blindness to obtain the roughness measurements in Table 17-1? What would be the advantage of using a blind approach?

10. What are the blocks in the lotion example?

11. To decrease the standard deviation of a proportion to one-third its value, how much must the sample size be increased?

12. Blocking and stratification are rather similar ideas. Explain.

17-4 ILLUSTRATIONS OF LOSS OF CONTROL

We can "lose control" in experiments for many reasons. Among these, the following factors may arise:

1. initial incomparability,
2. changing treatment,

3. confounding causes,
4. sampling bias,
5. biased measurement,
6. trends,
7. sampling unit,
8. variability.

EXAMPLE 5 *Lack of initial comparability of groups. Milk program.* Suppose that teachers have enough milk to give half of their pupils a glass each morning. Do pupils getting the milk gain more weight and height than the others? The teachers choose the smaller pupils to get the milk. Now the lack of control is not so much in the treatment as in initial lack of comparability of the experimental and control groups.

Remedy. We see that social pressures make experiments like this difficult to administer unless a whole class either does or does not get the treatment. Consequently, the unit of study might better be a classroom or a school rather than a child. We return to the issue of unit size later.

EXAMPLE 6 *Changing treatment. Sleep-learning.* In a sleep-learning experiment, two groups of soldiers were being trained in Morse code in supposedly identical manners, except that during sleep one group, the experimental group, got extra instruction through earphones, and the other group, the control group, got no extra training. The idea was to find out the value of the sleep-learning after a few weeks. When the noncommissioned officers in the control group heard about the sleep-learning, they instituted an extra 2 hours of wide-awake training each evening for their soldiers. Thus the investigators lost control of the treatment, and this experiment was spoiled beyond repair.

Remedy. Where possible, it is well to have an ineffective treatment in the control group parallel to the added treatment in the experimental group. For example, had the control group been given music through earphones while asleep, the treatments would have seemed more parallel. Of course, we need not have an ineffective treatment if we have an effective one to compare against.

EXAMPLE 7 *Possible misinterpretation of cause or confounding of causes. Speed limit.* The legislature reduces the speed limit to decrease the number of auto accidents. Simultaneously, gasoline and automobile prices increase, and a recession occurs. The number of accidents goes down. Lack of control of treatment makes it hard to tell what prevented the accidents: the reduced speed limit, the cost of driving, or fewer drivers, because they don't have work to drive to. And, of course, perhaps all of these matter. Here again we may easily misinterpret our findings. We say the effect of the new speed limit is confounded with those of the prices and the recession.

This sort of example is common when the legal situation seems all-or-none. That is, a whole political group must be subject to the same rule. We call such investigations **quasi experiments,** because if nothing changes but the treatment, then the situation before the speed limit change is introduced offers a control on the situation after. This is a very weak design, and its impressive name should not mislead us into trusting it. Sometimes, though, it is all we have.

Remedy. Society is beginning to develop better designs by being more flexible about their units. For example, perhaps a safety measure could be applied to some patches of roads and not to others. For speed limits, this might work; for a safety measure involving seat belts, probably not.

EXAMPLE 8 *Systematic error through sampling bias. TV viewing.* Suppose we knock on doors to determine if people are watching television. Can we afford to assume that the people who do not answer are watching at the same rate as those who do? Probably many are not home.

EXAMPLE 9 *Biased measurement. Slanted question.* When, in an opinion questionnaire, we ask "Wouldn't you say that it would be best if the United States did not get involved in ______?" the proportion answering yes is likely to be larger than if we merely ask "Do you think the United States should avoid participating in ______?" Not only does the first query push for a yes, but also it has the loaded word *involved*. Most of us don't want to be involved in anything, no matter how good. In such items as these, we speak of the bias of the measuring instrument.

Bias is not a simple idea, especially in the social area. We are likely to think of bias in a rather emotional way—our opponents are biased, our friends are unbiased (supporters). Let us get a bit technical about this. Earlier we said that the sample mean is an unbiased estimate of the population mean. The concept is based on averages. In order to get at the idea of bias, we need the idea of a true value. In the physical sciences, we often have measures that are essentially true values—the coefficient of gravitational attraction at sea level, the oral temperature of a human taken with a carefully calibrated thermometer, the amount of salt in 100 cubic centimeters of tap water. But if I take my temperature immediately after drinking a hot cup of coffee or after sucking an ice cube, I get a result different from the stable body temperature I want to measure, and we can say the result is biased.

In some problems it is difficult to discuss bias because there clearly is no true value. In the "Wouldn't you say . . ." question earlier, what is the true value? The percentage answering yes depends on the phrasing of the question. The "biased" phrasing might be one legitimate way to get the percentage who strongly feel that the United States should be involved. We then have to notice that often the size of a measurement depends on how the measurement is made. And our success with it may then depend on the consistency we use in taking the measurement in differing circumstances or with different groups.

EXAMPLE 10 *Trends. Spelling study.* A new method of teaching spelling to schoolchildren is proposed. We plan to try it out in a school with one class and compare the results with those for last year's class on the same standard spelling test. If the new class does better, we decide the new method of teaching is superior.

Here we must ask if this school is in a neighborhood where the scholarship of the pupils is increasing or decreasing because of population movement. If so, the difference in performance between this year's class and last year's class may be due to migration rather than the teaching method.

Remedy. To get better control, we need classes that are taught the old way at the same times and places where we also teach the new way. Accomplishing this may not be easy. If two classes in the same school are taught in different ways, unusual competition, discord from parental complaints, and leakage of information from one program to another all may destroy or damage the investigation. One might ask: Isn't competition good? It may be, for some students, if it can be controlled and maintained, and it may not be for others. Because our long-run plan is to teach by one method, not both, we should be measuring not only the effect of the program but also the competition and other confounding variables. Some of these conditions may not be reproducible in other schools. Even if we can reproduce conditions, perhaps we should explore the effects of competition and other factors in a separate study.

EXAMPLE 11 *Sampling unit. Spelling study continued.* We have agreed not to worry about the size of the class in Example 10. We pretend this class and last year's class are as large in numbers of pupils as need be. Still, in a different sense, the sample size is a problem. We have one class with one teacher in one school. In a sense, the sample size is 1, because the teacher and the students are all working together on spelling. If we want to say that the new method is good for the nation, or a well-specified group, not just that it is good in this school, observations of a single class will not be enough. The difficulty is like that of Lancaster's lemon juice experiment on the ships of the East India Company (Section 17-1).

Remedy. We need more schools in the study. Many school districts have several elementary schools. We may even be able to set up pairs or sets of schools somewhat comparable in year-to-year overall performance (matching or blocking).

Then, in school districts wishing to participate in this study, we can randomly assign one member of a pair to the new method, one to the standard. We prefer the random assignment to personal judgment (say that of the superintendent) because some secret bias may creep in—perhaps for some reason superintendents may tend to give new treatments to schools with better students, and this will create a bias in favor of the new treatment. The bias can also work the other way. Random assignment should reduce the chance of such a bias being large.

If we have trouble carrying out different treatments (the two methods of teaching, in our example) in the same school district, we may need to put treatments in a whole school district and consider matching school districts and randomizing these.

What if school districts are not willing to try both methods? That is, to be admitted to the study, a district has to be willing to use either method, even though they might insist on using the same method in all their schools. Suppose districts will not do this, some wanting the new method, the others wanting the old? Then we cannot carry out the randomized controlled field trial.

Observational study. We could, of course, carry out a weaker study. We could compare performance in those districts wanting and using the new method with that in districts wanting and using the standard method. The districts with the newer methods are likely to be those with better-educated parents, because better-educated people tend to volunteer more. Now we wonder how to adjust for this if the districts with the newer method do better than the others. Although we have statistical methods for making such adjustments, the strong inferences we hoped to make must be much watered down unless the resulting differences are huge. Unfortunately, in such investigations the differences are usually modest in size, although the gains are valuable if captured, and that is one reason we need very careful measurements: to detect improvements when they occur.

EXAMPLE 12 *Excessive variability.* In Section 17-3, Example 4 illustrates a situation in which the variability is too large to detect a small difference when a sample of 100 is used to nail down a difference of size 0.01.

PROBLEMS FOR SECTION 17-4

Match each of the following experimental situations with one or more of the sources of loss of control listed at the beginning of the section. Explain your reasoning.

1. An investigation of discrimination against minorities in employment in the building trades studies only males.
2. Suppose that to test a physical fitness program you apply it to either the Yale University or the Notre Dame University football team. The team that gets the new program will be determined at random.
3. A new fire alarm and response system is installed in a city, and the dollar claims for insurance rise compared with the previous 3 years. Opponents claim the new system is unsatisfactory.
4. A chain grocery store plans to compare the numbers of people using the fast checkout line when the sign says that the maximum number of items is 6 and when it says 10.
5. The city of St. Paul institutes a new team policing program on a citywide basis, and the number of crimes reported to the police drops by 10 percent.
6. To conserve energy during the winter, the local school district reduces the temperature in all of the classrooms by 10 degrees. The weather this winter is unusually severe, and the school district pays more for fuel than ever before.

17-5 THE PRINCIPAL BASIC DESIGNS

In the early part of this chapter we discussed several basic experimental designs. Sometimes only one group is tested, given a treatment, and then tested again. More often, two (or more) groups are given different treatments and then tested. The investigator may or may not choose to pretest the two groups. First, it costs something to take a measurement. Second, taking a measurement may have its own effect. For example, in a study of spelling, a psychologist found that giving a pretest before training in the spelling led ultimately to more errors. The direction need not have been negative; in some other areas of learning, a pretest can make pupils more aware and attentive to the training. The point is that the pretest can change the person or object about to be treated. Let us discuss these basic designs in a more organized fashion. Two main ideas sum up the distinctions:

1. Observations may be taken after the treatment period or both before and after.
2. One group may be followed, or two or more groups may be followed. In the latter case, we restrict our discussion to two groups.

Using these two distinctions, we have four kinds of experiments, as shown in Table 17-2.

TABLE 17-2

Four basic experimental designs

	Measurements	
	After only	*Before and after*
One group	(1)	(2)
Two or more groups	(3)	(4)

Design 1. One group (after only). For this to be useful, some outside standard of comparison must be available.

EXAMPLE 13 *Algebra.* Practically no one learns advanced mathematics on one's own; this offers an outside standard. Students do learn algebra in school. This shows that the treatment (schooling in algebra) works.

Design 2. One group (before and after).

EXAMPLE 14 *Arithmetic.* By measuring arithmetic performance before a semester begins and measuring it after a semester of training, we can compute the gain from the treatment (arithmetic training) for a semester by taking the difference in the scores. The weakness is that other features, such as maturation or experience with arithmetic outside class, may be contributing, and in this design we have no way to allow for this.

Design 3. Two groups (after only).

EXAMPLE 15 *Testing the polio vaccine.* Panic used to set in each summer at the onset of the poliomyelitis season, with polio's threat of death and paralysis. In 1954, considerable hope focused on the largest controlled public health experiment in the history of the United States. This field trial, which involved over one million children in public schools, was designed to determine the effectiveness of the Salk vaccine as a protection against polio.

The randomized experiment required such a large sample because of the rarity of polio, roughly 50 cases per 100,000, and because the incidence varied so much from year to year and from place to place.

The randomized experiment portion of the field trial involved over 400,000 children, half of whom were injected with the vaccine, while the other half were injected with a harmless salt solution, that is, placebo. The children were assigned to the treatment and placebo groups on the basis of random numbers, so that each child had an equal chance of being in either group. The experiment was double-blind; that is, neither the subjects nor those making the diagnoses knew who received the vaccine and who received the placebo. This feature of the experiment was used so that a physician's judgment in making a diagnosis would not be influenced by knowledge of whether or not the patient had received the vaccine.

The Salk polio vaccine trials used the two-group, after-only design. There was no need for pretreatment measurements, because the investigators assumed that all those children with polio would automatically be spotted and would receive special medical care.

TABLE 17-3
Results of the Salk polio vaccine trials

Group	*Study population*	*Number of cases*	*Rates (number of cases per 100,000)*
Randomized experiment			
Vaccinated	200,745	57	28.4
Placebo	201,229	142	70.6

Source: Adapted from Table 1 in Paul Meier (1978). The biggest public health experiment ever: the 1954 field trial of the Salk poliomyelitis vaccine. In *Statistics: A Guide to the Unknown*, second edition, p. 12. Holden-Day, San Francisco.

In Table 17-3 we give the details of the sample size and the numbers of confirmed cases of polio. Those children who were vaccinated were afflicted by paralytic polio at less than half the rate for those in the placebo group. The vaccine was a success!

Design 4. Two-groups (before and after).

EXAMPLE 16 *Psychotherapy for reformatory inmates.* A psychologist wanted to measure the effect of psychotherapy on inmates at a federal reformatory in Ohio. He gave all the inmates the Delinquency Scale test, and then he selected 52 of them to take part in the study. Thus, everyone received a pretest. From this group of 52, he randomly selected 12 inmates to receive 20 hours of psychotherapy counseling. The remaining 40 inmates made up the control group. After the therapy program was over, all 52 were tested again on the Delinquency Scale. Note that this study used a two-group, before-after design.

The group averages for the preliminary and final tests were as follows:

	Sample Size	**Before**	**After**
Therapy:	12	20.0	11.7
Control:	40	20.8	23.9

Those who underwent therapy showed a dramatic improvement; the control group actually got a little worse. The investigator found similar results using two other scales. Taken together, the results showed that the inmates who received therapy subsequently adapted to reformatory life far better than those who did not receive the therapy.

This design has the advantage over the one-group before-after design that maturation and outside effects can be measured in the control group.

PROBLEMS FOR SECTION 17-5

1. Choosing from the list at the beginning of Section 17-4, explain the weakness of the one-group before-after experiment (design 2).
2. What is the key need in the two-group after-only experiment (design 3) if bias is to be avoided?
3. How may the two-group before-after design (design 4) improve on the designs mentioned in Problems 1 and 2?
4. Describe an experiment other than any mentioned in the text where the presence or absence of a pretest may influence the outcome of the post-test.
5. Perform the following experiment to assess the biasing effect of preselection. Consider two adjacent columns in the table of random numbers, Table A-7 in Appendix III in the back of the book. For each row, record both numbers under "up" or "down." If the number in the first column is lower than that in the second, put both side by side in the "up" column; if the second is lower than the first, put both in the "down" column. If they are tied, discard them. Get the average first number in the "up" column, the average second number in the "up" column, the average first number in the "down" column, and the average second number in the "down" column. Thinking of the first column as a premeasurement and the second column as a postmeasurement within "up" or within "down," measure the selection effect. A random assignment would have given an average score of 4.5 to all columns. How large an effect does the selection have?

Class projects

6. *Effect of zip code.* Does using the zip code speed up the delivery of mail? Devise an experiment for the class to carry out to assess the difference in delivery time for mail addressed with and without zip codes. Remember that if two almost identical pieces of mail are mailed in the same box at the same time, they may well be sorted by the same person and at about the same time.
7. *Estimation of lengths.* Create transparencies for an overhead projector showing lines of three different lengths, one to a transparency. Each length will be shown

as in the following diagrams:

When a length is projected on the screen, it should be between 8 inches and 24 inches long. The subject sits a distance of 10 feet from the screen and is equipped with a yardstick. By looking at the yardstick and the screen, the subject estimates the length of the line segment in inches. You want to find the effects of the various treatments of the tips of the lines on the estimates.

8. *Word recognition.* Make a list of 40 three-letter words. Break the list at random into two lists of 20. Call them list *A* and list *B*. From each list of 20, draw a random 10 words. For list *A*, instruct your subject to remember the words on one of the lists of 10. Let the subject look at the list for 15 seconds. Now wait 2 minutes. Then show the subject the list of 20 from which you randomly drew the 10, and have the subject check off those that were on the short list. Compute the number correct. Now do the same thing with list *B*, except let the subject look at the list of 10 for 45 seconds, and then wait 2 minutes and have the subject check off on the list of 20. On half the subjects, use list *B* first. Now measure how much improvement in recognition the subject gains from the additional half minute of observation.

9. *Completing fragments of English.* From the sports pages (or the editorial page, or the front page) of a newspaper, obtain three 50-word passages; call them *A*, *B*, and *C*. Delete the spaces between words and the punctuation. From each passage, delete at random (using random number tables) 10, 30, and 50 percent of the letters. Form three tests using block capital letters:

Test 1		**Test 2**		**Test 3**	
50-Word Passage	**Deletion**	**50-Word Passage**	**Deletion**	**50-Word Passage**	**Deletion**
A	10%	*B*	10%	*C*	10%
B	30%	*C*	30%	*A*	30%
C	50%	*A*	50%	*B*	50%

Each subject takes one of the tests, being asked to fill in as many blanks as possible. Assess the percentage of blanks correctly filled in for each percentage of deleted letters, for example,

YOUC N ILLIN ISCA TYO

10. *Pulse rates.* (a) Compare the pulse rates of males and females in your class. (b) For each person in class, measure the resting pulse rate. Then have the person jog in place for 15 seconds, and again measure the pulse rate. What is the increase?

11. *A tasting problem.* Some people claim to be able to distinguish one soft drink from another. To do this experiment, you will need two bottles of cola, one from each of two different manufacturers, at the same temperature. A total of 16 subjects are required. The 16 subjects are to be randomly divided into two groups. Each subject in group 1 is to be given a sample of three colas, two from brand *A* and one from brand *B*. Each subject in group 2 is to be given a sample of three colas, one from brand *A* and two from brand *B*. The three colas will be presented to each subject in a random order, and the experiment will be double-blind—neither the subject nor the person supervising the experiment is to know the brand or order of presentation.

17-6 ANALYSIS OF VARIANCE AND EXPERIMENTATION

In the preceding section we discussed the basic designs of social and medical experiments, focusing on one-group and two-group problems. Often experiments have more than two treatment groups, and we naturally think about methods for analyzing the results that allow for several groups, that is, the analysis of variance. For a simple k-group randomized experiment, the basic one-way analysis of variance is the method of choice.

If a randomized experiment is strengthened by the organization of the experimental units into blocks, we need to use a two-way analysis of variance to adjust for blocking. In this section we describe how to do such adjustments.

CAUSATION AND EXPERIMENTATION

When we do an analysis of variance, the question arises whether the different categories or treatments in the rows and columns cause the different sizes of responses in the cells. When the rows and columns represent treatments in an experiment, we can be more comfortable about these variables contributing to the response. For example, if, in creating pottery, different clay bodies are fired to different temperatures chosen by the investigator, then the absorption (percentage weight of water taken up on immersion for a fixed period) and shrinkage may be affected by the variables in about the manner and magnitude implied by the analysis of variance.

If, on the other hand, the data do not come from experiments, then it is difficult to know what to conclude about causal effects. For example, Problem 4 at the end of Section 15-4 treats the cognitive performance of high school seniors as it is associated (a) with socioeconomic status of the senior's own family and (b) with high school climate, as measured by the percentage of high-status families. We see that a senior from a high-socioeconomic-status family and a high-climate school has a better chance of scoring above the median than others. What we do not know is whether moving a student from a lower-scoring school to a higher one or from a lower-status family to a higher one would change the student's chances, as a senior, of scoring above the median. This causal question is always in doubt in analysis of variance and regression studies when the data come from observational studies. The analysis of variance may be only a way of summarizing the data in such circumstances. That alone makes it a useful tool, but it does leave the causal question dangling. We have the same problem with regression. Recall that we can predict height from weight in adults, but gaining weight does not increase height.

BLOCKING AND ITS RELATIONSHIP TO TWO-WAY ANALYSIS OF VARIANCE

In Section 17-3 we discussed the merits of matching (for example, in twin studies) as a device for decreasing variability. Such matching, especially when applied to more than two individuals or items to be treated, is usually referred to as blocking. The idea is that by equating conditions in a group of treated individuals, we sharpen the comparison of their performances under the different treatments.

To take advantage in the analysis of such initial matching, we can use the analysis-of-variance methods of Chapter 15. We can think of laying out a two-way table with rows corresponding to treatments and columns corresponding to the blocks—the clumps of individuals (twins, triplets, family members, samples of liquid from the same mixture, cuts from a metal bar, and so on) chosen to be comparable. Within each of the blocks, the treatments are randomly assigned.

The two-way analysis of variance then gives us a way of assessing differences among treatments.

EXAMPLE 17 *Comparison of toothpastes.* Dentists graded children 11 to 13 years of age according to the state of their teeth (based on decay). Children with adjacent scores were grouped into blocks of three. Three toothpastes were randomly assigned to the three children. The experiment was double-blind. After 2 years, the change in state of teeth was measured for each child, and an analysis of variance was carried out, as shown in Table 17-4. The table shows a huge F ratio for treatments (dentifrices) and a modest one for blocks. We attend primarily to the differences in the effects of treatments. From Table A-4(a) in Appendix III in the back of the book we find that a value for $F_{2,842}$ lies between $F_{2,60}$ and $F_{2,\infty}$, that is, between 3.15 and 3.00 at the 0.05 level, and so the observed F value of 22.17 for treatments far exceeds the 5 percent level. The data suggest firmly that the toothpastes have performed differently.

TABLE 17-4

Analysis of changes in dental caries after 2 years with 422 blocks of three children each and three toothpastes

Source	*d.f.*	*SS*	*MS*	F
Treatments	2	1783.45	891.73	22.17
Blocks	421	26701.74	63.42	1.58
Error	842	33871.21	40.23	
Total	1265	62356.40		

Source: Adapted from K. M. Cellier, E. A. Fanning, T. Gotjamanos, and N. J. Vowels (1968). Some statistical aspects of a clinical study on dental caries in children. *Archives of Oral Biology* 13:483–508.

Although we are not so interested in the blocks, we can see from Table A-4(a) in Appendix III in the back of the book that a value of $F_{421,842}$ at the 5 percent level will be between $F_{60,60}$ and $F_{\infty,\infty}$, that is, between 1.53 and 1.00. Thus the blocks are also significantly different. And so the blocking does help the reliability.

The actual changes in average numbers of tooth surfaces newly decayed were as follows:

Toothpaste A:	9.67
Toothpaste B:	9.76
Toothpaste C:	12.23

Toothpaste *A* contained stannous fluoride, toothpaste *B* contained sodium monofluorophosphate, and toothpaste *C* was the control. The fluoride toothpastes were clearly superior to the control, leading to fewer cavities. There does not seem to be any substantial difference between the effects of the two different types of fluoride.

PROBLEMS FOR SECTION 17-6

1. Explain the distinction between good prediction in an observational study and the causal effects of changing the value of a variable.

Data for Problems 2, 3, and 4: In the Kansas City preventive patrol experiment, discussed in Example 11 in Section 15-3, one of the crimes examined was vandalism. There are 15 experimental units, arranged in five blocks of three. In each block, one unit received each of the three kinds of patrols. The analysis-of-variance table for vandalism is as follows:

Source	d.f.	SS	MS
Treatments	2	5.74	2.87
Blocks	4	10.76	2.69
Error	8	20.08	2.51
Total	14	36.58	

2. Compute the F ratio for treatments, and compare it with the 5 percent level from Table A-4(a) in Appendix III in the back of the book for the appropriate degrees of freedom. Do the effects of the different forms of police patrols seem to differ?
3. Was the blocking effective in this experiment?
4. Suppose there had been no blocking in this experiment, so that the within SS consisted of the sum of the block SS and the error SS. Would this have changed your conclusion in Problem 2? Explain your answer.

The data collected in the comparison of toothpastes experiment, discussed in Example 17, can be expressed in terms of the analysis of variance model—see Eq. (19) of Chapter 15—as

$$x_{ij} = \mu + \alpha_i + \beta_j + e_{ij}.$$

Here x_{ij} is the change in the state of teeth for the child in block *i* using toothpaste *j*, μ is the grand mean, α_i is the effect of block *i*, β_j is the effect of toothpaste *j*, and e_{ij} is the random error with mean 0 and variance σ^2. *For this model it can be shown that* x_{ij} *also has variance* σ^2.

5. Using this information and the formula for the variance of an average from Section 9-1 of Chapter 9, write the theoretical variance of the average change in the state of teeth for children who use toothpaste *j*. (This variance is the same for each toothpaste.)
6. Continuation. Using Table 17-4 and the result of Problem 5, compute the estimated standard error of the average change in the state of teeth for children using toothpaste *j*.
7. Continuation. Compute the difference between the average change in the state of teeth for children using toothpaste *A* and the average change for those using toothpaste *B*. Hint: Use the information given in Example 17.
8. Continuation. Using the formula for the variance of the difference from Chapter 9, Section 9-4, compute the estimated standard error for the difference between the average change in the state of teeth for children using toothpaste *A* and the average change for those using toothpaste *B*.
9. Continuation. Is there a significant difference in outcome for toothpaste *A* and toothpaste *B*?

17-7 SUMMARY OF CHAPTER 17

1. Controlled experiments comparing treatments or programs offer strong evidence in favor of the better-performing treatments and programs. The primary feature of the controlled experiment is that the investigator allocates the experimental units to treatments. An important requirement is comparability of other variables and conditions.
2. Experiments can be

Strengthened by:	Weakened by:
stratification	initial incomparability
randomization	changing treatment
matching	confounding causes
blocking	sampling bias
covariate analysis	biased measurement
blindness	trends
larger samples	sampling unit variability

3. The most basic designs of social and medical experiments are the following:

a) one group, measurement after only,
b) one group, measurement before and after,
c) two groups, measurement after only, one group given treatment 1, the other treatment 2,
d) two groups, measurement before and after, one group given treatment 1, the other treatment 2.

4. For analyzing data from experiments involving k groups, the analysis of variance is the method of choice.

5. When members of blocks are assigned randomly to treatments, with the number in the block equal to the number of treatments, the two-way analysis of variance gives a suitable way of taking advantage of this effective design in the analysis.

SUMMARY PROBLEMS FOR CHAPTER 17

Use the Salk polio vaccine experiment described in Section 17-5 as a basis for answering Problems 1 through 9. In each problem it may or may not be possible to use the method.

1. How could stratification have been used?

2. Was randomization used?

3. How might matching have been used?

4. How was "blinding" used?

5. What were the sample sizes?

6. What prevented initial incomparability?

7. Could the treatment change during the experiment?

8. What protection was there against trends in polio frequency?

9. How was the variability in susceptibility of children protected against?

10. In a study of reduction in deaths due to a change in speeding laws, the performance in a given state was compared with performance in other nearby states where no crackdown had occurred. What kind of study did this produce?

REFERENCES

G. E. P. Box, W. G. Hunter, and J. S. Hunter (1978). *Statistics for Experimenters*, Chapters 4 and 7. Wiley, New York.

T. D. Cook and D. T. Campbell (1979). *Quasi-Experimentation: Design and Analysis for Field Settings*. Rand McNally, Chicago.

F. Mosteller and W. Fairley (1977). *Statistics and Public Policy*. Addison-Wesley, Reading, Mass.

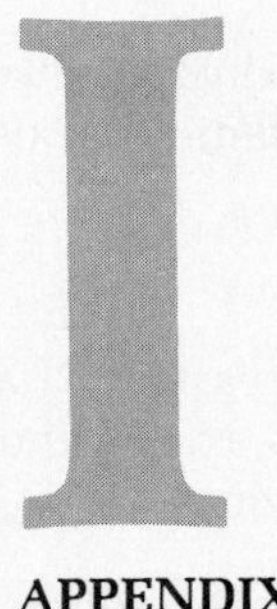

APPENDIX

Summations and Subscripts

We often wish to indicate the sum of several measurements or observations. For example: 30 students take a test, and we wish to know their average score, which is 1/30 of the sum of their scores. It is convenient to be able to express such a sum in compact form. The Greek letter Σ (capital *sigma*) is used for this purpose:

"Σ" denotes "summation of."

Let us arrange the names of the 30 students in alphabetical order; and let x_1 represent the test score of the first student, x_2 the score of the second student, and so on, with x_{30} representing the score of the final student on our alphabetical listing. The subscripts correspond to the positions of the students' names on the list. If the first three students received scores of 85, 79, and 94, in that order, then

$$x_1 = 85, \quad x_2 = 79, \quad x_3 = 94.$$

The sum of the 30 scores could be represented by

$$x_1 + x_2 + x_3 + \cdots + x_{30}, \tag{1}$$

where the three dots are used to indicate "and so on." Another way of representing the same sum using the symbol Σ is

$$\sum_{i=1}^{30} x_i. \tag{2}$$

We read expression (2) as "summation of x-sub-i from 1 through 30." Expression (2) has exactly the same meaning as expression (1); both indicate the sum of 30 scores x_1, x_2, and so on through x_{30}. In other words, the symbol

$$\sum_{i=1}^{30}$$

means "replace i by integers, one after another in ascending order, beginning at 1 and ending at 30, and add the results." The subscript may be any convenient letter, although i, j, k, and n are most frequently used.

Study the following examples along with the attached notes.

EXAMPLE 1 If $x_1 = -3$, $x_2 = 5$, $x_3 = 7$, and $x_4 = 6$, find

(a) $\sum_{i=1}^{4} x_i$, (b) $\sum_{i=2}^{4} x_i$, (c) $\sum_{j=1}^{3} x_j$,

(d) $\sum_{k=1}^{4} 5x_k$, (e) $\sum_{i=1}^{4} x_i^2$, (f) $\sum_{i=1}^{3} 5$, (g) $\sum_{i=1}^{3} x_i y_i$.

SOLUTIONS.

(a) $\sum_{i=1}^{4} x_i = x_1 + x_2 + x_3 + x_4 = -3 + 5 + 7 + 6 = 15.$

(b) $\sum_{i=2}^{4} x_i = x_2 + x_3 + x_4 = 5 + 7 + 6 = 18.$

Note: The notation $i = 2$ written beneath the summation sign tells us where the sum begins; the integer 4 at the top of the summation sign tell us where the sum ends. The subscript i on x is first to be replaced by 2. We then proceed through the integers from the starting place (in this case, integer 2) until we reach the integer written above the summation sign (in this case, integer 4), and add the results.

(c) $\sum_{j=1}^{3} x_j = x_1 + x_2 + x_3 = -3 + 5 + 7 = 9.$

Note: Here we use the letter j instead of i for the subscript on x and for the corresponding **index of summation.** The notation $j = 1$ beneath the summation sign tells us the first value to substitute for j, and this substitution converts x_j into x_1. We then proceed, as before, one by one through the integers until we reach the upper limit of summation, in this case 3. Then we add results and get $x_1 + x_2 + x_3$.

(d) $\sum_{k=1}^{4} 5x_k = 5x_1 + 5x_2 + 5x_3 + 5x_4 = 5(x_1 + x_2 + x_3 + x_4) = 5(15) = 75.$

Note: We replace the subscript k by 1, 2, 3, and 4, in that order, but we have a common factor 5 in each term. In fact, we see that

$$\sum_{k=1}^{4} 5x_k = 5\sum_{k=1}^{4} x_k,$$

and this result can be easily generalized.

Note: Examples (c) and (d) illustrate that the letter used for the index of summation is immaterial. This also explains why that index is often called a dummy index.

(e) $\sum_{i=1}^{4} x_i^2 = x_1^2 + x_2^2 + x_3^2 + x_4^2 = (-3)^2 + 5^2 + 7^2 + 6^2 = 119.$

(f) $\sum_{i=1}^{3} 5 = 5 + 5 + 5 = 15.$

$$\sum_{i=1}^{n} c = c + c + \ldots + c = nc.$$

(g) $\sum_{i=1}^{3} x_iy_i = x_1y_1 + x_2y_2 + x_3y_3.$

Omission of limits of summation. We frequently lighten the notation by omitting the limits of summation and writing simply Σx_i. This notation means that the summation is to extend over all values of x_i under discussion, unless something is said to the contrary. For instance, if the only values in a particular discussion are x_1, x_2, x_3, and x_4, then Σx_i means $x_1 + x_2 + x_3 + x_4$.

PROBLEMS FOR APPENDIX I

Let $x_1 = 3$, $x_2 = -1$, $x_3 = 0$, $x_4 = 5$.

1. Find

$$\sum_{i=1}^{3} x_i.$$

2. Find

$$\sum_{i=1}^{4} 2x_i.$$

3. Find

$$\sum_{i=1}^{4} x_i^2.$$

4. Find

$$\sum_{j=1}^{4} (2x_j + 1).$$

5. Find

$$\sum_{k=1}^{5} 4.$$

6. Find

$$\sum x_i.$$

Formulas for Least-Squares Coefficients for Regression with Two Predictors

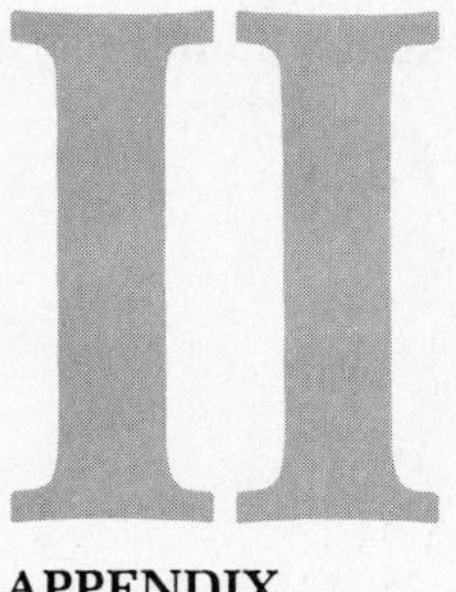

APPENDIX

To calculate the least-squares coefficients for the equation

$$y = b_0 + b_1 u + b_2 w$$

based on n observations for y, u, and w, we first need to compute the sample averages,

$$\bar{y} = \Sigma y_i / n,$$
$$\bar{u} = \Sigma u_i / n,$$
$$\bar{w} = \Sigma w_i / n,$$

and the sums of squared deviations,

$$SSu = \Sigma u_i^2 - n\bar{u}^2,$$
$$SSw = \Sigma w_i^2 - n\bar{w}^2,$$
$$SSy = \Sigma y_i^2 - n\bar{y}^2,$$

and the sums of products of deviations from the means,

$$Suy = \Sigma u_i y_i - n\bar{u}\bar{y},$$
$$Swy = \Sigma w_i y_i - n\bar{w}\bar{y},$$
$$Suw = \Sigma u_i w_i - n\bar{u}\bar{w}.$$

Then the least-squares coefficients are

$$\hat{b}_1 = \frac{(SSw)(Suy) - (Suw)(Swy)}{(SSu)(SSw) - (Suw)^2}, \tag{1}$$

$$\hat{b}_2 = \frac{(SSu)(Swy) - (Suw)(Suy)}{(SSu)(SSw) - (Suw)^2}, \tag{2}$$

and

$$\hat{b}_0 = \bar{y} - \hat{b}_1\bar{u} - \hat{b}_2\bar{w}. \tag{3}$$

Note that the denominators for $\hat{b}_1$ and $\hat{b}_2$ are the same.

EXAMPLE 1 Compute the least-squares coefficients using the ski-trip data in Table 13-1.

SOLUTION. First we compute the various summary statistics:

$$\bar{y} = \frac{1640}{5} = 328,$$

$$\bar{u} = \frac{20}{5} = 4,$$

$$\bar{w} = \frac{500}{5} = 100,$$

and

$$SSu = 90 - 5(4)^2 = 10,$$

$$SSw = 62{,}000 - 5(100)^2 = 12{,}000,$$

$$SSy = 598{,}450 - 5(328)^2 = 60{,}530,$$

$$Suy = 7335 - 5(4)(328) = 775,$$

$$Swy = 159{,}800 - 5(100)(328) = -4200,$$

$$Suw = 1920 - 5(4)(100) = -80.$$

Then we substitute these values in equations (1), (2), and (3):

$$\hat{b}_1 = \frac{(12{,}000)(775) - (-80)(-4200)}{(10)(12{,}000) - (-80)^2} = \frac{8{,}964{,}000}{113{,}600} = 78.9,$$

$$\hat{b}_2 = \frac{(10)(-4200) - (-80)(775)}{113{,}600} = \frac{20{,}000}{113{,}600} = 0.176,$$

and

$$\hat{b}_0 = 328 - (78.9)(4) - (0.176)(100) = -5.2.$$

Tables

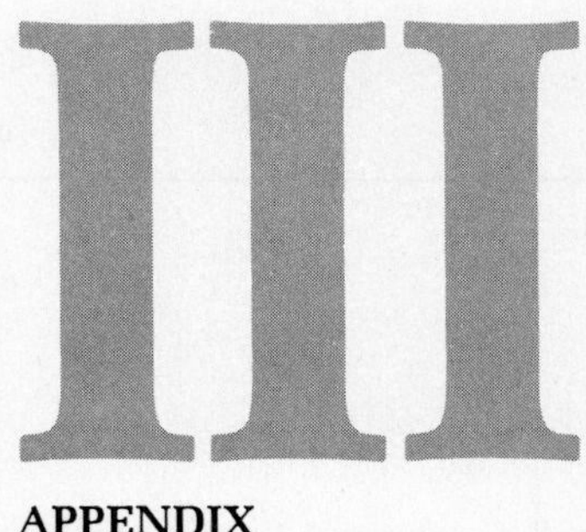

APPENDIX

Table A-1 Normal Curve Areas *page* **538**
Table A-2 Values of t that Give Two-Sided Probability Levels for Student t Distributions **540**
Table A-3 Values of Chi-Square for Various Degrees of Freedom and for Various Probability Levels **542**
Table A-4 Tables of the F Distribution **544**
Table A-5 Three-Place Probabilities of the Cumulative Binomial Distribution for $p = \frac{1}{2}$ **547**
Table A-6 Exact Cumulative Distribution for the Mann-Whitney-Wilcoxon Test **554**
Table A-7 One Thousand Random Digits **556**

TABLE A-1
Normal curve areas

The area under the standard normal curve from 0 to a, shown shaded, is A.

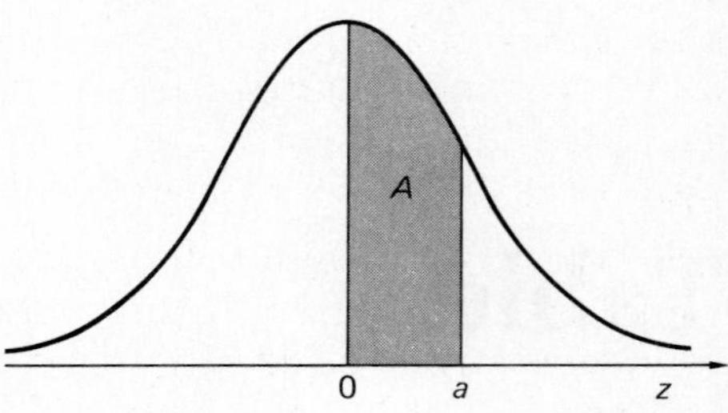

Illustrations. The probability of a value of z:

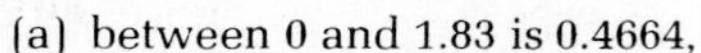

(a) between 0 and 1.83 is 0.4664,

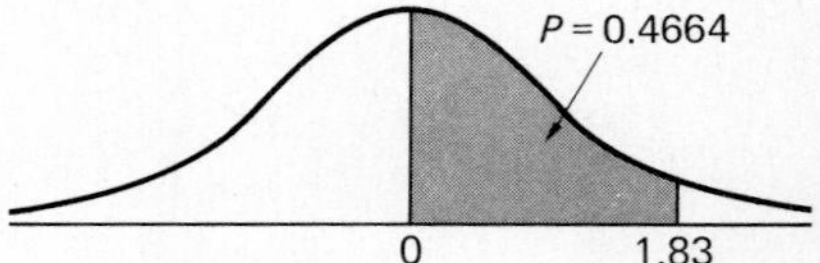

(b) greater than 1.83 is
0.5000 − 0.4664 = 0.0336,

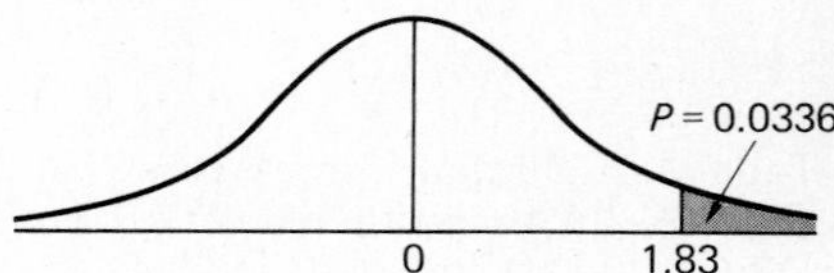

(c) less than 1.83 is
0.4664 + 0.5000 = 0.9664,

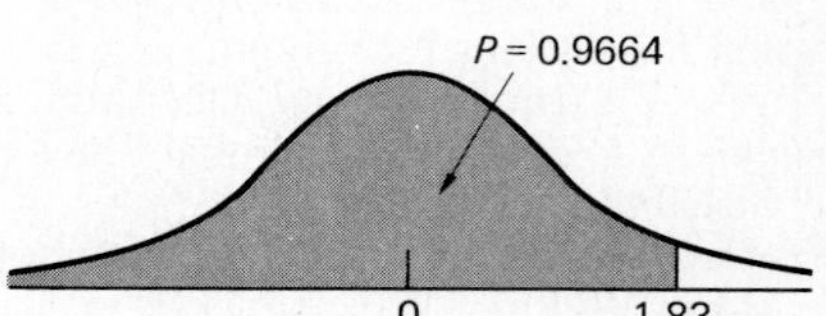

(d) between −1.83 and +1.83 is
2(0.4664) = 0.9328,

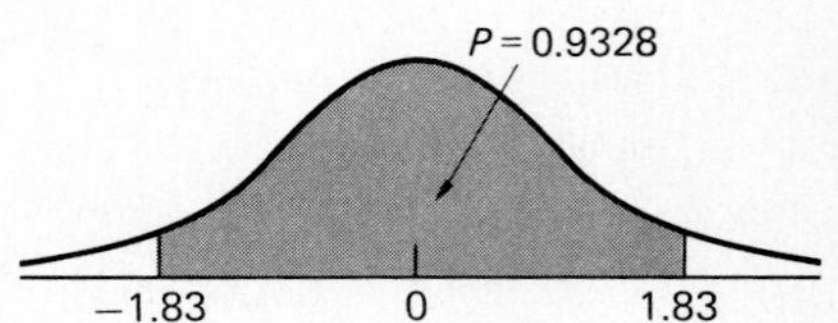

(e) less than −1.83 and greater than
+1.83 is 1 − 0.9328 = 0.0672.

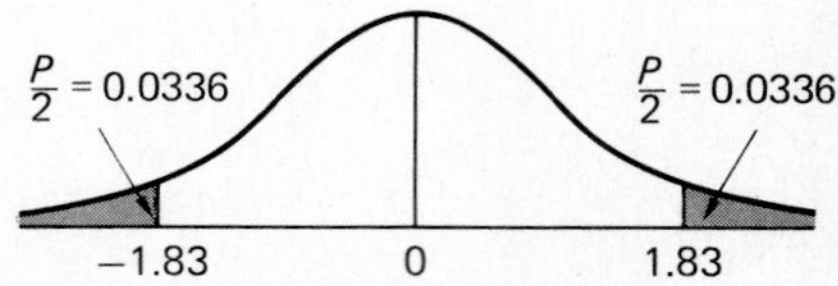

TABLE A-1 *(Cont.)*

a	.00	.01	.02	.03	.04	.05	.06	.07	.08	.09
0.0	.0000	.0040	.0080	.0120	.0160	.0199	.0239	.0279	.0319	.0359
0.1	.0398	.0438	.0478	.0517	.0557	.0596	.0636	.0675	.0714	.0753
0.2	.0793	.0832	.0871	.0910	.0948	.0987	.1026	.1064	.1103	.1141
0.3	.1179	.1217	.1255	.1293	.1331	.1368	.1406	.1443	.1480	.1517
0.4	.1554	.1591	.1628	.1664	.1700	.1736	.1772	.1808	.1844	.1879
0.5	.1915	.1950	.1985	.2019	.2054	.2088	.2123	.2157	.2190	.2224
0.6	.2257	.2291	.2324	.2357	.2389	.2422	.2454	.2486	.2517	.2549
0.7	.2580	.2611	.2642	.2673	.2704	.2734	.2764	.2794	.2823	.2852
0.8	.2881	.2910	.2939	.2967	.2995	.3023	.3051	.3078	.3106	.3133
0.9	.3159	.3186	.3212	.3238	.3264	.3289	.3315	.3340	.3365	.3389
1.0	.3413	.3438	.3461	.3485	.3508	.3531	.3554	.3577	.3599	.3621
1.1	.3643	.3665	.3686	.3708	.3729	.3749	.3770	.3790	.3810	.3830
1.2	.3849	.3869	.3888	.3907	.3925	.3944	.3962	.3980	.3997	.4015
1.3	.4032	.4049	.4066	.4082	.4099	.4115	.4131	.4147	.4162	.4177
1.4	.4192	.4207	.4222	.4236	.4251	.4265	.4279	.4292	.4306	.4319
1.5	.4332	.4345	.4357	.4370	.4382	.4394	.4406	.4418	.4429	.4441
1.6	.4452	.4463	.4474	.4484	.4495	.4505	.4515	.4525	.4535	.4545
1.7	.4554	.4564	.4573	.4582	.4591	.4599	.4608	.4616	.4625	.4633
1.8	.4641	.4649	.4656	.4664	.4671	.4678	.4686	.4693	.4699	.4706
1.9	.4713	.4719	.4726	.4732	.4738	.4744	.4750	.4756	.4761	.4767
2.0	.4772	.4778	.4783	.4788	.4793	.4798	.4803	.4808	.4812	.4817
2.1	.4821	.4826	.4830	.4834	.4838	.4842	.4846	.4850	.4854	.4857
2.2	.4861	.4864	.4868	.4871	.4875	.4878	.4881	.4884	.4887	.4890
2.3	.4893	.4896	.4898	.4901	.4904	.4906	.4909	.4911	.4913	.4916
2.4	.4918	.4920	.4922	.4925	.4927	.4929	.4931	.4932	.4934	.4936
2.5	.4938	.4940	.4941	.4943	.4945	.4946	.4948	.4949	.4951	.4952
2.6	.4953	.4955	.4956	.4957	.4959	.4960	.4961	.4962	.4963	.4964
2.7	.4965	.4966	.4967	.4968	.4969	.4970	.4971	.4972	.4973	.4974
2.8	.4974	.4975	.4976	.4977	.4977	.4978	.4979	.4979	.4980	.4981
2.9	.4981	.4982	.4982	.4983	.4984	.4984	.4985	.4985	.4986	.4986
3.0	.4987	.4987	.4987	.4988	.4988	.4989	.4989	.4989	.4990	.4990
3.1	.4990	.4991	.4991	.4991	.4992	.4992	.4992	.4992	.4993	.4993
3.2	.4993	.4993	.4994	.4994	.4994	.4994	.4994	.4995	.4995	.4995
3.3	.4995	.4995	.4995	.4996	.4996	.4996	.4996	.4996	.4996	.4997
3.4	.4997	.4997	.4997	.4997	.4997	.4997	.4997	.4997	.4997	.4998

TABLE A-2
Values of t that give two-sided probability levels for student t distributions

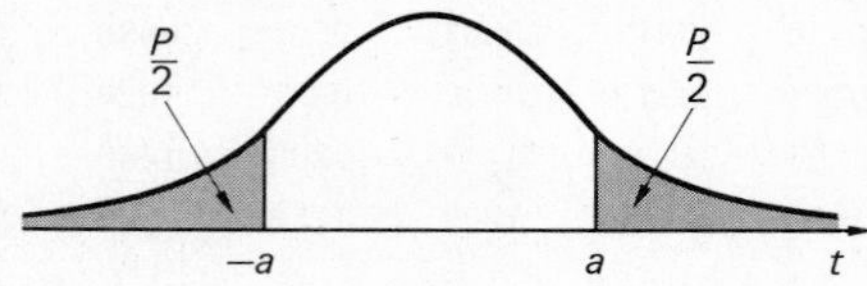

Illustrations. For 8 degrees of freedom (d.f.), the probability of a value of t:

(a) less than −2.90 or greater than +2.90 is 0.02,

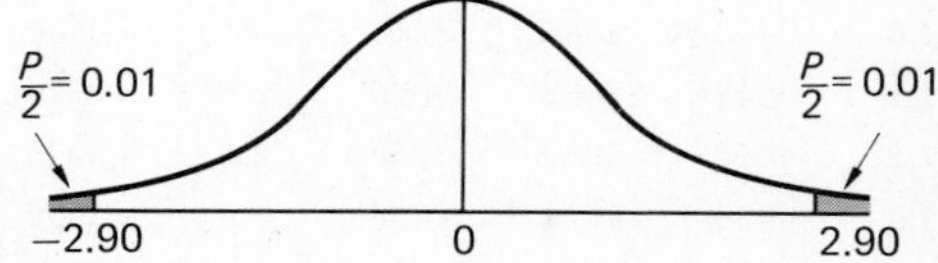

(b) greater than 2.90 is 0.02/2 = 0.01,

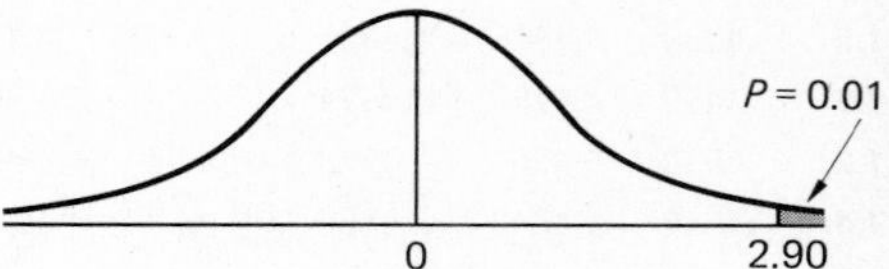

(c) between −2.90 and +2.90 is 1 − 0.02 = 0.98,

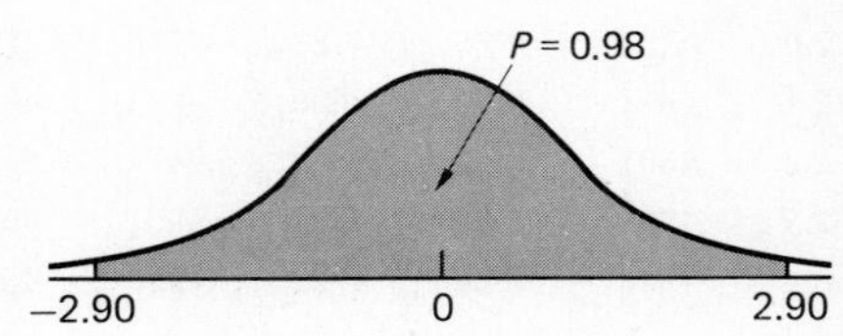

(d) less than −1.50 or greater than +1.50 is between 0.10 and 0.20 (simple interpolation gives 0.18).

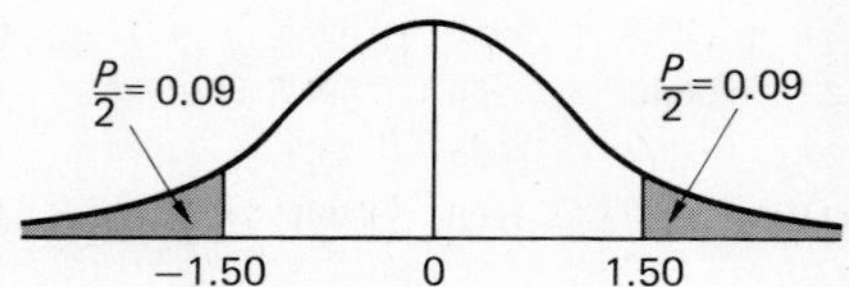

TABLE A-2 (*Cont.*)

Degrees of freedom	*Two-sided probability level, P*							
	.80	*.60*	*.50*	*.20*	*.10*	*.05*	*.02*	*.01*
1	.32	.73	1.00	3.08	6.31	12.71	31.82	63.66
2	.29	.62	.82	1.89	2.92	4.30	6.96	9.92
3	.28	.58	.76	1.64	2.35	3.18	4.54	5.84
4	.27	.57	.74	1.53	2.13	2.78	3.75	4.60
5	.27	.56	.73	1.48	2.02	2.57	3.36	4.03
6	.26	.55	.72	1.44	1.94	2.45	3.14	3.71
7	.26	.55	.71	1.41	1.89	2.36	3.00	3.50
8	.26	.55	.71	1.40	1.86	2.31	2.90	3.36
9	.26	.54	.70	1.38	1.83	2.26	2.82	3.25
10	.26	.54	.70	1.37	1.81	2.23	2.76	3.17
11	.26	.54	.70	1.36	1.80	2.20	2.72	3.11
12	.26	.54	.70	1.36	1.78	2.18	2.68	3.05
13	.26	.54	.69	1.35	1.77	2.16	2.65	3.01
14	.26	.54	.69	1.34	1.76	2.14	2.62	2.98
15	.26	.54	.69	1.34	1.75	2.13	2.60	2.95
16	.26	.54	.69	1.34	1.75	2.12	2.58	2.92
17	.26	.53	.69	1.33	1.74	2.11	2.57	2.90
18	.26	.53	.69	1.33	1.73	2.10	2.55	2.88
19	.26	.53	.69	1.33	1.73	2.09	2.54	2.86
20	.26	.53	.69	1.33	1.72	2.09	2.53	2.85
21	.26	.53	.69	1.32	1.72	2.08	2.52	2.83
22	.26	.53	.69	1.32	1.72	2.07	2.51	2.82
23	.26	.53	.69	1.32	1.71	2.07	2.50	2.81
24	.26	.53	.68	1.32	1.71	2.06	2.49	2.80
25	.26	.53	.68	1.32	1.71	2.06	2.49	2.79
26	.26	.53	.68	1.32	1.71	2.06	2.48	2.78
27	.26	.53	.68	1.31	1.70	2.05	2.47	2.77
28	.26	.53	.68	1.31	1.70	2.05	2.47	2.76
29	.26	.53	.68	1.31	1.70	2.05	2.46	2.76
30	.26	.53	.68	1.31	1.70	2.04	2.46	2.75
50	.25	.53	.68	1.30	1.68	2.01	2.40	2.68
100	.25	.53	.68	1.29	1.66	1.98	2.36	2.63
∞†	.25	.52	.67	1.28	1.64	1.96	2.33	2.58

† Standard normal distribution.

TABLE A-3

Values of chi-square for various degrees of freedom and for various probability levels

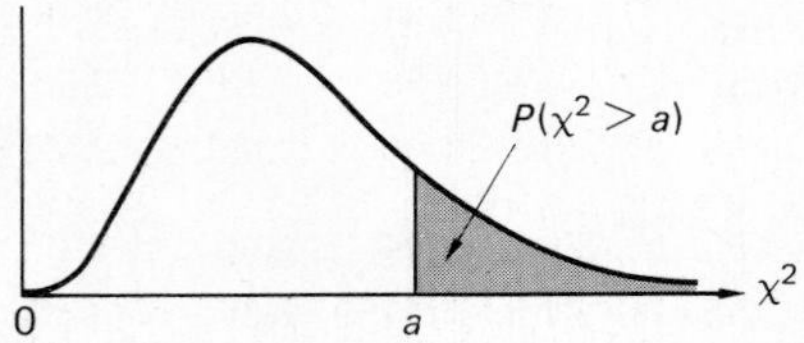

Illustrations. For 10 degrees of freedom (d.f.), the probability of a value of χ^2:

(a) greater than 18.31 is
$P(\chi^2 > 18.31) = 0.05$,

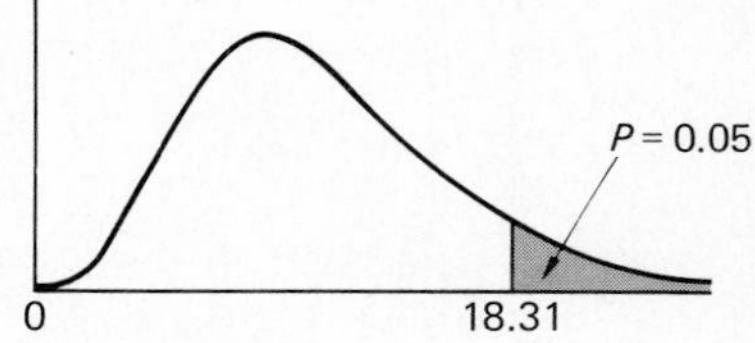

(b) greater than 14.00, which lies between 0.20 and 0.10 is (simple interpolation gives the approximation)
$P\chi^2 < 14.00) = 0.18$,

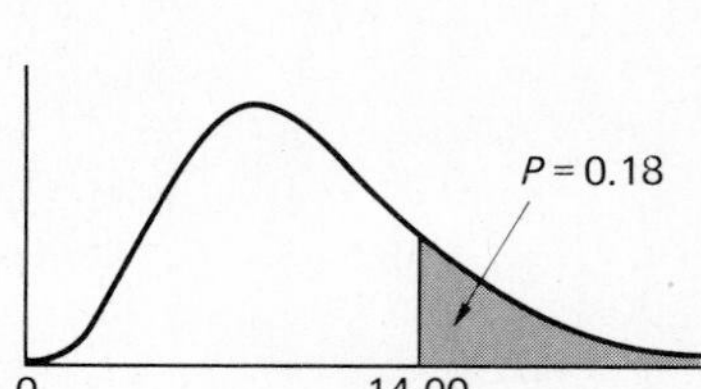

(c) less than 2.56 is
$P(\chi^2 < 2.56) = 1 - 0.99 = 0.01$.

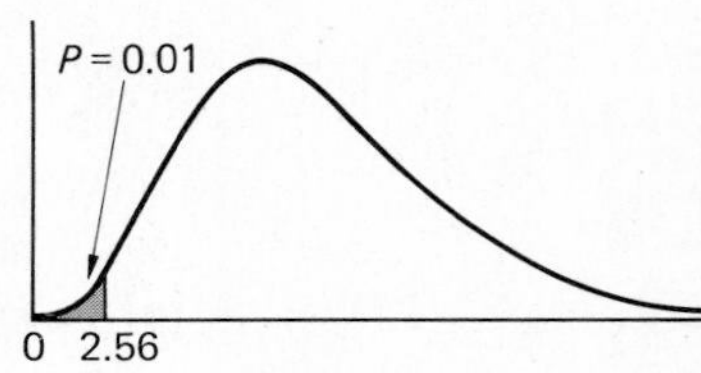

TABLE A-3 (Cont.)

Degrees of freedom	*Probability levels*								
	0.99	***0.95***	***0.90***	***0.50***	***0.20***	***0.10***	***0.05***	***0.01***	***0.001***
1	0.00	0.00	0.02	0.45	1.64	2.71	3.84	6.63	10.83
2	0.02	0.10	0.21	1.39	3.22	4.61	5.99	9.21	13.82
3	0.11	0.35	0.58	2.37	4.64	6.25	7.81	11.34	16.27
4	0.30	0.71	1.06	3.36	5.99	7.78	9.49	13.28	18.47
5	0.55	1.15	1.61	4.35	7.29	9.24	11.07	15.09	20.52
6	0.87	1.64	2.20	5.35	8.56	10.64	12.59	16.81	22.46
7	1.24	2.17	2.83	6.35	9.80	12.02	14.07	18.48	24.32
8	1.65	2.73	3.49	7.34	11.03	13.36	15.51	20.09	26.12
9	2.09	3.33	4.17	8.34	12.24	14.68	16.92	21.67	27.88
10	2.56	3.94	4.87	9.34	13.44	15.99	18.31	23.21	29.59
11	3.05	4.57	5.58	10.34	14.63	17.28	19.68	24.72	31.26
12	3.57	5.23	6.30	11.34	15.81	18.55	21.03	26.22	32.91
13	4.11	5.89	7.04	12.34	16.98	19.81	22.36	27.69	34.53
14	4.66	6.57	7.79	13.34	18.15	21.06	23.68	29.14	36.12
15	5.23	7.26	8.55	14.34	19.31	22.31	25.00	30.58	37.70
16	5.81	7.96	9.31	15.34	20.47	23.54	26.30	32.00	39.25
17	6.41	8.67	10.09	16.34	21.61	24.77	27.59	33.41	40.79
18	7.01	9.39	10.86	17.34	22.76	25.99	28.87	34.81	42.31
19	7.63	10.12	11.65	18.34	23.90	27.20	30.14	36.19	43.82
20	8.26	10.85	12.44	19.34	25.04	28.41	31.41	37.57	45.31
21	8.90	11.59	13.24	20.34	26.17	29.62	32.67	38.93	46.80
22	9.54	12.34	14.04	21.34	27.30	30.81	33.92	40.29	48.27
23	10.20	13.09	14.85	22.34	28.43	32.01	35.17	41.64	49.73
24	10.86	13.85	15.66	23.34	29.55	33.20	36.42	42.98	51.18
25	11.52	14.61	16.47	24.34	30.68	34.38	37.65	44.31	52.62
26	12.20	15.38	17.29	25.34	31.79	35.56	38.89	45.64	54.05
27	12.88	16.15	18.11	26.34	32.91	36.74	40.11	46.96	55.48
28	13.56	16.93	18.94	27.34	34.03	37.92	41.34	48.28	56.89
29	14.26	17.71	19.77	28.34	35.14	39.09	42.56	49.59	58.30
30	14.95	18.49	20.60	29.34	36.25	40.26	43.77	50.89	59.70
50	29.71	34.76	37.69	49.33	58.16	63.17	67.50	76.15	86.66

TABLE A-4
Tables of the *F* distribution

Illustrations. The probability of a value of F, for 8 d.f. for the *BMS* and 10 d.f. for the *WMS*:

(a) greater than 3.07 is $P(F_{8,10} > 3.07) = 0.05$, from Table A-4(a),

(b) greater than 2.50 is $P(F_{8,10} > 2.50)$ which is greater than 0.05, from Table A-4(a),

(c) greater than 5.06 is $P(F_{8,10} > 5.06) = 0.01$, from Table A-4(b),

(d) greater than 5.00 is $P(F_{8,10} > 5.00)$ is between 0.01 and 0.05, from Table A-4(a) and Table A-4(b).

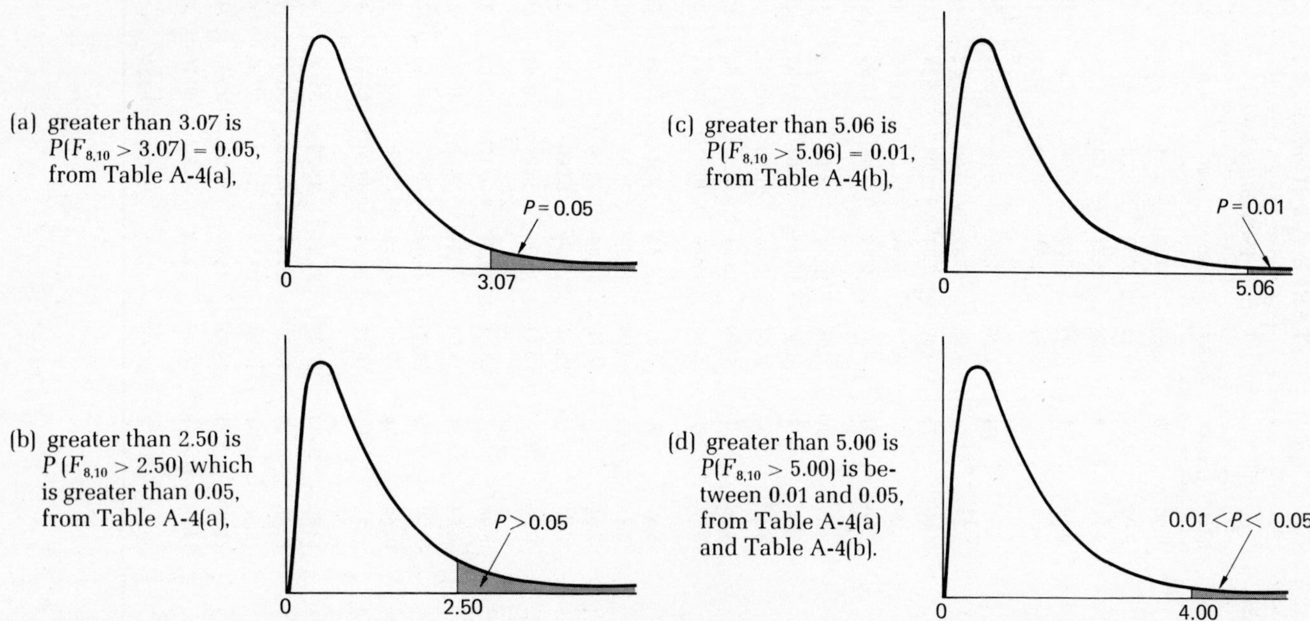

TABLE A-4(a): Tables of the F distribution (*Cont.*)

Values of a for which the probability is 0.05 that a value of $F_{u,v}$ is greater than a, where u is the number of degrees of freedom for the numerator and v is the number of degrees of freedom for the denominator in the F ratio BMS/WMS (see Section 15-3)

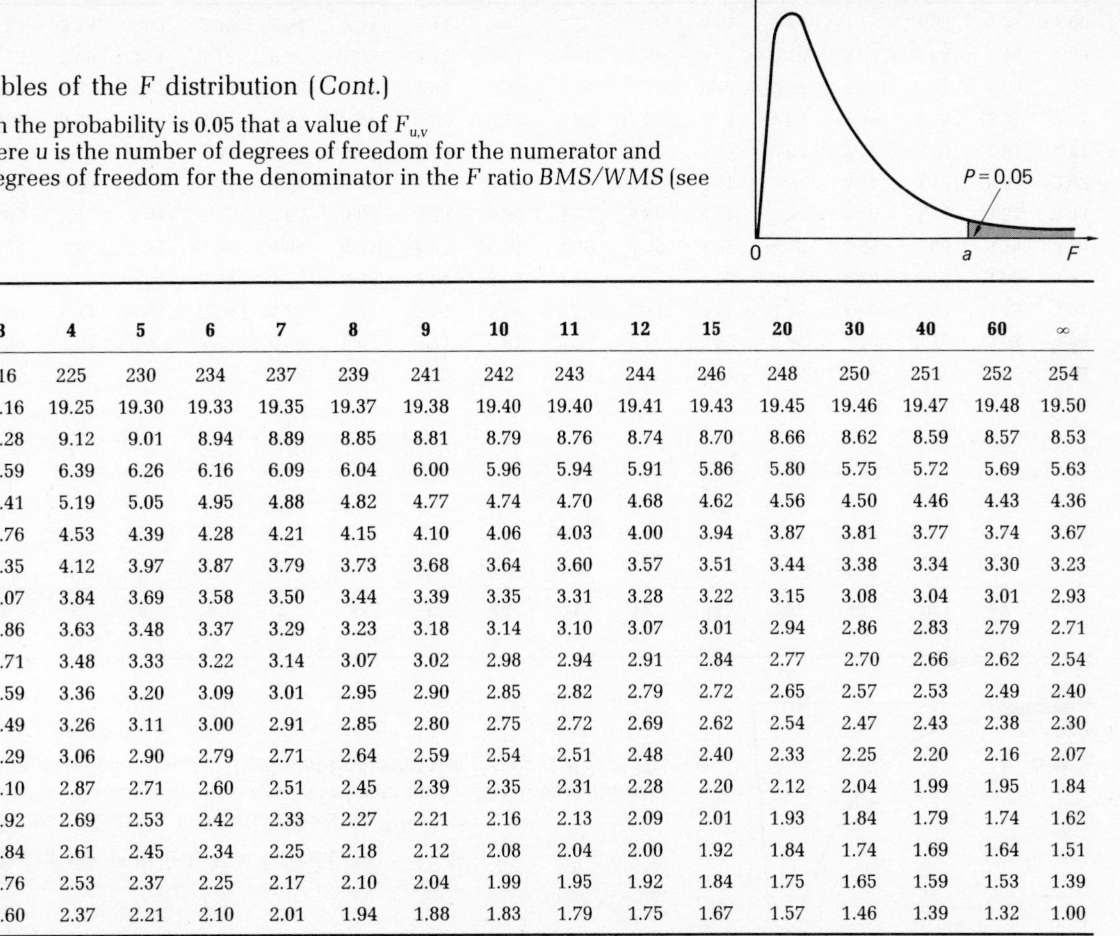

v \ u	1	2	3	4	5	6	7	8	9	10	11	12	15	20	30	40	60	∞
1	161	200	216	225	230	234	237	239	241	242	243	244	246	248	250	251	252	254
2	18.51	19.00	19.16	19.25	19.30	19.33	19.35	19.37	19.38	19.40	19.40	19.41	19.43	19.45	19.46	19.47	19.48	19.50
3	10.13	9.55	9.28	9.12	9.01	8.94	8.89	8.85	8.81	8.79	8.76	8.74	8.70	8.66	8.62	8.59	8.57	8.53
4	7.71	6.94	6.59	6.39	6.26	6.16	6.09	6.04	6.00	5.96	5.94	5.91	5.86	5.80	5.75	5.72	5.69	5.63
5	6.61	5.79	5.41	5.19	5.05	4.95	4.88	4.82	4.77	4.74	4.70	4.68	4.62	4.56	4.50	4.46	4.43	4.36
6	5.99	5.14	4.76	4.53	4.39	4.28	4.21	4.15	4.10	4.06	4.03	4.00	3.94	3.87	3.81	3.77	3.74	3.67
7	5.59	4.74	4.35	4.12	3.97	3.87	3.79	3.73	3.68	3.64	3.60	3.57	3.51	3.44	3.38	3.34	3.30	3.23
8	5.32	4.46	4.07	3.84	3.69	3.58	3.50	3.44	3.39	3.35	3.31	3.28	3.22	3.15	3.08	3.04	3.01	2.93
9	5.12	4.26	3.86	3.63	3.48	3.37	3.29	3.23	3.18	3.14	3.10	3.07	3.01	2.94	2.86	2.83	2.79	2.71
10	4.96	4.10	3.71	3.48	3.33	3.22	3.14	3.07	3.02	2.98	2.94	2.91	2.84	2.77	2.70	2.66	2.62	2.54
11	4.84	3.98	3.59	3.36	3.20	3.09	3.01	2.95	2.90	2.85	2.82	2.79	2.72	2.65	2.57	2.53	2.49	2.40
12	4.75	3.89	3.49	3.26	3.11	3.00	2.91	2.85	2.80	2.75	2.72	2.69	2.62	2.54	2.47	2.43	2.38	2.30
15	4.54	3.68	3.29	3.06	2.90	2.79	2.71	2.64	2.59	2.54	2.51	2.48	2.40	2.33	2.25	2.20	2.16	2.07
20	4.35	3.49	3.10	2.87	2.71	2.60	2.51	2.45	2.39	2.35	2.31	2.28	2.20	2.12	2.04	1.99	1.95	1.84
30	4.17	3.32	2.92	2.69	2.53	2.42	2.33	2.27	2.21	2.16	2.13	2.09	2.01	1.93	1.84	1.79	1.74	1.62
40	4.08	3.23	2.84	2.61	2.45	2.34	2.25	2.18	2.12	2.08	2.04	2.00	1.92	1.84	1.74	1.69	1.64	1.51
60	4.00	3.15	2.76	2.53	2.37	2.25	2.17	2.10	2.04	1.99	1.95	1.92	1.84	1.75	1.65	1.59	1.53	1.39
∞	3.84	3.00	2.60	2.37	2.21	2.10	2.01	1.94	1.88	1.83	1.79	1.75	1.67	1.57	1.46	1.39	1.32	1.00

TABLE A-4(b): Tables of the F distribution (*Cont.*)

Values of a for which the probability is 0.01 that a value of $F_{u,v}$ is greater than a, where u is the number of degrees of freedom for the numerator and v is the number of degrees of freedom for the denominator in the F ratio *BMS/WMS* (see Section 15-3)

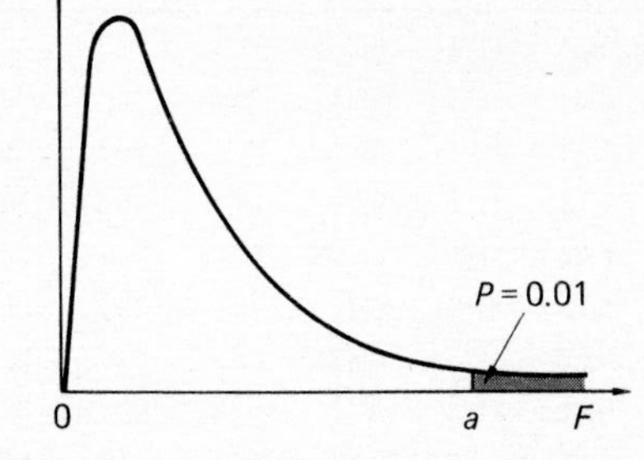

v \ u	1	2	3	4	5	6	7	8	9	10	11	12	15	20	30	40	60	∞
1	4052	5000	5403	5625	5764	5859	5928	5981	6022	6056	6083	6106	6157	6209	6261	6287	6313	6366
2	98.50	99.00	99.17	99.25	99.30	99.33	99.36	99.37	99.39	99.40	99.41	99.42	99.43	99.45	99.47	99.47	99.48	99.50
3	34.12	30.82	29.46	28.71	28.24	27.91	27.67	27.49	27.34	27.23	27.13	27.05	26.87	26.69	26.50	26.41	26.32	26.12
4	21.20	18.00	16.69	15.98	15.52	15.21	14.98	14.80	14.66	14.55	14.45	14.37	14.20	14.02	13.84	13.74	13.65	13.46
5	16.26	13.27	12.06	11.39	10.97	10.67	10.46	10.29	10.16	10.05	9.96	9.89	9.72	9.55	9.38	9.29	9.20	9.02
6	13.74	10.92	9.78	9.15	8.75	8.47	8.26	8.10	7.98	7.87	7.79	7.72	7.56	7.40	7.23	7.14	7.06	6.88
7	12.25	9.55	8.45	7.85	7.46	7.19	6.99	6.84	6.72	6.62	6.54	6.47	6.31	6.16	5.99	5.91	5.82	5.65
8	11.26	8.65	7.59	7.01	6.63	6.37	6.18	6.03	5.91	5.81	5.73	5.67	5.52	5.36	5.20	5.12	5.03	4.86
9	10.56	8.02	6.99	6.42	6.06	5.80	5.61	5.47	5.35	5.26	5.18	5.11	4.96	4.81	4.65	4.57	4.48	4.31
10	10.04	7.56	6.55	5.99	5.64	5.39	5.20	5.06	4.94	4.85	4.77	4.71	4.56	4.41	4.25	4.17	4.08	3.91
11	9.65	7.21	6.22	5.67	5.32	5.07	4.89	4.74	4.63	4.54	4.46	4.40	4.25	4.10	3.94	3.86	3.78	3.60
12	9.33	6.93	5.95	5.41	5.06	4.82	4.64	4.50	4.39	4.30	4.22	4.16	4.01	3.86	3.70	3.62	3.54	3.36
15	8.68	6.36	5.42	4.89	4.56	4.32	4.14	4.00	3.89	3.80	3.73	3.67	3.52	3.37	3.21	3.13	3.05	2.87
20	8.10	5.85	4.94	4.43	4.10	3.87	3.70	3.56	3.46	3.37	3.29	3.23	3.09	2.94	2.78	2.69	2.61	2.42
30	7.56	5.39	4.51	4.02	3.70	3.47	3.30	3.17	3.07	2.98	2.90	2.84	2.70	2.55	2.39	2.30	2.21	2.01
40	7.31	5.18	4.31	3.83	3.51	3.29	3.12	2.99	2.89	2.80	2.73	2.66	2.52	2.37	2.20	2.11	2.02	1.80
60	7.08	4.98	4.13	3.65	3.34	3.12	2.95	2.82	2.72	2.63	2.56	2.50	2.35	2.20	2.03	1.94	1.84	1.60
∞	6.63	4.61	3.78	3.32	3.02	2.80	2.64	2.51	2.41	2.32	2.25	2.18	2.04	1.88	1.70	1.59	1.47	1.00

TABLE A-5
Three-place probabilities, $P(a \le r)$, of the cumulative binomial distribution for $p = \frac{1}{2}$

r \ n†	1	2	3	4	5	6	7	8	9	10
10										1
9									1	999
8								1	998	989
7							1	996	980	945
6						1	992	965	910	828
5					1	984	938	855	746	623
4				1	969	891	773	637	500	377
3			1	938	812	656	500	363	254	172
2		1	875	688	500	344	227	145	090	055
1	1	750	500	312	188	109	062	035	020	011
0	500	250	125	062	031	016	008	004	002	001

† n = sample size (number of trials); r = number of successes; P is the probability of r or fewer successes in n trials.

Note 1: Each three-digit entry should be read with a decimal point preceding it. For entries 1−, the probability is larger than 0.9995 but less than 1. For entries 0+, the probability is less than 0.0005 but greater than 0. For entries 1, the probability is exactly 1.

Note 2: The probability given is the probability of a value a equal to or less than r, $P(a \le r)$. Illustration: For $n = 10$, $p = \frac{1}{2}$, $r = 3$, the probability of a value of a less than or equal to 3 is $P(a \le 3) = 0.172$.

Note 3: To get a probability for a value of a greater than r, use the relation $P(a > r) = 1 - P(a \le r)$. Illustration: For $n = 10$, $p = \frac{1}{2}$, $r = 3$, the probability of a value of a greater than 3 is $P(a > 3) = 1 - 0.172 = 0.828$.

Note 4: $P(a \ge r) = P(a \le n - r)$.

Note 5: The probability for an individual term $P(a = r)$ can be found by using the relation

$$P(a = r) = P(a \le r) - P(a \le r - 1).$$

Illustration: For $n = 10$, $p = \frac{1}{2}$.

$$P(a = 3) = P(a \le 3) - P(a \le 2) = 0.172 - 0.055 = 0.117.$$

Note 6: To find the probability of a value between a and b inclusive, use the relation

$$P(x \le b) - P(x \le a - 1).$$

Illustration: For $n = 10$, $p = \frac{1}{2}$, $P(x = 3, 4, \text{or } 5)$ is

$$P(x \le 5) - P(x \le 2) = 0.623 - 0.055 = 0.568.$$

TABLE A-5 (Cont.)

r \ n	*11*	*12*	*13*	*14*	*15*	*16*	*17*	*18*	*19*	*20*
20										1
19									1	1−
18								1	1−	1−
17							1	1−	1−	1−
16						1	1−	1−	1−	999
15					1	1−	1−	999	998	994
14				1	1−	1−	999	996	990	979
13			1	1−	1−	998	994	985	968	942
12		1	1−	999	996	989	975	952	916	868
11	1	1−	998	994	982	962	928	881	820	748
10	1−	997	989	971	941	895	834	760	676	588
9	994	981	954	910	849	773	685	593	500	412
8	967	927	867	788	696	598	500	407	324	252
7	887	806	709	605	500	402	315	240	180	132
6	726	613	500	395	304	227	166	119	084	058
5	500	387	291	212	151	105	072	048	032	021
4	274	194	133	090	059	038	025	015	010	006
3	113	073	046	029	018	011	006	004	002	001
2	033	019	011	006	004	002	001	001	0+	0+
1	006	003	002	001	0+	0+	0+	0+	0+	0+
0	0+	0+	0+	0+	0+	0+	0+	0+	0+	0+

TABLE A-5 *(Cont.)*

r \ n	21	22	23	24	25	26	27	28	29	30
30										1
29									1	1−
28								1	1−	1−
27							1	1−	1−	1−
26						1	1−	1−	1−	1−
25					1	1−	1−	1−	1−	1−
24				1	1−	1−	1−	1−	1−	1−
23			1	1−	1−	1−	1−	1−	1−	999
22		1	1−	1−	1−	1−	1−	1−	999	997
21	1	1−	1−	1−	1−	1−	999	998	996	992
20	1−	1−	1−	1−	1−	999	997	994	988	979
19	1−	1−	1−	999	998	995	990	982	969	951
18	1−	1−	999	997	993	986	974	956	932	900
17	999	998	995	989	978	962	939	908	868	819
16	996	992	983	968	946	916	876	828	771	708
15	987	974	953	924	885	837	779	714	644	572
14	961	933	895	846	788	721	649	575	500	428
13	905	857	798	729	655	577	500	425	356	292
12	808	738	661	581	500	423	351	286	229	181
11	668	584	500	419	345	279	221	172	132	100
10	500	416	339	271	212	163	124	092	068	049
9	332	262	202	154	115	084	061	044	031	021
8	192	143	105	076	054	038	026	018	012	008
7	095	067	047	032	022	014	010	006	004	003
6	039	026	017	011	007	005	003	002	001	001
5	013	008	005	003	002	001	001	0+	0+	0+
4	004	002	001	001	0+	0+	0+	0+	0+	0+
3	001	0+	0+	0+	0+	0+	0+	0+	0+	0+
2	0+	0+	0+	0+	0+	0+	0+	0+	0+	0+
1	0+	0+	0+	0+	0+	0+	0+	0+	0+	0+
0	0+	0+	0+	0+	0+	0+	0+	0+	0+	0+

TABLE A-5 (Cont.)

r \ n	31	32	33	34	35	36	37	38	39	40
40										1
39									1	1–
38								1	1–	1–
37							1	1–	1–	1–
36						1	1–	1–	1–	1–
35					1	1–	1–	1–	1–	1–
34				1	1–	1–	1–	1–	1–	1–
33			1	1–	1–	1–	1–	1–	1–	1–
32		1	1–	1–	1–	1–	1–	1–	1–	1–
31	1	1–	1–	1–	1–	1–	1–	1–	1–	1–
30	1–	1–	1–	1–	1–	1–	1–	1–	1–	1–
29	1–	1–	1–	1–	1–	1–	1–	1–	999	999
28	1–	1–	1–	1–	1–	1–	1–	999	998	997
27	1–	1–	1–	1–	1–	999	999	997	995	992
26	1–	1–	1–	1–	999	998	996	993	988	981
25	1–	1–	999	999	997	994	990	983	973	960
24	1–	999	998	995	992	986	976	964	946	923
23	998	996	993	988	980	967	951	928	900	866
22	995	990	982	971	955	934	906	872	832	785
21	985	975	960	939	912	879	838	791	739	682
20	965	945	919	885	845	797	744	686	625	563

TABLE A-5 (Cont.)

r \ n	31	32	33	34	35	36	37	38	39	40
19	925	892	852	804	750	691	629	564	500	437
18	859	811	757	696	632	566	500	436	375	318
17	763	702	636	568	500	434	371	314	261	215
16	640	570	500	432	368	309	256	209	168	134
15	500	430	364	304	250	203	162	128	100	077
14	360	298	243	196	155	121	094	072	054	040
13	237	189	148	115	088	066	049	036	027	019
12	141	108	081	061	045	033	024	017	012	008
11	075	055	040	029	020	014	010	007	005	003
10	035	025	018	012	008	006	004	003	002	001
9	015	010	007	005	003	002	001	001	001	0+
8	005	004	002	001	001	001	0+	0+	0+	0+
7	002	001	001	0+	0+	0+	0+	0+	0+	0+
6	0+	0+	0+	0+	0+	0+	0+	0+	0+	0+
5	0+	0+	0+	0+	0+	0+	0+	0+	0+	0+
4	0+	0+	0+	0+	0+	0+	0+	0+	0+	0+
3	0+	0+	0+	0+	0+	0+	0+	0+	0+	0+
2	0+	0+	0+	0+	0+	0+	0+	0+	0+	0+
1	0+	0+	0+	0+	0+	0+	0+	0+	0+	0+
0	0+	0+	0+	0+	0+	0+	0+	0+	0+	0+

TABLE A-5 (Cont.)

r \ n	41	42	43	44	45	46	47	48	49	50
50										1
49									1	1–
48								1	1–	1–
47							1	1–	1–	1–
46						1	1–	1–	1–	1–
45					1	1–	1–	1–	1–	1–
44				1	1–	1–	1–	1–	1–	1–
43			1	1–	1–	1–	1–	1–	1–	1–
42		1	1–	1–	1–	1–	1–	1–	1–	1–
41	1	1–	1–	1–	1–	1–	1–	1–	1–	1–
40	1–	1–	1–	1–	1–	1–	1–	1–	1–	1–
39	1–	1–	1–	1–	1–	1–	1–	1–	1–	1–
38	1–	1–	1–	1–	1–	1–	1–	1–	1–	1–
37	1–	1–	1–	1–	1–	1–	1–	1–	1–	1–
36	1–	1–	1–	1–	1–	1–	1–	1–	1–	1–
35	1–	1–	1–	1–	1–	1–	1–	1–	999	999
34	1–	1–	1–	1–	1–	1–	999	999	998	997
33	1–	1–	1–	1–	1–	999	998	997	995	992
32	1–	1–	1–	999	999	998	996	993	989	984
31	1–	1–	999	998	997	994	991	985	978	968
30	999	999	997	995	992	987	980	970	957	941
29	998	996	993	989	982	973	961	944	924	899
28	994	990	984	976	964	948	928	903	874	839
27	986	978	967	952	932	908	879	844	804	760
26	970	956	937	913	884	849	809	765	716	664
25	941	918	889	854	814	769	720	667	612	556

TABLE A-5 (Cont.)

r \ n	41	42	43	44	45	46	47	48	49	50
24	894	860	820	774	724	671	615	557	500	444
23	826	780	729	674	617	559	500	443	388	336
22	734	678	620	560	500	441	385	333	284	240
21	622	561	500	440	383	329	280	235	196	161
20	500	439	380	326	276	231	191	156	126	101
19	378	322	271	226	186	151	121	097	076	059
18	266	220	180	146	116	092	072	056	043	032
17	174	140	111	087	068	052	039	030	022	016
16	106	082	063	048	036	027	020	015	011	008
15	059	044	033	024	018	013	009	007	005	003
14	030	022	016	011	008	006	004	003	002	001
13	014	010	007	005	003	002	002	001	001	0+
12	006	004	003	002	001	001	001	0+	0+	0+
11	002	001	001	001	0+	0+	0+	0+	0+	0+
10	001	0+	0+	0+	0+	0+	0+	0+	0+	0+
9	0+	0+	0+	0+	0+	0+	0+	0+	0+	0+
8	0+	0+	0+	0+	0+	0+	0+	0+	0+	0+
7	0+	0+	0+	0+	0+	0+	0+	0+	0+	0+
6	0+	0+	0+	0+	0+	0+	0+	0+	0+	0+
5	0+	0+	0+	0+	0+	0+	0+	0+	0+	0+
4	0+	0+	0+	0+	0+	0+	0+	0+	0+	0+
3	0+	0+	0+	0+	0+	0+	0+	0+	0+	0+
2	0+	0+	0+	0+	0+	0+	0+	0+	0+	0+
1	0+	0+	0+	0+	0+	0+	0+	0+	0+	0+
0	0+	0+	0+	0+	0+	0+	0+	0+	0+	0+

TABLE A-6
Exact cumulative distribution for the Mann-Whitney-Wilcoxon test†

	m = 8		m = 7		m = 6		m = 5		m = 4		m = 3		m = 2		m = 1	
t	P	t	P	t	P	t	P	t	P	t	P	t	P	t	P	t
$n = 1$																
1	111	9	125	8	143	7	167	6	200	5	250	4	333	3	500	2
2	222	8	250	7	286	6	333	5	400	4	500	3				
3	333	7	375	6	429	5	500	4								
4	444	6	500	5												
$n = 2$																
3	022	19	028	17	036	15	048	13	067	11	100	9	167	7		
4	044	18	056	16	071	14	095	12	133	10	200	8	333	6		
5	089	17	111	15	143	13	190	11	267	9	400	7				
6	133	16	167	14	214	12	286	10	400	8						
7	200	15	250	13	321	11	429	9								
8	267	14	333	12	429	10										
9	356	13	444	11												
10	444	12														
$n = 3$																
6	006	30	008	27	012	24	018	21	029	18	050	15				
7	012	29	017	26	024	23	036	20	057	17	100	14				
8	024	28	033	25	048	22	071	19	114	16	200	13				
9	042	27	058	24	083	21	125	18	200	15	350	12				
10	067	26	092	23	131	20	196	17	314	14	500	11				
11	097	25	133	22	190	19	286	16	429	13						
12	139	24	192	21	274	18	393	15								
13	188	23	258	20	357	17	500	14								
14	248	22	333	19	452	16										
15	315	21	417	18												
16	388	20	500	17												
17	461	19														

TABLE A-6 (Cont.)

	m = 8		m = 7		m = 6		m = 5		m = 4		m = 3		m = 2		m = 1	
t	P	t	P	t	P	t	P	t	P	t	P	t	P	t	P	t
n = 4																
10	002	42	003	38	005	34	008	30	014	26						
11	004	41	006	37	010	33	016	29	029	25						
12	008	40	012	36	019	32	032	28	057	24						
13	014	39	021	35	033	31	056	27	100	23						
14	024	38	036	34	057	30	095	26	171	22						
15	036	37	055	33	086	29	143	25	243	21						
16	055	36	082	32	129	28	206	24	343	20						
17	077	35	115	31	176	27	278	23	443	19						
18	107	34	158	30	238	26	365	22								
19	141	33	206	29	305	25	452	21								
20	184	32	264	28	381	24										
21	230	31	324	27	457	23										
22	285	30	394	26												
23	341	29	464	25												
24	404	28														
25	467	27														

† Test: Sample A has n measurements and sample B has m measurements ($n \leq m$). The $m + n$ measurements are ranked 1, 2, . . . , $m + n$, and t is the total of the ranks of the n measurements of sample A.

Note 1: Each three-digit entry is the probability and should be read with a decimal point preceding it.

Note 2: The integers at the extreme left are values of t to be used in finding $P(T \leq t)$ on that line.

Note 3: The integer on the right of the probability is the value of t for which the probability is $P(T \geq t)$.

Illustrations: (a) For $m = 5$, $n = 3$, and $t = 6$, $P(T \leq 6) = 0.018$. (b) For $m = 5$, $n = 4$, and $t = 25$, $P(T \geq 25) = 0.143$.

TABLE A-7
One thousand random digits†

Line number	*Column number*									
	1–10		*11–20*		*21–30*		*31–40*		*41–50*	
1	15544	80712	97742	21500	97081	42451	50623	56071	28882	28739
2	01011	21285	04729	39986	73150	31548	30168	76189	56996	19210
3	47435	53308	40718	29050	74858	64517	93573	51058	68501	42723
4	91312	75137	86274	59834	69844	19853	06917	17413	44474	86530
5	12775	08768	80791	16298	22934	09630	98862	39746	64623	32768
6	31466	43761	94872	92230	52367	13205	38634	55882	77518	36252
7	09300	43847	40881	51243	97810	18903	53914	31688	06220	40422
8	73582	13810	57784	72454	68997	72229	30340	08844	53924	89630
9	11092	81392	58189	22697	41063	09451	09789	00637	06450	85990
10	93322	98567	00116	35605	66790	52965	62877	21740	56476	49296
11	80134	12484	67089	08674	70753	90959	45842	59844	45214	36505
12	97888	31797	95037	84400	76041	96668	75920	68482	56855	97417
13	92612	27082	59459	69380	98654	20407	88151	56263	27126	63797
14	72744	45586	43279	44218	83638	05422	00995	70217	78925	39097
15	96256	70653	45285	26293	78305	80252	03625	40159	68760	84716
16	07851	47452	66742	83331	54701	06573	98169	37499	67756	68301
17	25594	41552	96475	56151	02089	33748	65239	89956	89559	33687
18	65358	15155	59374	80940	03411	94656	69440	47156	77115	99463
19	09402	31008	53424	21928	02198	61201	02457	87214	59750	51330
20	97424	90765	01634	37328	41243	33564	17884	94747	93650	77668

† A good source of random digits and random normal deviates is *A Million Random Digits with 100,000 Normal Deviates*, The RAND Corporation, Glencoe, Ill., Free Press of Glencoe, 1955, although these do not come from that source.

Short Answers to Selected Odd-Numbered Problems

CHAPTER 1

Section 1-1

1. about 4800 **3.** about 6950

5. 7700 from midnight to 8 A.M., 6175 from 8 A.M. to 4 P.M., 11,125 from 4 P.M. to midnight

7. 40 **9.** 80–84 years, 17.3% **11.** 96.26%

Section 1-2

3. 1, 7, 14, 21, 28 **5.** no **9.** 70–74 years, 13.51%

Section 1-3

1. 90%, 10% **3.** a) $M = 4.3$, b) $Q_1 = 3.75$, $Q_3 = 4.75$, $IQR = 1.00$

5. partial solution:

Measurements Less Than	Cumulative Frequency	Cumulative Percent
15	0	0
20	1	1
25	7	8
...	...	...
65	90	100

9.

Difference in Scores	Frequency	Percent	Cumulative Frequency	Cumulative Percent
1	282	35	282	35
2	165	20	447	55
3	124	15	571	70
4	89	11	660	81
5	58	7	718	89
6	28	3	746	92
7	24	3	770	95
8	15	2	785	97
9	11	1	796	98
10	5	1	801	99
11	3	0	804	99
12	4	0	808	100
13	1	0	809	100
14	1	0	810	100
Total	810			

15. larger

Chapter 1 summary problems

1. partial solution:

Length	Cumulative Frequency	Cumulative Percent
1	0	0
2	0	0
3	2	2
4	6	5
...	...	...
15	125	99
≥ 16	126	100

5. $M = 6.7$, $Q_1 = 5.5$, $Q_3 = 8.2$, $IQR = 2.7$ **7.** yes

13. very similar to right of Fig. 1-4

CHAPTER 2

Section 2-1

1. a) 24.5 megohms, b) 46.4, 47.9, 49.0, c) 46.0, d) isolated groups at both ends

3. a) 21.4 megohms, b) 43.8, 44.5, 47.0, 50.0, c) 44.0, d) isolated groups at both ends

5. 1 unit = $10

Stem	Leaf
0	4
1	6
2	4488
3	26
4	044488
5	2266
6	00488
7	2

7. 1 unit = $10

Stem	Leaf
3	6
4	8
5	2666
6	00000444
7	666
8	088
9	26
10	04
11	2

9.

Stem	Leaf
0	889
1	00144455556778999
2	01112244444456888899
3	02455667
4	1
5	
6	
7	2

11. a) 64, b) 24, c) 22, d) one isolated measurement (for Problem 9 data)

13. a) 33, b) 4, c) 6, d) no isolated measurements

Section 2-3

1. $Q_1 = 59.5$, $M = 63.4$, $Q_3 = 68.1$, $IQR = 8.6$

3. summer temperatures about 39° higher than winter temperatures; winter temperatures slightly more variable

13. less variable

15. 140, 11, 90, 77, 65, 96, 106; 83, 365, 163, 203, 106, 162, 144, 327, 310, 74, 99, 87, 276, 194, 75, 96, 70, 799, 279, 167, 108, 324

17. $M = 90$ **19.** $Q_1 = \$840$, $M = \$1120$, $Q_3 = \$1620$, $IQR = \$780$

21. $Q_1 = \$820$, $M = \$1000$, $Q_3 = \$1060$, $IQR = \$240$

23. yes, except possibly for very large observation in Northeast

Section 2-4

1. 0.93, 1.93, −0.07

3.

Stem	Leaf
0	9
1	6778888888899

Chapter 2 summary problems

3. highest and most variable in Fox Chapel; lowest and least variable in 19th Ward; 14th Ward and Wilkinsburg intermediate

5. partial solution:

Fox Chapel
5.64
5.46
5.53
5.29
5.30
5.45

7. 14th Ward: $Q_1 = 4.78$, $M = 4.89$, $Q_3 = 5.04$, $IQR = 0.26$; 19th Ward: $Q_1 = 4.44$, $M = 4.57$, $Q_3 = 4.65$, $IQR = 0.21$; Fox Chapel: $Q_1 = 5.30$, $M = 5.46$, $Q_3 = 5.56$, $IQR = 0.26$; Wilkinsburg: $Q_1 = 4.40$, $M = 4.62$, $Q_3 = 4.79$, $IQR = 0.39$

CHAPTER 3

Section 3-1

3. $X = \sqrt{x}$, $y = 1 \cdot X + 0$ **7.** $y = (3/4)x + 0$

Section 3-2

1. (b), (d), (g) **3.** $X = x^2$, $B = 1 \cdot X + 2$

5. $y = \log D$, $y = (-\log 2)x + 11 \log 2$

7. $F = 3.14x^2$ **9.** $X = x^3$, $y = 5X - 10$ **11.** $Y = \log y$, $X = 10^x$, $Y = X + \log 5$

Section 3-3

1. $T \approx 11$, $T \approx 55$, $T \approx 1.1R - 22$ **3.** $T \approx 44$, $T = 42$

Section 3-4

1. partial solution: as age increases, variability in height increases

13. large residuals at 1912, 1920, and 1948

Chapter 3 summary problems

1.

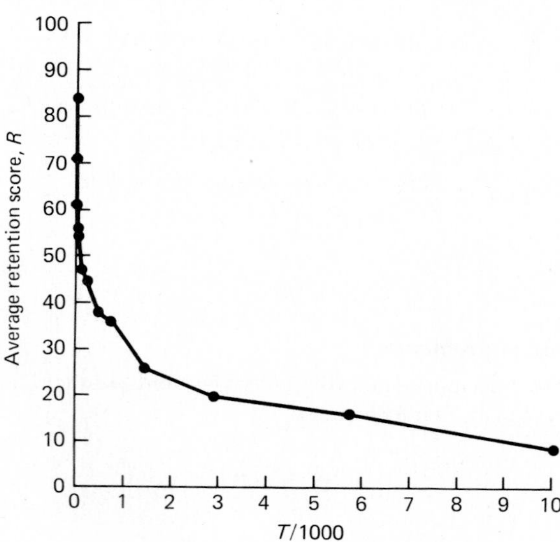

5. 0, 0.70, 1.18, 1.48, 1.78, 2.08, 2.38, 2.68, 2.86, 3.16, 3.46, 3.76, 4.00

7. $R \approx -20t + 89$ **9.** yes

CHAPTER 4

Section 4-1

1. 15°, 79°, 82.3° **3.** a) 9.6, b) 44.8 **5.** yes, 28.6

7.

0	112334444456667889
1	024688
2	5
3	4

9. 34 **11.** yes, 20 **13.** 5.5, 4.5

15. N: 21,080; 1240; S: 10,680; 445; C: 21,680; 985; W: 17,840; 714; grand mean = 810

21. 780 **23.** 240 **25.** they are the same

Section 4-2

1.

			Perfect pitch			
			Present	**Absent**	**Row Mean**	**Training Effects (row)**
			−20	**−20**		
Previous training	Present	30	190	110	150	40
	Absent	−30	90	50	70	−40
Column Mean			140	80	110	0
Pitch effects (column)			30	−30	0	

Estimates

180	120
100	40

Residuals

10	−10
−10	10

3. Average of column means $= [(a + c)/2 + (b + d)/2]/2 = (a + b + c + d)/4$

Average of row means $= [(a + b)/2 + (c + d)/2]/2 = (a + b + c + d)/4$

5. 83.30, 81.80, 76.00, 64.43, 52.80, 45.87, 43.43

7. Flagstaff −18.06, Phoenix 6.35, Yuma 11.71

9. 65.24, 63.74, 57.94, 46.37, 34.74, 27.81, 25.37

11. 95.01, 93.51, 87.71, 76.14, 64.51, 57.58, 55.14

Section 4-3

1. yes,

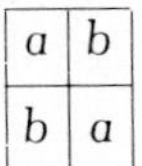

a	b
b	a

13. largest residuals at lowest scholastic aptitude quartile for grades 7 and 8

15. extreme age groups and activity groups have effects of similar magnitudes

19. yes, residuals are mostly positive in upper left and lower right, and they are mostly negative in lower left and upper right

Chapter 4 summary problems

1. brand means: 1.0, 3.2, 3.6, 3.0, 3.8, rater means: 2.2, 3.2, 2.8, 3.0, 3.4, grand mean: 2.92

3. −0.72, 0.28, −0.12, 0.08, 0.48

5.

		Rater				
		1	**2**	**3**	**4**	**5**
Brand	1	.72	−.28	.12	−.08	−.48
	2	1.52	.52	−1.08	−.28	−.68
	3	−.88	−.88	.52	.32	.92
	4	−1.28	.72	.12	−.08	.52
	5	−.08	−.08	.32	.12	−.28

7. party effects: 17.2, −2.1, −2.8, −3.1, −4.1, −5.4, column effects: 6.7, 1.6, −8.4

CHAPTER 5

Section 5-4

1. If we wanted to collect detailed information about each individual, it might be too difficult or too expensive to do so for the entire population.

Section 5-7

1. F **3.** F **5.** T **7.** F **9.** F

11. F **13.** T **15.** T

Chapter 5 summary problems

7. because numbers of people doing grocery shopping vary from day to day, it would be difficult to make comparisons

9. the information tells you what treatments to use

CHAPTER 6

Section 6-1

1. $\frac{2(2/3) + 2(1/3)}{4} = \frac{1}{2} = p$ **3.** $\frac{3(1/2) + 3(0)}{6} = \frac{1}{4} = p$

5. $\dfrac{1(0) + 2(1)}{3} = \dfrac{2}{3} = p$ **7.** $\dfrac{3(1/2) + 2(3/4)}{5} = 0.6 = p$

Section 6-2

3. 1.2 **5.** 0.040 **7.** variance decreases as sample size increases

9. 1/8

Section 6-3

1. N is the population size, and n is the sample size

3. bias is not a property of a single sample

5. $N = 12$, $n = 2$, $p = 1/6$ **7.** 0.152

Section 6-4

1. 0.04, 0.015 **3.** 0.015 **5.** 29 **7.** 1433

Section 6-5

1. 0.0037 **5.** 60

9. Var $\overline{p} = (p - p^2)/n$, and for p near 0, $p - p^2$ is approximately p

Chapter 6 summary problems

1. 0.4, 0.027, 0.164 **3.** 0.4, 0.048, 0.219 **5.** they do not change

7. 0.901 **9.** 25%

CHAPTER 7

Section 7-2

1. 21.2% **3.** 99.8%, 99.8% **5.** more likely that $x = 9$

7. A **9.** yes

Section 7-3

1. as areas between the curve and the horizontal axis

3. mean 0 and standard deviation 1

5. a) 0.683, b) 0.954, c) 0.496 **7.** 0.383 **9.** 0.159 **11.** 0.657

Section 7-4

3. partial solution:

digit:	0	1 . . . 9
frequency:	25	19 . . . 21

Chapter 7 summary problems

3. 0.976 **5.** 69.9% **7.** answer b) **9.** 0.014

CHAPTER 8

Section 8-2

1. a) 1, 2, 3, 4, 5, 6 **3.** *ab*, *ac*, *bc* **5.** 1/5, 2/5

7. 5/6 **9.** 3/4 **11.** 1/4 + 3/4 + 0 = 1 **13.** 0.303

15. 90/132 + 40/132 + 2/132 = 1

Section 8-3

1. 0.045 **3.** 0.197 **5.** 0.0625, 0.0059 **7.** 0.833 **9.** 0.031, 0.031

11. 0.092 **13.** 0.00024

Section 8-4

1. 5/6 **3.** 0.382 **5.** 5/33 + 5/33 = 10/33 = 0.303 **7.** 0.667

9. 4/15 + 4/15 = 8/15 = 0.533

Section 8-5

1. 1/6, 5/6 **3.** 4/9 **5.** 5/9 **7.** 4/9 **9.** 0.4 **11.** 11/18

Section 8-6

1. 0.375 **3.** 3/52 **5.** 0.133 **7.** 0.576 **9.** 0.111

Section 8-7

1. a) 0.714, b) 0.666, c) 0.643 **3.** 0.428 **5.** 0.110, 0.082

7. 0.545 **9.** 0.008

Section 8-8

3. 3.4, 0.897 **5.** 2/3, 0.471 **7.** a) 3.52 cents, b) $352

9. 2.889, 0.737 **11.** 1

Chapter 8 summary problems

3. a) *art*, *atr*, *rat*, *rta*, *tar*, *tra*, b) 1/2 **5.** 0.651 **7.** 0.0016

9. 0.127 **11.** 1/3 **13.** 0 **15.** 0.28, 0.262, 0.511

CHAPTER 9

Section 9-1

1. F **3.** last year's crime rate **5.** the standard deviation

7. $\dfrac{\text{statistic} - \text{standard}}{\text{standard deviation of statistic}}$

9. statistic ± tabled value (standard deviation of statistic) **11.** T

Section 9-2

1. 0.0005, 0.0005 **3.** 0.1314 **5.** P value = 0.058

7. $p = 1/2$ **9.** 0.058

Section 9-3

1. P value = 0.101 **3.** P value = 0.344

Section 9-4

1. P value $\approx$ 0, have strong grounds **3.** P value = 0.2389, not strong evidence

5. P value = 0.0154

Section 9-5

5. **Expected**

22	11
8	4

Residuals

1	−1
−1	1

7. **Expected**

11.5	11.5
3.5	3.5

Residuals

2.5	−2.5
−2.5	2.5

Section 9-6

1. $\chi^2 = 45.87$, P value = ≈ 0 **3.** $\chi^2 = 0.18$, P value = 0.75 (by interpolation), conclude height and diameter independent

5. $\chi^2 = 0$, independent **7.** $|z|$ large

9. $P(|z| > 1) = 0.3174$, $P(\chi^2 > 1) \approx 0.3613$ (by interpolation). A more detailed χ^2 table would make the answers look much closer.

Section 9-7

1. $\chi^2 = 4.8$, P value = 0.09 **3.** $\chi^2 = 9.28$, P value = 0.056

5. a)

3.8	18.0
14.6	69.2

b) $P(\chi^2_{16} > 562) \approx 0$

7. a) 0.05, b) 0.0367 **9.** As degrees of freedom increase, median increases.

Chapter 9 summary problems

1. $\dfrac{\text{statistic} - \text{standard}}{\text{standard deviation of statistic}}$

3. 0.3085 **5.** (−9.6, 29.6) **7.** P value = 0.1587

9. $z^2 = (-1.005)^2 = 1.01 = \chi^2$ **11.** 0.5

CHAPTER 10

Section 10-1

1. reduced by 29.3%

3.

		Second Observation			
		0	**2**	**4**	**6**
	0	0	1	2	3
First observation	2	1	2	3	4
	4	2	3	4	5
	6	3	4	5	6

$\text{Var}\ \overline{X} = 2.5$

9. $\dfrac{s^2}{n} = \dfrac{\Sigma(x - \overline{x})^2}{n(n-1)}$,

when sampling with replacement;

$$\frac{s^2}{n}\left(\frac{N-n}{N-1}\right) = \frac{\Sigma(x - \overline{x})^2}{n(n-1)}\left(\frac{N-n}{N-1}\right),$$

when sampling without replacement

11. 10

13. $\sqrt{12\left(\dfrac{N-1}{N+1}\right)}\,\sigma$; $\sqrt{12}\,\sigma$ for N large

Section 10-2

3. 99.92%

5. partial solution:

sum:	0	1	2 ...	9 ...	18
frequency:	1/100	2/100	3/100 ...	10/100 ...	1/100

Section 10-3

1. $t_5 = 2.21$, $P(t > 2.21) \approx 0.0415$, by interpolation **3.** [4.38, 7.62]

5. $t_9 = 3$, $P(t > 3) \approx 0.008$, the treatment appears to reduce roughness

7. [−8.24, 4.24]

9. $t_6 = -1.32$, $P(|t| > 1.32) \approx 0.25$, little evidence to suggest that the mean difference is different from 0

11. [2.08, 4.43]

Section 10-4

1. [−5.16, 44.83] **3.** yes, 0.16 **5.** 88%

7. [−11.38, 81.38], little evidence to suggest that the true mean scores of the two districts differ

Section 10-5

1. [−17.85, −2.15] **3.** $t_{33} = -19.05$, $P(t < -19.05) \approx 0$, evidence suggests that mean score for second set of groups may be higher

5. $t_{18} = 1.16$, $P(t > 1.16) \approx 0.14$, little evidence to suggest that treatment is effective

Section 10-6

1. $t = 2.46$, $P(t > 2.46) \approx 0.02$, by interpolation

3. $t = 6.88$, $P(t > 6.88) \approx 0$, evidence suggests that mean score for practice group may be higher

5. $t_5 = 2.93$, $P(t > 2.93) \approx 0.02$

Chapter 10 summary problems

1. the Central Limit Theorem

3. σ refers to the population standard deviation; s is an estimate of σ calculated from a sample

5. use the standard normal table when the standard deviation is known; use the t table when the standard deviation is unknown

9. no, only that the data do not give strong evidence aganist μ being equal to 0.

CHAPTER 11

Section 11-1

3. 164.62 mi/hr, observed − predicted = −6.03 mi/hr **5.** 1888 or 1889

Section 11-2

3. $A = \pi r^2$; no; yes; $\log A = \log \pi + 2 \log r$

Section 11-3

1. 31.25, 31.25 **9.** more reliable and objective

Section 11-4

1. $y = 2.8x + 6$ **3.** $\hat{S} = 3.6$ **5.** $y = 12.9x - 67.3$ **7.** $\hat{S} = 6831.75$

9. $\hat{S} = 1025.60$

Chapter 11 summary problems

1. $R = -0.0056T + 52.6$ **3.** $\hat{S} = 2550.9$

5.

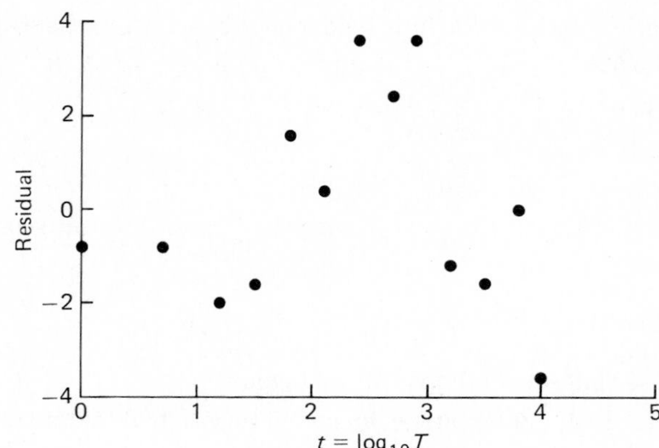

yes, but we notice that now large and small values of t correspond to negative residuals, while moderate values of t correspond to positive residuals

7. $y = 60.5x - 82.6$ **9.** $y = 75.9x - 730$ **11.** $\hat{S} = 195{,}633$

13. $y = 63.0x - 228$ **15.** $y = 4.03x + 61.0$ **17.** $y = 2.27x + 157$

19. $\hat{S} = 1310.1$ **21.** $y = 4.91x + 10.8$

CHAPTER 12

Section 12-2

1. 0.6826, 0.9544 **3.** a) 12.71, b) 2.92, c) 9.92, d) 2.57, e) 2.33, f) 2.58

5.

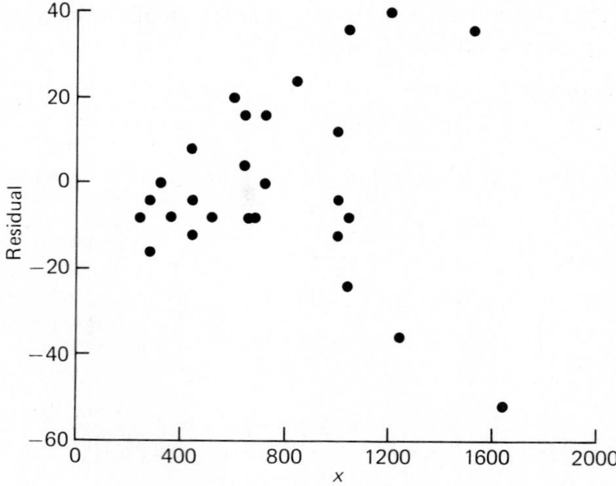

The variance does not seem to be constant. It tends to increase as x increases.

7. observational

Section 12-3

1. $\hat{m} = 0.5, \hat{b} = 0.833$ **3.** 5 of 6 are within 1 s_e, and all are within 2 s_e

5. $\hat{S} = 10124.8, s_e^2 = 562.49$ **7.** $\hat{S} = 12084, s_e = 22$ using $\hat{m} = 0.105$

Section 12-4

1. $r = +1$

3. a) $\Sigma x^2 = 141, SSx = 60$, b) $\Sigma y^2 = 984, SSy = 308$, c) $\Sigma xy = 234, Sxy = 0$

5. 9%, 9%

7. 0%, November snowfall is not a good linear predictor of snowfall for the rest of the year

9. 0.806

Section 12-5

3. 6 **5.** 2.271×10^{-5}

7. $t = -6.9$, evidence suggests that 0.75 is not a very plausible value of m

Section 12-6

3. $[-1.738, -0.442]$

Section 12-7

1. $t = 1.12$, $P(t_8 > 1.12) \approx 0.16$ by interpolation, the prediction seems to be consistent with the actual winning speed

3. winning speed does not fall in the 95% confidence interval

5. 124.54 pounds

7. 2.895 in units of 1/10 times the earth's distance from the sun

9. [424.6, 1452.1] **11.** [206.65, 218.80]

Chapter 12 summary problems

1. [10.09, 15.79] **3.** 77.75% **5.** 84.81% **7.** [6.71, 7.70]

9. [14.95, 16.05]

CHAPTER 13

Section 13-1

1. y is the dependent variable, u and w are the independent variables **3.** yes

5. average wind velocity **7.** yes

9. for example age, high school GPA

Section 13-2

1. 1.06

3. for the least-squares estimator $\hat{S} = 0.015$, while for the best predictor from Problem 2 $S = 1.06$

5. S from (b) is the lowest **7.** 0.103

Section 13-3

1.

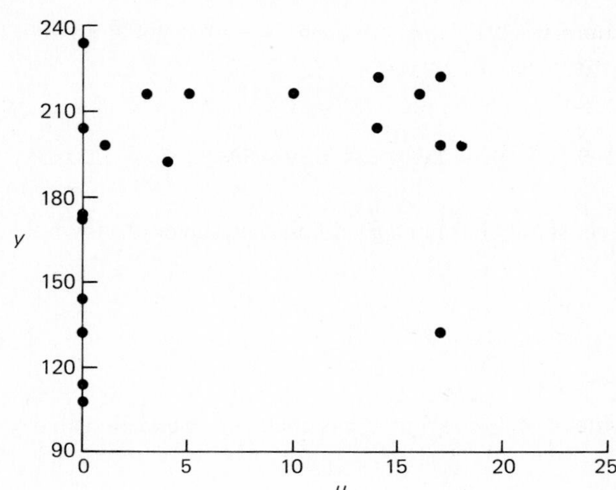

3. a) y seems to have only a slight linear relation to u, while there is a definite linear relation between y and w

5. $y = 43.18 + 13.7\,w$

7. for the regression of y on u, $R^2 = 12.6\%$, for the regression of y on w, $R^2 = 73.9\%$

9. $\hat{S} = 19044.9$, $R^2 = 74.9\%$

Section 13-4

1. 4.0453, 3.9003, 3.9526, 3.9950, 4.0363, 3.9239, 3.9718, 4.0033, 3.9851, 3.9127, 3.9584, 3.9481

3. 0.0118 5. R^2 is increased 3.6%

Chapter 13 summary problems

1. negative 3. a) $t = -2.2$, $P(t_{45} < -22) \approx 0.018$ by interpolation, b) $t = -2.8$, $P(t_{45} < -2.8) < 0.005$

5. $\hat{S} = 6.204$, $R^2 = 96.3\%$ 7. $y = -1.61 - 85.1u + 0.0439w$

9. $\hat{S} = 5824195$, $R^2 = 96.9\%$

CHAPTER 14

Section 14-1

3. 6 5. 0 7. $S = \Sigma e_i^2$

Section 14-2

1. a) TEMP, RAIN, b) LAT and LONG, RAIN and LONG, RAIN and LAT

3. LENG = 37.113 − 0.345 LAT − 0.175 LONG + 0.001 ALT + 0.050 RAIN + 0.174 TEMP

5. 13 − 6 = 7   **7.** RAIN, LAT, LONG

11. LENG = 9.786 + 0.002 ALT + 0.217 TEMP

Chapter 14 summary problems

1. a) DLIC, b) ROAD and TAX, DLIC and FUEL

3. FUEL = 377.291 − 34.790 TAX + 1336.449 DLIC − 0.067 INC − 0.002 ROAD

5. R^2 is increased from 49% to 68%

9.

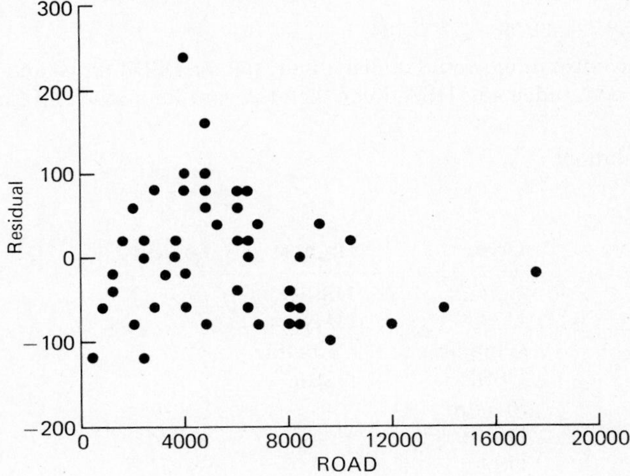

The variance seems to decrease as ROAD increases; and Wyoming, with a residual of 242.6, seems to be an outlier.

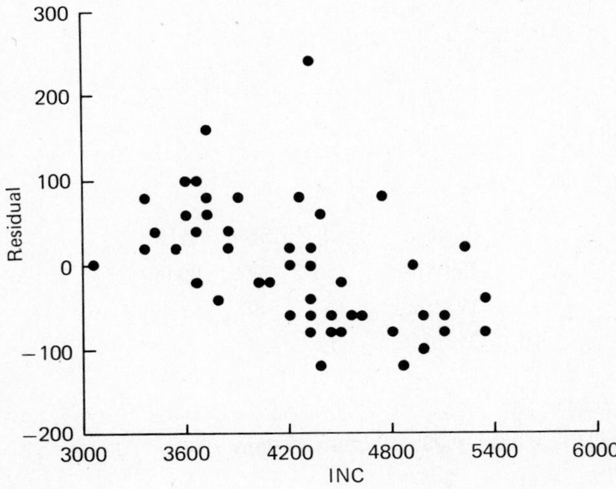

The residuals seem to decrease slightly as INC increases, and Wyoming again seems to be an outlier.

CHAPTER 15

Section 15-1

3. Table 15-1, mean = 3.58, Table 15-2, mean = 2.69, the means for the real data vary more, even with the removal of Nevada.

5. 0.5, −1.5, −0.5, −0.7, 0.3, 1.9 **7.** because of rounding

9. −8, 45, −58, 39, 39, 99, −57, −104

11. 120 for Swedish judge and Swedish diver; 108 for USSR judge and USSR diver; 95 for USA judge and USA diver; 95 for Australian judge and Canadian diver

13. partial solution:

Diver	Highest	Lowest
Canada	USSR	—
USA	USA	—
Canada	Colombia	—
USSR	USSR	—
Sweden	—	USSR
USSR	—	Italy
USA	—	Sweden
E. Germany	—	Australia

Section 15-2

1. $\overline{x}_1 = 4$, $\overline{x}_2 = 2$, $\overline{x}_3 = 5$, $\overline{x} = 4$ **3.** 5 **5.** 42

Section 15-3

1. partial solution:

Source	df	SS	MS
Between	2	160	80.00
Within	—	86	7.17
Total	14	246	

3. no **5.** $P(F \leq 0.06) \approx 0.033$ by interpolation

7. $BSS = 18$, $TSS = 44$ **9.** 3.71

Section 15-4

1.

Source	df	SS	MS	F-Ratio
Between heights	2	$BHSS$	$BHSS/2$	$3BHSS/RSS$
Between pressures	3	$BPSS$	$BPSS/3$	$2BPSS/RSS$
Residual	6	RSS	$RSS/6$	
Total	11	TSS		

where $BHSS$ and $BPSS$ represent the between heights sum of squares and the between pressures sum of squares, respectively

3. partial solution:

Source	df	SS	MS
Stoves	1	216.00	216.00
Pots	2	86.33	43.17
Residual	—	—	9.50
Total	5	321.33	

5. residuals for first row are 0.44, 4.64, 1.44, −2.16, −4.36; negative residuals tend to lie near the diagonal

Chapter 15 summary problems

1.

Due to	df	SS	MS	F-Ratio
Factor	4	89.037	22.259	33.074
Error	15	10.102	0.673	
Total	19	99.139		

$P(F_{4,15} > 33.074)$ is much less than 0.01.

3. partial solution:

Source	df	SS	MS	F-Ratio
Factor	—	26235	—	12.11
Error	16	—	722	
Total	19	37794		

5. partial solution:

Source	*df*	*SS*	*MS*	*F*-Ratio
Judge	6	44864	7477	—
Diver	7	—	32588	10.71
Error	—	127819	3043	
Total	55	400800		

7.

Source	*df*
Popper	1
Corn	2
Error	2
Total	5

9.

Source	*df*
Popper	2
Corn	2
Error	4
Total	8

CHAPTER 16

Section 16-2

1. 0.002 **3.** 0.238

5. we are trying to decide if the evidence suggests that p is far from ½

7. $t = 1.38$, $P(t_{11} > 1.38) \approx 0.1$ **9.** no effect since they are ties

Section 16-3

1. 0.2483 **3.** $P(T \leq 5) = 0.143$ **5.** $P(Z \leq -2.80) = 0.0026$

7. $P(Z \leq -1.54) = 0.0618$ **9.** $P(Z \leq -0.81) = 0.3090$

Section 16-4

1. $P(H \geq 9.2) \approx 0.10$ **3.** $P(\chi_2^2 \geq 0.368) > 0.5$

5. $P(\chi_4^2 \geq 10.072) \approx 0.044$ by interpolation **7.** $P(\chi_2^2 \geq 2.379) > 0.2$

Chapter 16 summary problems

3. Mann-Whitney-Wilcoxon and Kruskal-Wallis, respectively

5. Using a one-sided test, the probability of 7 or more successes when $p = 0.3$ is 0.011. Thus we reject the null hypothesis of $p = 0.3$ at the 5 percent level, or any level down to 0.011.

7. both give 27/14

CHAPTER 17

Section 17-1

1. a) control group consists of the other hostages; experimental group consists of Daniel and friends, b) experimental treatment was vegetable and water diet, control treatment was king's rich food and wine, c) that it might cause Daniel and his friends to be less healthy than the other hostages

7. yes, there may be an effect due to warming up or an effect due to tiring or becoming bored

Section 17-2

1. a) location, time of year, day of week, b) rainy weather, c) yes

Section 17-3

11. 9 times

Section 17-4

1. sampling bias, sampling unit

3. confounding causes, trends, variability

5. confounding causes, trends, variability

Section 17-5

1. confounding causes, trends

3. Maturation and outside effects can be measured in the control group, and the groups can be adjusted for initial comparability.

Section 17-6

3. $P(F_{4,8} > 1.072) > 0.05$, blocking does not appear to have been very effective

5. $\sigma^2/422$

7. 0.09

9. no

Chapter 17 summary problems

1. block children by geographic area, age, sex, etc.

3. match on basis of such variables as age, sex, and school attended, and give one of each pair the vaccine and the other the placebo

5. over 200,000 received the vaccine and over 200,000 received the placebo

7. no

9. by randomly selecting those who received the vaccine and those who received the placebo

APPENDIX I

1. 2 **3.** 35 **5.** 20

Index to Data Sets and Examples

Age and activity for poor adults, 110–111
Air Force pilot accidents, 256–257
Allocation of police, 2–4, 7–8
Ambiguity and speech, 287–288
Amoebas and intestinal disease, 253–254
Antelope bands, 34, 96
Area of Vermont counties, 28–30, 40, 55
Assessing the demand for fuel, 426–427
Awards for school district projects in California, 38
Awards for school district projects in Tennessee, 36–37, 50–51, 55

Baseball scores, 20–21
Bedwetting therapies, 291–292
Birthday-deathday, 485–486

Cancer mortality and fluoridation, 389–391
College football scores: 1967, 10–11
Cooling down Arizona, 102–103

Daniel and the diets, 500–501, 503
Deterrent effects of punishment on homicide rates, 399–400
Diabetic mice weights, 81–82, 324–325
Diameters of steel rods, 4–5, 8
Distances of planets from sun (Bode's Law), 70–72, 369
Dr. Spock and the women jurors, 260
Draft lotteries and random selection, 183–184

Electrical resistances of insulation, 30–32
Employment, price deflator, and GNP data, 401–402

Feed corn demand in the U.S., 321–324, 372
Fuel consumption data, 430–433

Geographic distribution of divorced men, 436–438, 440, 449–451, 456, 490–491, 494
Grade-point average predictions, 380, 384, 400–401
Graduate admissions at Berkeley, 125, 126
Graduate admissions at Harvard, 487–488
Gravitational attraction, 77–78
Group therapy for the aged, 290–291
Growth of Japanese larch trees, 427

Hail suppression, 255
Hamilton and Madison word usage, 289–290
Handedness in baseball, 284–285
Heights and weights for high school boys, 309–311, 369
Hen's eggs measurements, 397–398
Heroin purchases, 47–49
Hunting caps, red and yellow, 240–241, 248, 251, 255
Hydrolysis completion measurements, 111–112

Indianapolis 500 winning speeds, 293–301, 318–319, 343–344, 351, 366–369

Kansas City preventive patrol experiment, 460, 525

Length of life for U.S. females, 5–7, 8, 16–19, 20
Length of life for U.S. males, 11–12
Lind on scurvy, 502–503
Living histograms, 25–26
Lizards of Bimini, 255

Marijuana and intellectual performance, 276–278, 290
Mathematics programs for sixth grade classes, 456–458
Metabolic rate and body weight for university women, 305–306, 308, 359–360
Midterm congressional elections, 427–428
Minnesota snowfall data, 304–305, 345–346, 351, 358–359

Numbers of industrial workers and supervisors, 338–340

Occupation and aptitude for Army Air Corps volunteers, 258–259
Ohm's law experiment data, 302–304, 359
Olympic platform diving, 441–442, 470

Party preference in French survey, 114–115
Pennsylvania Daily Lottery, 201–203, 215–216
pH for cigarettes and cigars by puff, 97–98, 103–106
Picture frame measurements, 376–377, 383–384
Population distribution of Ghana, 81–83
Population of states in 1970, 54–55
Property tax assessments in Ramsey County, Minnesota, 428–429
Property taxes for 88 U.S. cities, 32–34, 49, 97

Rating of typewriters, 113–115
Real estate prices in Pittsburgh, 55–57
Realistic career choices, 108–110
Rosier futures, 481, 482–483
Roughness scores for hands, 282, 508–509, 511

Salk polio vaccine experiment, 518–519, 527
Samples from the Williams family, 138–144, 147, 158–159
Size of cricket frogs, 420–425
Ski trip expense, 377–379, 382–383, 385–387, 388, 534–535
Small-world problem, 21–22, 23–24
Social mobility in Great Britain, 260–261
Social status in jury deliberations, 259–260
Steam-consumption data, 379–380
Sterile bandages, 247, 252, 255
Stopping distances and car speed, 330–331, 332–336, 363–364, 372
Strong's retention data, 85–87, 324
Student achievement-score predictions, 408–420
Submarine sinkings in World War II, 78–80, 319–321, 352, 355–356

Temperatures at Boston airport, 12, 20, 42–47, 96
Temperatures below the earth's surface, 73–75, 362–363
Temperatures for five northeast cities, 438–440, 459, 465–466, 470
Toothpaste comparisons, 524–526

Vitamin C and the cold, 242–245, 250–251, 252–253
Volume of trees, 392–397
Voting registration and turnout, 75–76, 77

Winning times for Olympic metric mile, 84–85, 323, 369–370

General Index

Additive model, 90, 99, 103, 439
  column effect, 99–100
  grand mean, 91–92
  group effect, 91–92
  interaction effect, 105
  outliers in, 106–108
  residuals from, 105, 439, 462
  row effect, 99–100
Alternative hypothesis, 281, 292
American Statistical Association, 135, 165, 183
Analysis of covariance, 510
Analysis of variance, 436
  between-means sum of squares, 444
  blocking and two-way, 523
  computer output for, 448, 465
  one-way model, 444
  for ranks, 486
  residual variation, 465–466
  table, 447, 464
  total sum of squares, 445
  two-way breakdown, 463
  two-way model, 464
  within-group sum of squares, 445
ANOVA; *see* Analysis of variance
Association, 245

Bayes' formula, 214
Beaton, A. E., 259
Bias, 147, 514
Biased estimate, 147, 149
Bickel, P. J., 125, 134
Bingham, C., 432
Binomial distribution, 168
  approximation by normal, 173
  histogram of, 171
  and sign test, 475
  standard deviation of, 170
  symmetry for $p = \frac{1}{2}$, 168
  table for $p = \frac{1}{2}$, 547
Binomial proportion, 151
Binomial tables, 168, 547
Black-thread method, 63, 69, 73–76, 298, 309
Bliss, C. I., 427, 433
Blocking, 459, 509
  and two-way analysis of variance, 523
Box, G. E. P., 115, 471, 528
Brown, G. F., Jr., 48

Brown, K. S., 441
Bureau of Labor Statistics, 122
Bush, R. R., 489

Calculators, 50, 383
Campbell, D. T., 528
Campbell, E. Q., 409
Carlson, R., 253
Case studies, 118, 120–121, 131
Causal relations, 123, 129, 348, 522
  improper, 420
Cellier, K. M., 524
Censuses, 121–123
Central limit theorem, 274
Central values; *see* Mean; Median
Chain rule, 219–220
Chatterjee, S., 339, 373, 402
Chi-square distribution, 253
  normal approximation to, 258
  mean and variance of, 258
  tables for, 253, 542
Cochran, W. G., 164, 454, 471
Cohen, J. E., 253
Coleman, J. S., 408, 409
Collinearity, 411
Comparisons
  for measurements
    independent samples, 284, 286, 289
    matched pairs, 276
  need for, 234
  for proportions and counts, 234
    independent samples, 242
    matched pairs, 237, 240
Computer output
  from ANOVA, 449–450, 465
  from regression analysis, 411–425
Confidence interval, 235
  see also Analysis of variance; Linear regression model
Confidence limits, 235, 278
Contingency table
  association and independence in, 245
  chi-square tests for, 252
  expected counts, 249
  independence, 248
  r × c, 256
  2 × 2, 245
Control groups, 126, 501, 505
Controlled field studies; *see* Experiments
Cook, T. D., 528

Correlation coefficient, 348, 389
  among regression predictors, 410
  and test for slope, 357
Covariance, 510
Covariates, 510
Critical ratio, 235, 337
Critical region, 281
Cumulative frequency distribution, 13–20, 45–46, 90
  and reverse cumulative distribution, 15
  steps in construction of, 18

Data collection methods; *see* Case studies; Experiments; Observational studies; Sample surveys
Data displays; *see* Frequency distribution; Frequency histogram; Stem-and-leaf display; Tally
Davies, A. P., 492, 496
Degrees of freedom, 253, 257, 270, 341, 396, 406, 446, 452
Delury, G. E., 29, 54, 85
Deming, W. E., 4, 165
Dempster, A. P., 397
Dependent variables in regression, 376, 381, 410
Descriptive significance level, 238, 245
Design of experiments, 517
Distribution; *see* Cumulative frequency distribution; Frequency histogram; Probability distribution
Drake, A. W., 3
Draper, N. R., 380, 402

Effect; *see* Additive model; Causal relations
Elementary event, 192
Events, 192
  complementary, 210
  dependent, 203
  independent, 196
  mutually exclusive, 207
Expectation, 225
Expected counts, 249
Expected value, 149, 224
Experimental control, 504, 507
  loss of, 511
Experimental group, 501
Experiments, 118, 120, 127–129, 131, 500
  and causation, 522
  devices for strengthening, 507

double-blind, 510
loss of control in, 511
principal basic designs, 517
Extreme values; *see* Outliers

F distribution, 452–454
tables of, 455, 544
F ratio, 448, 451
small values of, 458
Fairley, W. B., 115, 134, 528
Fanning, E. A., 524
Ferber, R., 135, 165
Ferrar, T. A., 426, 433
Fienberg, S. E., 183, 184, 187
Fisher, R. A., 450
Fitting linear relations, 298
aims of, 302–307
choosing a criterion, 310
least squares, 312, 314–316
residuals from, 312
Flieger, W., 6, 11
Frequency distribution, 2
see also Cumulative frequency distribution; Frequency histogram
Frequency histogram, 2, 18–19, 28
gaps, 4, 10
holes, 9
outliers in, 10
peak, 5, 9
pit, 5, 9
spike, 6, 11
steps in construction of, 8–9

Galton, F., 329, 332, 337
Good, I. J., 87
Gosset, W. S., 278
Gotjamanos, T., 524
Graphs; *see* Plotting data; Multiple regression analysis, residual plots; Straight line
Group effect; *see* Additive model

Hamilton, A., 289
Hammel, E. A., 134
Harmon, K. M., 112
Hartley, H. O., 454, 471
Hawkins, R., 502
Histogram; *see* Frequency histogram
Hoaglin, D. C., 57, 87, 135
Hobson, C. J., 409
Hoffman, D., 98
Hollender, J. W., 109
Hunter, J. S., 115, 471, 528
Hunter, W. G., 115, 471, 528
Hypothesis; *see* Alternative hypothesis; Null hypothesis; Test of significance

Independence, 196, 221, 245, 247
Independent variable in regression, 376, 381
Interaction, 105, 466
Interquartile range, 20, 39–42, 95
Izenman, A. J., 421

Joiner, B. L., 19, 25, 471
Jones, E. E., 82
Juster, F. T., 259

Keeney, R. L., 3
Keyfitz, N., 6, 11
Korte, C., 23, 24
Kruskal, W. H., 4, 26, 76, 87, 114, 165, 253, 497
Kruskal-Wallis test, 486
chi-square approximation for, 487
ties, 488

Lancaster, J., 502, 503, 515
Larson, R. C., 3
Least-squares, 66, 312
advantages of, 314
formulas for two regression predictors, 533–535
by high-speed computer, 383, 405
regression coefficients, 315, 329, 383, 404
sensitivity to outliers, 314
Lefcowitz, M. J., 110
Lehmann, E. L., 4, 76, 497
Light, R. J., 135
Lind, J., 501, 502, 503
Lindzey, G., 489
Linear regression analysis; *see* Multiple regression analysis; Regression analysis
Linear regression model, 329
control-knob, 332
error variance estimate, 341
features of, 329
interval for intercept, 360
interval for prediction, 364
interval for slope, 352

Linear regression model *(continued)*
  normal error assumption, 331
  observational situation and, 332
  predicting new y values, 364
  residual analysis to check, 331
  test for intercept, 360
  test for prediction, 364
  test for slope, 352
Link, R. F., 4, 26, 76, 114, 253, 497
Logarithms, 49–50, 93
  base, 49
  common, 50, 52, 70
  and multiplicative formulas, 305, 391, 426
  natural, 50
Longley, J. W., 401

MacFarlane, P., 21
McKittrick, E. J., 306
McNeil, D. R., 57
McPartland, J., 409
McPeek, B., 135
Madison, J., 289
Mann-Whitney-Wilcoxon two-sample test, 478
  normal approximation for, 481
  tables for, 554
  ties, 483
Mason, W., 391
Matched pairs
  measurements, 276
  proportions, 237
  0, 1 outcomes, 240
Matching in experiments, 508
Mean, sample, 39–40, 90–91, 475
Mean absolute deviation, 142
Mean of a random variable, 224
Mean square, 412, 447–448, 464
Measures of location; *see* Mean; Median
Measures of variation; *see* Mean absolute deviation; Variance
Median, sample, 20, 39–41, 90–91, 475
Median of a distribution, 474
Meier, P., 519
Milgram, S., 21, 22, 23, 24
Mnemonics, 410
Model; *see* Additive model; Analysis of variance; Linear regression model; Multiple regression model
Mood, A. M., 409
Moore, G. H., 246
Morse, P. M., 3
Mosteller, F., 4, 10, 26, 57, 74, 76, 87, 114, 115, 134, 135, 165, 187, 232, 253, 264, 289, 290, 295, 315, 325, 350, 373, 402, 433, 489, 497, 528
Multiple correlation coefficient, 389
Multiple regression analysis, 404
  choosing an equation, 406
  computer output and, 404
  estimated residual variance, 408
  least squares coefficients, 405
  residual plots, 405, 417, 418
  residual sum of squares, 405, 408
  strength of, 429
  studentized residuals, 413
  subsets of predictors, 407
  t tests for coefficients, 405
Multiple regression model, 404
  estimated standard errors, 396, 405
  for two predictors, 377
  least squares coefficients, 379, 382, 404

National Bureau of Standards, 183
Nelsen, J. M., 277
Nelson, J. P., 426, 433
Niemann, C., 112
Nonparametric methods, 474
Normal distribution, 168, 172
  analysis of variance model errors, 444, 464
  approximation to binomial, 173
  importance of, 274
  for regression model errors, 331
  standard, 172
  tables of, 173, 177, 538
Null hypothesis, 238, 281
  accepting or rejecting, 292

Observational studies, 118, 119, 124–126, 131
O'Connell, J. O., 134
One-sided tests, 281
Outliers, 10, 31, 40, 63, 78
  from additive model, 94–95, 106
    rule of thumb, 95
  in regression analysis, 413
  from straight lines, 63, 310, 314
Owen, D. B., 454, 471

P value; *see* Descriptive significance level
Parameter, 150
Pauling, L., 242

Pearson, E. S., 454, 471
Phillips, D. P., 485, 497
Pieters, R. S., 4, 26, 76, 114, 253, 497
Pittsburgh Post-Gazette, 56
Placebo, 242, 502, 505
Plotting data, 298
Pocket calculators, 50, 383
Population, 138
Population frequencies, 162, 190
Population mean, 266
Population variance, 153, 267
Price, B., 339, 373, 402
Probability, 162
  of A and B, 206
  of A or B, 208
  conditional, 212, 214
  distribution, 162, 190
    binomial, 168
    chi-square, 253
    discrete uniform, 272, 274
    F, 452
    normal, 168, 172
    Mann-Whitney-Wilcoxon, 480
    t, 278
  equally likely, 192, 212
  of an event, 192
  experiments, 190
  as long-run frequency, 190
  marginal, 216
  multiplication rule, 197, 200, 205
  rules, 229
  sample space, 190
Proportion of variability explained, 346, 387, 405
Public opinion polling, 159

Quadratic function, 66
Quality control, 30, 159, 273
Quartiles, 19, 39, 41–42, 95
Quasi-experiments, 513

r^2, 348
R^2, 387, 405
RAND Corporation, 556
Random numbers, 168, 179, 556
Random samples, drawing, 168, 178–182
Random variables, 223
  differences, 243
  mean, 224
  normal, 168, 172
  notation, 239, 270
  sums, 275
  variance, 225
Randomization, 507
Range, 39
Ranks of measurements, 474, 479
Real number line, 13
Regression analysis
  intervals, 337
  least squares coefficients, 329
  one versus two predictors, 384
  residual sum of squares, 341
  tests, 337
  with two predictors, 376
Relationships; *see* Straight line
Residuals
  from independence, 250
  from mean, 90
  from median, 90
  minimum sum of squared, 316
  from multiple regression analysis, 405, 412
  from straight lines, 63, 77–81, 312
  from two-way additive model, 100–101
Resistant methods, 474, 489
Rising, G. R., 4, 26, 76, 114, 253, 497
Robust methods, 474
Roewe, C., 21
Rourke, R. E. K., 74, 165, 187, 232, 264, 295, 325, 373, 497
Ryan, B. F., 471
Ryan, T. A., Jr., 471

Sample proportions, 139, 149
  mean of, 141
  unbiasedness of, 141, 149
  variability of, 151
  variance with replacement, 157
  variance without replacement, 151
Sample space, 190, 191
Sample surveys, 118, 119, 121–123, 131, 178, 182
Sample variance, 269
  computation of, 271
Sampling, 138–141, 195
  with replacement, 145, 147–150, 156–157
  without replacement, 138–141, 147–151, 205
    improvement factor for, 151
Sampling fraction, 154
Schoener, T., 255

Scientific notation, 410
Sears, A. W., 492, 496
Seton, E. T., 34, 96
Sheatsley, P., 135, 165
Shewhart, W. A., 30, 32
Sign test, 475
Significance level
  choice of, 245
  descriptive, 238
  P value, 238
Significance test; *see* Test of significance
Silverman, L., 48
Simulations, 184
Slam-bang effect, 502
Slonim, M. J., 165
Smith, H., 380, 402
Snedecor, G. W., 454, 471
Spare-tire principle, 319
Spock, B., 260
Spread; *see* Interquartile range; Range; Variance
Standard deviation, 144, 170, 226
Standard error; *see* Standard deviation
Standard normal distribution; *see* Normal distribution
Standard score, 173
Statistic, 150
Stem-and-leaf display, 29–31, 90, 94, 106
  cycles, 35
  lack of symmetry, 30
  logarithms in, 49
  outliers in, 31
  for several groups, 438
  spike, 36
Stoetzel, J., 114
Stoto, M. A., 135
Straight line
  equation of, 60
  intercept of, 60
  predictions from, 298
  relating variables, 60
  slope of, 61
  straightening, 62
  see also Black-thread method; Fitting linear relations; Least-squares
Stratification in experiments, 507
Stuart, A., 165
Student, 278
Studentized residuals; *see* Multiple regression analysis
Student's distribution; *see* *t* distribution
Sum of squares; *see* Analysis of variance; Least squares; Residuals
Summary statistics; *see* Interquartile range; Mean; Median; Quartiles; Range; Standard deviation; Variance
Summation notation, 529
Symmetry, 9, 11
  lack of, 452

t distribution, 278
  degrees of freedom for, 278, 287
  limiting normal distribution, 279
  for regression coefficients, 337
  tables for, 279, 540
Table of counts; *see* Contingency table
Tables for probability distributions; *see* Binomial distribution; Chi-square distribution; *F* distribution; Mann-Whitney-Wilcoxon two-sample test; Normal distribution; *t* distribution
Tables of random numbers, 179, 556
Table of measurements
  one-way, 90–93, 436
  two-way, 90, 97–108, 439
Tally, 28–31
Tanur, J. M., 4, 76, 87, 165, 497
Target population, 122
Test of hypothesis; see Test of significance
Test of significance, 235, 245
  see also Analysis of variance; Comparisons; Kruskal-Wallis test; Linear regression model; Mann-Whitney-Wilcoxon test; Multiple regression analysis; Sign test; Tests for measurements
Tests for measurements
  differences, 276
  independent samples
    known variances, 284
    unknown equal variances, 286
    unknown unequal variances, 289
  matched pairs, 276
Theory of measurement, 266
Thomas, G. B., Jr., 74, 165, 187, 232, 264, 295, 325, 373
Transformations of data, 51–53, 62, 69–70, 93, 307, 330, 336
  see also Logarithms
Travers, J., 21, 22
True value, 138
Tufte, E. R., 76, 427, 433

Tukey, J. W., 57, 87, 102, 108, 115, 402, 433
Turner, A., 135, 165
Two-sided tests, 281

Unbiasedness, 141, 149, 268, 270
U.S. Department of Commerce, 12, 33, 43
U.S. News and World Report, 33

Variability, sources of, 436, 440
Variance, 143
  of binomial observation, 152
  of difference, 243, 285
  of mean, 267
  of sample proportion, 143
  sample size and, 154
  of sum, 275
Velleman, P. F., 57
Vowles, N.J., 524

Waksberg, J., 135, 165
Wallace, D., 289, 290
Watson, B. F., 305, 345
Weil, A. T., 277
Weinfeld, F. D., 409
Weisberg, S., 373, 431, 433
White, H. C., 24
Who's Who in America, 182, 185
Womack, A. W., 322
Wynder, L., 98

York, R. L.

z score, 173
Zelinka, M., 114, 253
Zinberg, N. E., 277